THE PLANTS OF
MEFOU PROPOSED NATIONAL PARK
YAOUNDÉ, CAMEROON

His Excellency SYLVESTRE NAAH ONDOAH (centre), then Minister for Environment and Forest of the Government of Cameroon, during his visit to the Royal Botanic Gardens, Kew on 22nd February 2001.

During his visit he requested that RBG, Kew work with Cameroonian botanists to produce a Conservation Checklist for the Plants of the Mefou National Park, the creation of which he has championed. This book is the result.

THE PLANTS OF
MEFOU PROPOSED NATIONAL PARK
YAOUNDÉ, CAMEROON

A CONSERVATION CHECKLIST

Martin Cheek, Yvette Harvey and Jean Michel Onana

Royal Botanic Gardens, Kew
IRAD-National Herbarium of Cameroon

Kew Publishing
Royal Botanic Gardens, Kew

ROYAL BOTANIC GARDENS

First published in 2011 by
Royal Botanic Gardens, Kew,
Richmond, Surrey, TW9 3AB, UK
www.kew.org

ISBN 978-1-84246-400-7

British Library Cataloguing in Publication Data
A catalogue record for this book is available from the British Library

Design and typesetting by Christine Beard
Publishing, Design & Photography
Royal Botanic Gardens, Kew

Front cover: *Clerodendrum splendens.* Photo taken by Martin Cheek, March 2004.

Printed in the UK by Hobbs the Printers

For information or to purchase all Kew titles please visit
www.kewbooks.com or email publishing@kew.org

Kew's mission is to inspire and deliver science-based plant conservation worldwide, enhancing the quality of life.

Kew receives half of its running costs from Government through the Department for Environment, Food and Rural Affairs (Defra). All other funding needed to support Kew's work comes from members, foundations and commercial activities including book sales.

CONTENTS

In Memoriam Dr Louis Elwyn Fay III (9 July 1928 – 6 May 2009)

Lou Fay initially joined our botanical visits to Cameroon under the Earthwatch programme by which time he was already over 70 years of age and knew what he wanted to do: collect herbarium specimens of ferns. His wife Alice had developed a passion for fern collecting in their years in Trinidad and he had continued the interest. He enjoyed the climate and culture of rural Cameroon since he had spent two happy years in Liberia as Fulbright Professor when his five children were small and later, while also teaching, in Trinidad and Kenya. In Cameroon he never missed an expedition with us. His pattern was to hire a field assistant go collecting at dawn; return for an early lunch, cold beer and nap, then to prepare his specimens. That almost every one of the 42 species of pteridophyte recorded in this book was among the 82 collections that Lou made is a tribute to his dedication. He was born in Tucson, Arizona. After school he joined the US Air Force to benefit from their education programme. He trained as an engineer and later went into industry, developing among other things magnetic tape for recording sound. He then went into his teaching career.

Lou was a strong man in many ways. Despite his age he delighted in arriving in Cameroon at Douala airport by himself, and finding his own way to join the team. He would pack a large amount of baggage and carry most of it himself — he was certainly physically strong in his early 70s. He always brought his own tent and equipment, which he would donate to good causes at the end of his visit. His personality, humour, and stories are missed by those of us who knew him in Cameroon. He was a devout Episcopalian Christian.

LIST OF FIGURES

CAMEROON PLANT CONSERVATION CHECKLISTS, THE SERIES SO FAR

Martin Cheek

Herbarium, Royal Botanic Gardens, Kew, Richmond, Surrey, TW9 3AE, UK

This book is the seventh so far, in a series of Conservation Checklists for different natural areas in Cameroon. These books have each been produced at the request of different conservation NGOs that are seeking to protect and manage the area concerned.

What is a conservation checklist?

'Normal' plant checklists seek to catalogue all the plant species present in an area, citing specimens as evidence, often with introductory information detailing the history, vegetation types etc. A conservation checklist is so named because it includes a Red Data chapter resulting from screening all the species present for their conservation status. We have used the IUCN (1994) standard in our first two checklists, and the IUCN 2001 standard for later volumes.

Cable, S. & Cheek, M. (eds) (1998). The Plants of Mount Cameroon: A Conservation Checklist. Royal Botanic Gardens, Kew, UK. lxxix + 198 pp.

Cheek, M., Onana, J.-M. & Pollard, B.J. (eds) (2000). The Plants of Mount Oku and the Ijim Ridge, Cameroon: A Conservation Checklist. Royal Botanic Gardens, Kew, UK. iv + 211 pp.

Harvey, Y., Pollard, B.J., Darbyshire, I., Onana, J.-M. & Cheek, M. (eds) (2004). The Plants of Bali Ngemba Forest Reserve, Cameroon: A Conservation Checklist. Royal Botanic Gardens, Kew, UK. iv + 154 pp.

Cheek, M., Pollard, B.J., Darbyshire, I., Onana, J.-M. & Wild, C. (eds) (2004). The Plants of Kupe, Mwanenguba and the Bakossi Mountains, Cameroon: A Conservation Checklist. Royal Botanic Gardens, Kew, UK. iv + 508 pp.

Cheek. M., Harvey, Y. & Onana, J.-M. (eds) (2010). The Plants of Dom, Bamenda Highlands, Cameroon: A Conservation Checklist. Royal Botanic Gardens, Kew, UK. iv + 162 pp.

Harvey, Y., Tchiengué, B. & Cheek, M. (eds) (2010). The Plants of Lebialem Highlands (Bechati-Fosimondi-Besali), Cameroon: A Conservation Checklist. Royal Botanic Gardens, Kew, UK. 170 pp.

Cheek. M., Harvey, Y. & Onana, J.-M. (2011). The Plants of Mefou proposed National Park, Yaoundé, Cameroon: A Conservation Checklist. Royal Botanic Gardens, Kew, UK. 252 pp.

Each entry for a species in our conservation checklists is designed to include information useful to those managing the survival of the species. This includes the range of the species as evidenced by specimens, general information including the assessment, often including a short description to facilitate identification. Habitat and threats are also detailed. Management suggestions are the final entry. Here we list recommended actions to facilitate management of the species for its long-term survival.

Over the years improvements have been made. In the first volume, species descriptions in the main checklist were usually reduced to the habit, or absent, as is usual in a checklist. However, beginning with the second volume, we included brief, hopefully diagnostic descriptions, of one to several lines. In the first two volumes only species assessed as threatened were distinguished. In later volumes all species were assessed for IUCN status.

IUCN (2001). IUCN Red List Categories and Criteria. Version 3.1. IUCN, Gland, Switzerland. 30 pp.

PREFACE

The forest of Mefou is a special place, the only protected area of forest close to Cameroon's capital city, Yaoundé, and this book marks an important step on a journey that has really only just begun, although that journey is already 10 years old.

Mefou came to my attention in 2000 when it was offered by the Government of Cameroon's Ministry of Forestry and Wildlife to Cameroon Wildlife Aid Fund, a British NGO now known as Ape Action Africa (AAA), as a site for a rescue centre for confiscated apes and other primates. Bristol Zoo Gardens has worked in Cameroon, supporting Ape Action Africa, since 1998, and continues to work closely with AAA both in Cameroon and in the UK through our Bristol Conservation and Science Foundation.

Ape Action Africa's story begins with the commercial bushmeat trade; the hunting and trade in wild animals for food. A by-product of the trade for meat is the trade in live animals as pets, particularly young primates. The trade in great apes is illegal and many animals are confiscated by the authorities. These then need care, rehabilitation, and eventually perhaps, for some, reintroduction back to the wild. A British zoologist, Chris Mitchell, travelling in Cameroon, decided in 1997 to set up Cameroon Wildlife Aid Fund, with several objectives; to provide a home for confiscated primates thus supporting important conservation law enforcement, to improve the living conditions of primates held in Cameroon, and to provide an education centre for visitors to learn about the bushmeat trade and to understand the consequences of unsustainable and illegal hunting. Originally based at the Mvog Betsi Zoo in Yaoundé, it was clear that the scale of operation needed was much greater than the zoo could hold, so the charity was offered Mefou, with a governmental commitment to the area being gazetted as a National Park.

The site is just over 10 km^2 and comprises a mosaic of logged forest, fields, forests under-planted with cocoa trees and riverine swamp forest. The wildlife of the forest has been heavily hunted, so that now only little of the original fauna remains. The Mefou National Park, however, represents a unique resource for the Government of Cameroon and the city of Yaoundé. It is a site where some of the original flora remains; an example, albeit a little impoverished, of the ancient forests of the area. In a country with a rapidly growing urban population, increasing habitat loss and unsustainable hunting of wildlife for food, Mefou is a piece of forest to which inhabitants and visitors to Yaoundé can come to rediscover part of the natural heritage of Cameroon.

The site has now developed into a remarkable centre for confiscated primates and for conservation education focusing on the bushmeat trade and primate care and welfare. This survey, conducted by the expert scientists of the Royal Botanic Gardens, Kew, in partnership with those at IRAD-Herbier National Camerounais, reveals as well the high importance of Mefou as an area for plants. It shows that a surprisingly diverse representation of important vegetation types remain here, as well as containing some of the unique plant species of the Yaoundé area. The team have discovered new species and rediscovered others, which underlines the importance of the flora of Mefou and the protection of this forest.

The process of gazetting the National Park is still not complete, but is now in its final stages. When that happens, the next stage of regeneration and restoration of this unique forest resource should start, providing Cameroon with a living example of it natural heritage. This book provides the basis from which that long-term project can be developed.

Dr Bryan Carroll
Director
Bristol Zoo Gardens & Bristol Conservation and Science Foundation

VISITING MEFOU PROPOSED NATIONAL PARK

Martin Cheek

Herbarium, Royal Botanic Gardens, Kew, Richmond, Surrey, TW9 3AE, UK

Reaching the proposed National Park is now easily achieved by travelling about 20 km S on the main Yaoundé–Mbalmayo road (surfaced) from the SW of Yaoundé, passing Nsimalen International Airport on the way. The town of Bikok is indicated on the right, while on the left, a sign for the proposed Mefou National Park should be followed through the village of Ekali, over the river Mefou by the new bridge (opened c. 2005) to Metet, then heading S to the Park Headquarters at Ndanan 1. Both Bikok and Ekali are featured on the 1:200 000 and 1:500 000 maps of Yaoundé, 3rd edition 1972, NA-32-XXIV. Those travelling by public transport should leave Yaoundé from the Mbalmayo gare routiere, and ask to drop at the carrefour for Bikok, or better, Ekali, and then walk the final two to three kilometres. For those with cars the roads are now good in all seasons.

Before the new bridge opened over the Mefou, access was via M'fou town and a maze of country roads making the journey lengthy and uncertain. Thankfully this is no longer necessary.

A charge is made for entry to the proposed National Park. Please contact Ape Action Africa (AAA) for the current rates.

It is important not to assume that Mefou proposed National Park is connected with or adjacent to, the major town of Mefou, to the SE of Yaoundé. Rather, the National Park is named for the River Mefou which forms its boundary. Confusingly, the largest adminstrative centre near to the Park is the town named M'fou.

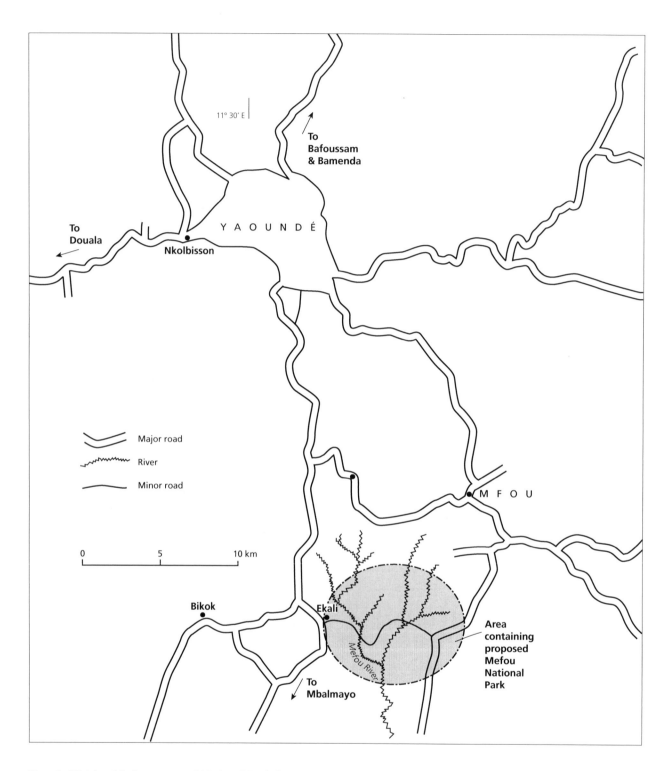

Fɪɢ. 1 Visiting Mefou proposed National Park from Yaoundé

NEW NAME

Martin Cheek

Herbarium, Royal Botanic Gardens, Kew, Richmond, Surrey, TW9 3AE, UK

The following is published in this volume for the first time:

RUBIACEAE

FOREWORD

Martin Cheek

Herbarium, Royal Botanic Gardens, Kew, Richmond, Surrey, TW9 3AE, UK

Mefou proposed National Park is easily reached being situated just south of Yaoundé, on the road to Mbalmayo, near the western edge of the S Cameroon plateau. With its many large enclosure of primates that are being rehabilitated in a forest setting, the Mefou proposed National Park is an attraction to visitors from Yaoundé. But it is more than that. It offers great potential as a centre for environmental education on the forests of Cameroon and their conservation. Mefou proposed National Park is representative of one of the better preserved areas of surviving forest vegetation in the greater Yaoundé area. It is also the only site where the vegetation types are protected. Although much of Mefou's semi-deciduous forest was once logged and farmed, important fragments remain with Red Data species and there are hopes to restore them. Large areas of intact evergreen *Gilbertiodendron dewevrei* forest with even more Red Data species exist in great swathes along the Park's Rivers which feed into the Nyong.

This book describes the 863 species and varieties of flowering plant and ferns discovered so far in the forests of the Mefou proposed National Park. These were found over several visits to the park in 2001–2006, but mainly during two weeks in October 2002 and four weeks of March 2004, by teams of mainly British and Cameroonian research botanists, supported by teams of Earthwatch volunteers from all over the world, when 2405 specimen, sight and photographic records were made. Following study of these specimens at the herbarium of the Royal Botanic Gardens, Kew in London by the authors, and by numerous international specialists, 26 globally threatened plant species were found to be present. These are discussed in detail in the chapter on Red Data.

Surprisingly, 12 plant species new to science were found in the study (see Endemic, Near Endemic and New Taxa), one of which, *Morinda mefou* is formally published in this book. Five of these 12 new species are, so far, known only from Mefou. While it is likely that with further research they will be found elsewhere in the area, we cannot be certain of this and so Mefou is vital if these species are to avoid extinction. It is to be hoped that full National Park status will be obtained soon.

A major purpose of this book is to enable identification of the plant species within the checklist area, in particular those threatened with extinction - the highest priorities for conservation. To this end, details to aid the management of each of the threatened taxa are given in the Red Data chapter, which includes illustrations for as many of the species involved as could be obtained. It is hoped that this information will aid the long-term survival of these threatened taxa.

We believe that this book is the most complete account yet of the plant species and forests, (both semi-deciduous and evergreen *Gilbertiodendron dewevrei* types) of an area around Yaoundé. However, it is far from exhaustive since it results from only two major study periods. It is certain that additional studies in November to February, and April to September inclusive, would reveal new aspects, so far unknown. Such studies would certainly result in the addition of further species to the list.

Our hope is that this book is just the beginning of a long series of more intense studies of the plants of the National Park, and will form a useful baseline for that purpose.

ACKNOWLEDGEMENTS

The following abbreviations are to be found in the text: HNC (IRAD-National Herbarium of Cameroon); IUCN (International Union for Conservation of Nature and Natural Resources); K (Herbarium, Royal Botanic Gardens, Kew); Kew (Royal Botanic Gardens, Kew).

Firstly we wish to thank those who facilitated our fieldwork at the village of Ndanan 1 at the edge of the proposed Mefou National Park, both the father of the village ("Le Père") and his family, especially Belinga, and the staff of the conservation NGO CWAF (Cameroon Wildlife Aid Fund), now AAA (Ape Action Africa), especially Rachel Hogan and Sarah Ndi.

At the National Herbarium of Cameroon, Yaoundé, apart from our co-author, Jean Michel Onana, the present leader, we thank the following, who have joined us in fieldwork at Mefou and/or identified specimens: Fulbert Tadjouteu (Technician); Victor Nana (Technician); Felicité Nana (Accountant). Their biographies as plant collectors are featured in the chapter on the History of Botanical Collection at Mefou.

We would like to thank Julie Mafanny of Limbe Botanic Garden who facilitated Elias Ndive's Darwin Initiative sponsored visit to Kew in Spring of 2008. Elias, amongst a myriad of other work, helped identify Mefou specimens for this checklist.

The Darwin Initiative grant to Kew has been the main single source of funds to our 'Red List Plants of Cameroon' Project (http://darwin.defra.gov.uk/project/15034/), of which this volume is one of the outputs. We are extremely grateful for this assistance, without which specimen identifications and species descriptions which form the basis of this book would not have been completed so swiftly. A substantial part of the publication costs of this book were met by this grant. In particular we thank Ruth Palmer and Helen Beech for their administration of this grant at DEFRA, as well as George Sarkis at Kew for assistance in managing the finances. The Edinburgh Centre for Tropical Forestry, particularly Eilidh Young, contracted by DEFRA to monitor the grant, is thanked for their constructive reviews of our project throughout its life.

The Kew Publications Committee agreed to contribute towards the publication costs and arranged review of the manuscript. Lydia White and Lloyd Kirton are thanked for guidance and for arranging scanning of images and design work by Chris Beard.

At Kew we also thank David Mabberley, Keeper of the Herbarium, Library, Art and Archives (HLAA) and Simon Owens, former Keeper of the Herbarium, who has long supported our work in Cameroon, as has Rogier de Kok, Assistant Keeper for Regional teams, who has also championed our work. Eimear Nic Lughadha has also been consistently supportive of our efforts in Cameroon. We also thank Daniela Zappi and her support and guidance in her role as the previous Assistant Keeper under whom this project was initially run. Helen Fortune-Hopkins, Laura Pearce and Xander van der Burgt have provided valuable sectional support within our team and were key in arranging the Mefou specimens and naming several important families. Laura Pearce is also to be thanked for standardising the economic uses of the plants featured in the main checklist.

Determinations of the specimens gathered in the course of our fieldwork were made by ourselves and by the following botanists who are credited in the individual family accounts. These botanists are often world experts in their fields. We sincerely thank them all and their institutions, for their work, often done without reservation, towards producing the names which are used in this book.

Gratitude is to be given to The Andrew J. Mellon foundation for its continued support to http://www.aluka.org/. Type material stored around the world is instantly accessible as is important literature, speeding up identifications and finding references. In fact, scanning and computing equipment, purchased with a grant from the foundation has proven very useful for naming the unicates stored at HNC, along with exchange of data and accounts for editing.

Thanks goes to Stuart Cable of Kew. For the DFID-supported Mount Cameroon Project in 1997 and 1998, he originated the original species database arrangement from which the checklist part of this book was produced. George Gosline subsequently developed the 'Cameroon specimen' database, enabling us to print off data labels whilst still in the field, a real time-saver. He also developed the system by which data could be exported from the Access database using XML and XSLT into its Microsoft Word format. For meticulous data entry into this database over many years, Karen Sidwell, Suzanne White, Julian Stratton, Benedict Pollard and Harry de Voil are also to be thanked. Without the database this volume would have taken considerably longer.

An enormous debt of gratitude goes to Janis Shillito, one of Kew's volunteers, for typing most of the Introductory Chapters. Ian Turner kindly reviewed the book before it was published.

Our fieldwork would not have been possible without the sponsorship and assistance from Earthwatch. While the Darwin grant and Royal Botanic Gardens, Kew paid for National Herbarium per diem expenses, 2000–2004, Earthwatch provided all the other fieldwork funding for Mefou, apart from an RBG, Kew Overseas Fieldwork Committee grant in 2004. We are grateful to all the Earthwatch volunteers that gave their time to help us at Mefou – these are listed in the chapter on The History of Botanical Collection. Shigeo Yasuda is singled out for special praise since apart from being the best and most diligent specimen-presser at Mefou, he also joined us for weeks at the herbarium, Kew to help with the last stages of completing this book.

Julian Laird, Gill Barker, Lucy Beresford-Stooke and Pamela Mackney of Earthwatch all helped greatly in recruiting various sorts of volunteers for us in Europe, as has Tania Taranovski in the U.S.A.

Emma Fenton, our sandwich student in 2005–2006, is singled out for special praise and thanks. During her year with the Wet Tropics Africa team she did much of the data-basing. She extracted many of the species descriptions, helped with identifications and even visited Mefou briefly in 2006. For these reasons she is co-author on 34 of the family accounts in this book — a major contribution.

THE CHECKLIST AREA OF MEFOU PROPOSED NATIONAL PARK

Martin Cheek

Herbarium, Royal Botanic Gardens, Kew, Richmond, Surrey, TW9 3AE, UK

The checklist area is the boundary of the proposed Mefou National Park as it had been drawn by the AAA/CWAF/MINEF, the co-managers of the area. The accompanying map (Fig. 2) is based on that provided by Chris Mitchell, formerly of Bristol Zoo and then director of CWAF, in 2001. In the meantime it is possible that some boundary changes have occurred.

The Park boundary encompasses an area of about 10 km², extending about 5 km from W to E, and a maximum of 4 km N to S. The Park falls in two main blocks, W and E, separated by a narrowed portion where the N and S boundaries are less than 1 km apart. The western boundary, and part of the southern, is formed by a major branch of the Mefou River. A second branch forms part of the eastern and southern. These two branches unite at the S boundary of the Park and the river flows S for 12 km, joining the Nyong river as it flows west to the Atlantic. The main access to the proposed Park is from the north, via the Yaoundé-Mbalmayo road and Ekali and Metet (see Visiting Mefou Proposed National Park). From Metet it is little more than 1 km to the Park HQ at Ndanan 1, on the northern boundary of the W parcel. Ndanan 1 is a crossoads, with four main tracks meeting, the most important of which is the road which heads SE and bisects the Park, reaching the village of Edzassana and a network of roads beyond. This must be followed to Ndangan 1 (not to be confused with Ndanan 1) if the eastern parcel of the Park is to be reached by old logging roads which were becoming overgrown when we used them in 2004. This eastern parcel, being less accessible to our botanical teams was consequently much less well-surveyed and undoubtedly additional species to those listed in this volume await discovery there.

From our base at Ndanan 1, we primarily explored down the two footpaths to Ndanan 2, the second settlement within the Park boundary, 2 km SE of Ndanan 1, which does not have motorable access. Later we surveyed along the Mefou River itself (the western Park boundary) and also along the bisecting road. The Mbetemanger river (See Fig. 2) was followed upstream from its junction with the bisecting road, into the E parcel. Some collections were also made along the road from Ndanan 1 to Metet, and on towards Ekali.

The general terrain is dominated by what was once semi-deciduous forest rich in timber species but which has been heavily logged in recent decades and which has now been replaced in many areas by fields (and farm fallow) of cassava (*Manihot esculenta*) and other crops. Cocoa (*Theobroma cacao*) and pineapple (*Ananas comusus*) are also planted. By contrast forest along the rivers is evergreen, dominated by *Gilbertiodendron dewevrei*, and is often completely intact since it is poor in commercial timber species, partly or completely inundated in the wet season, and since logging regulations stipulate that forest within some tens of metres of water-courses should be left intact to avoid their damage.

Where the rivers are wide enough (the main branches of the Mefou river) to avoid forest shade, interesting aquatic plant communitites occur, that are also found in *Raphia* swamp-grassland areas such as that found N of Ndanan 2. Vegetation types are discussed at length in the chapter of that name.

All three tracks from Ndanan 1 (but not the access road from Metet) have large enclosures in which orphaned primates rescued from hunters by MINEF patrols elsewhere in Cameroon, and former household pets donated by mainly expatriate families, are being rehabilitated with a view to future reintroduction to the wild. Thus the Park represented much-needed expansion and development space for CWAF/MINEF outside of their Mvog-Betsi zoo in Yaoundé. However it is also the most accessible and nearest natural protected area to Yaoundé, capital of Cameroon and has great potential as a centre for environmental education.

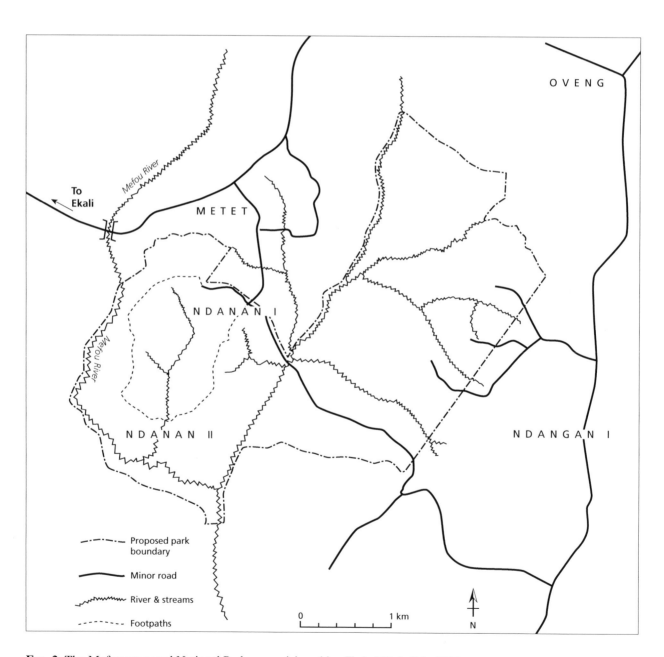

FIG. 2 The Mefou proposed National Park, as envisioned by Chris Mitchell in 2001.

GEOLOGY, GEOMORPHOLOGY, SOILS AND CLIMATE

Martin Cheek

Herbarium, Royal Botanic Gardens, Kew, Richmond, Surrey, TW9 3AE, UK

Mefou proposed National Park lies near the western edge of the South Cameroon Plateau, an area of flat or rolling terrain 600–900 m alt. which occupies most of Cameroon S of the Adamawa Mts and E of the Cameroon Highlands occupying c. 200,000 km^2. To the W altitude drops as the coastal plain is reached. In the SE corner of Cameroon, around the towns of Yokadouma and Moloundou the plateau also gives way to lowlands. The geology of the Plateau is mixed, but that part underlying the S of Yaoundé is ancient Pre-Cambrian base complex of migmatites. These are crystalline acidic rocks. However in the Park area, surface rock is not seen. Instead, deep red ferralitic soils overlay them. When wet, these are soft and clayey. The terrain has a gently rolling aspect and steep slopes are rare. During the dry season when the river bottoms are exposed, they are seen to be of white sand, probably produced by decomposition of the migmatite.

The annual rainfall of Yaoundé (20 km to the N) is 1.529 m per annum, explaining the prevailing semi-deciduous forest, since evergreen forest, as a rule-of-thumb, requires about 2 m per annum unless it is supplied by ground water. There are two main wet seasons, when rainfall exceeds 100 mm per month. The first, March to June inclusive, has its peak in May (183 mm). The second is September to November, with October the highest (288 mm).

Curiously the river beds, either dry (Mbetemanger) or non-flowing, with only pools (Mefou near Ndanan 2) refill in the wet season by water rising from the mother river, the Nyong. For this reason fish traps are pointed downstream, to catch the fish that swim upstream with the refilling rivers.

The data above is mainly derived from Letouzey's Étude Phytogéographique du Cameroun, Edition Lechevalier, Paris (1968), and from personal observations.

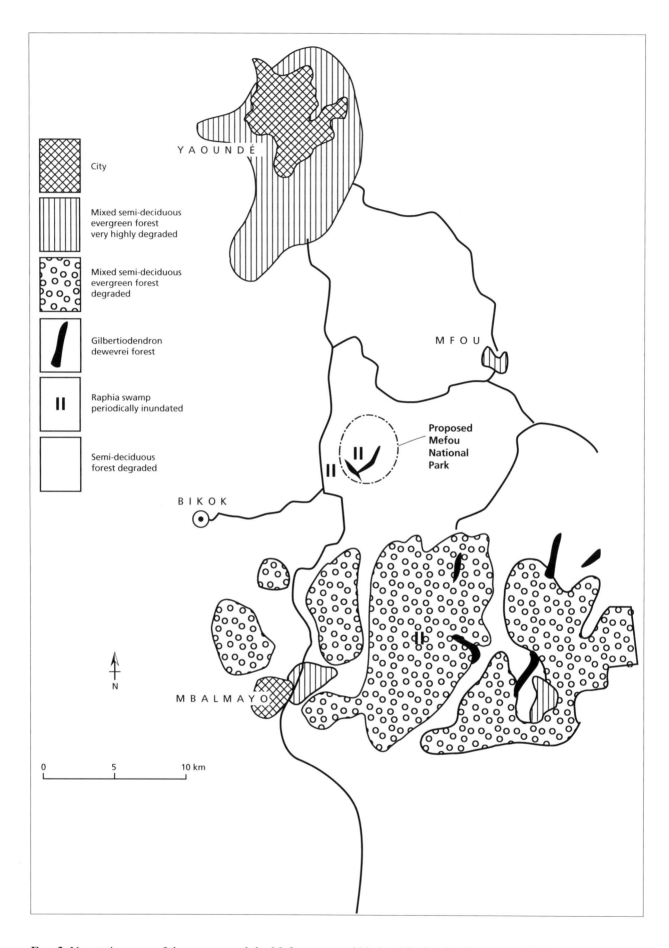

Fig. 3 Vegetation map of the area around the Mefou proposed National Park (after Letouzey, 1985).

VEGETATION

Martin Cheek

Herbarium, Royal Botanic Gardens, Kew, Richmond, Surrey, TW9 3AE, UK

In this chapter the three main vegetation types recognised at the proposed Mefou National Park are characterised:

1) semi-deciduous forest mosaic;

2) evergreen *Gilbertiodendron dewevrei* forest; and

3) open swamp grassland and river-edge communities.

For each of these three, sampling by way of specimen collecting levels is gauged, the physiognomy and species composition of the vegetation is examined, reference is made to endemic and rare taxa, and threats to the vegetation types are listed, and phytogeographical links are discussed. There is also a discussion on the submontane element of the South Cameroon Plateau. Red data taxa for each vegetation type are listed in the introduction to the Red Data chapter. Notes on exotic, cultivated and weed species are given in the chapter on Human Impacts. Links with the Congo basin are discussed in the chapter on the Nyong-Congo hypothesis.

PREVIOUS STUDIES

Prior to the studies presented in this book, no previous studies had been conducted on the vegetation of the proposed Mefou National Park itself. Indeed in the greater Yaoundé area as a whole, remarkably little previous work could be found, apart from that by Dr Achoundong on the submontane forest of the mountains in Yaoundé. Since Mefou proposed National Park does not contain submontane forest per se, not having sufficient altitude, Dr Achoundong's excellent work need not be considered here.

Letouzey (1985) produced the definitive vegetation map of Cameroon. That part corresponding to the Yaoundé-Mbalmayo area is the basis for our Fig. 3. Aerial photography and extrapolation from field observations made elsewhere are thought to be the basis for this superb study which succeeds and supports Letouzey's immense Étude Phytogéographique du Cameroun (1968) which is a lengthy descriptive memoire. Clearly Letouzey did not visit Mefou proposed National Park since although *Gilbertiodendron dewevrei* monotypic stands are present along rivers, he has not mapped it for the area although this vegetation type is shown further south towards Mbalmayo. Nevertheless his work is highly useful for general context and so is referred to in the vegetation treatments below.

MAPPING UNIT 1. DEGRADED SEMI-DECIDUOUS FOREST MOSAIC (c. 600–700 m alt.)

Letouzey (1985) gives his mapping unit 169, equating to degraded semi-deciduous forest, as the dominant vegetation type for the area in and around Mefou. Indeed much of the Park does conform to this definition although there is an evergreen forest element, and certainly at the bottom of the shallow valleys and along valleys, evergreen forest is dominant. So comprehensively degraded is the semi-deciduous forest that it is difficult to delineate its physiognomy and structure with any confidence. Despite this, we did place our 25 m × 25 m Hall & Swaine plot MF2 in this vegetation type immediately N of the village of Ndanan 1 (vouchers were Cheek in MF 202–412). Sampling levels in this vegetation type were high since access was made easy by the larger number of old logging roads and the open and damaged canopy assisted collections.

Semi-deciduous forest takes its name from the phenomenon that, in the dry season, as many as half the trees present might be leafless. This is presumably a device to conserve moisture, needed since rainfall (upon which available moisture forest is often dependent for survival) is lower than in evergreen forest areas. Whereas, as a rule of thumb, 2 m per annum is held to be the minimum to sustain evergreen forest, Yaoundé receives only 1.5 m rainfall per annum. The absence of shade, or at least a great reduction in shade, from the canopy for several months of the year, means that vastly more light penetrates to the forest understorey and floor. As a result differences in both the species composition and frequency can be expected if semi-deciduous and evergreen forest are compared.

The most striking marker for semi-deciduous forest is the presence of a diversity of canopy trees of two families which have been used to name this vegetation type, hence, the forêt semi-décidue à Sterculiacées et Ulmacées of Letouzey (1968: 191). These indicators present at Mefou are:

Celtis adolfi-friderici	(Ulmaceae)
Celtis gomphophylla	(Ulmaceae)
Celtis mildbraedii	(Ulmaceae)
Celtis tessmannii	(Ulmaceae)
Celtis zenkeri	(Ulmaceae)
Holoptelea grandis	(Ulmaceae)
Sterculia oblonga	(Sterculiaceae)
Sterculia rhinopetala	(Sterculiaceae)
Triplochiton scleroxylon	(Sterculiaceae)

These species are totally absent in evergreen forest such as obtains towards the coast, e.g. in Littoral, S and SW Regions. *Triplochiton* is a particularly good indicator since it is frequent and has a distinctive, palmately 5-lobed leaf which is especially easily recognised. Almost all the 16 timber tree species previously logged at Mefou are also semi-deciduous forest indicators (listed in the chapter on Human Impacts) with the exception of *Pycnanthus angolensis*, *Baillonella toxisperma* and *Diospyros crassiflora* which also occur in evergreen forest. Species of *Entandophragma* and *Khaya* can also be found in drier types of evergreen forest, but much more sparsely than in semi-deciduous forest. In some ways 'Sterculiaceae – Ulmaceae' as a label misrepresents the situation since in the understorey of evergreen forest, diverse species of *Cola* (Sterculiaceae) occur, and it is not unknown for the odd species of *Celtis* to be found. However if it is clear that this label is applied to the canopy layer alone, it is succinct and accurate.

The most extensive and profound characterisation of semi-deciduous forest in Cameroon is that by Letouzey (1968: 181–237). Here he documents the many 0.5 ha transects that he conducted in semi-deciduous forest near Deng Deng, Nanga Eboko, Yokadouma-Batouri and Médoum. The similarity in species composition between his findings of this vegetation type and that which we encountered at Mefou is demonstrated by the table below.

TABLE 1. Most frequent 20 semi-deciduous forest tree species per plot of Letouzey (1968: 211) and their absence or presence at Mefou Proposed National Park

Species name used by Letouzey and our probable equivalent, if applicable	Proportion of Letouzey's plots in which species occurred	Present in semi-deciduous forest at Mefou?
Alstonia boonei = A. congensis	9/12	✓
Terminalia superba	8/12	✓
Entandrophragma cylindricum = E. sp.	8/12	✓
Strombosia glaucescens = S. pustulata?	8/12	✓?
Funtumia elastica	8/12	✓
Corynanthe pachyceras	8/12	✓
Celtis mildbraedii	7/12	✓
Erythrophleum guineense = E. suaveolens?	7/12	✓?
Keayodendron bridelioides	7/12	✗
Trichilia zenkeri = T. monadelpha	7/12	✓
Polyalthia suaveolens = Greenwayodendron suaveolens	7/12	✓
Tabernaemontana crassa	7/12	✓
Triplochiton scleroxylon	6/12	✓
Combretodendron macrocarpum = Petersianthus macrocarpus		
Rauvolfia macrophylla	6/12	✓
Mansonia altissima	6/12	✗
Angylocalyx zenkeri = A. oligophyllus?	6/12	✓?
Celtis adolfi-friderici	6/12	✓
Trichilia rubescens	6/12	✓
Enantia chlorantha = Annickia affinis	6/12	✓?

The top 20 species listed above were all recorded as trees >20 cm diam. by Letouzey (1968) in intact semi-deciduous forest, mainly in E Region Cameroon. Name changes in subsequent decades have been indicated for several species (the *Alstonia, Polyalthia, Combretodendron* and *Enantia*). In the case of several others, e.g. the *Erythrophleum, Strombosia* and *Angylocalyx*, the probable equivalents in our checklist have been given, but more research is needed to confirm, hence the insertion of interrogation marks.

In summary it seems likely that 16/20 of the most frequent larger tree species per transect of Letouzey occur at Mefou, although five of these need further resolution and three species appear to be absent or unrecorded from Mefou:

Keayodendron bridelioides

Petersianthus macrocarpus

Mansonia altissima

This absence is probably due to the relatively low plot sampling at Mefou (only one 25 × 25 m plot in semi-deciduous forest) or in the case of *Keayodendron*, possible regional differences between the Mefou area and that in which most of Letouzey's plots were placed.

Mixed evergreen and semi-deciduous forest mosaic

Semi-deciduous forest often occurs mixed with evergreen forest as a mosaic. In practise, in rolling terrain, the tops and slopes of rises often have semi-deciduous trees, while the bottom of the slopes, where more soil moisture can be found, may have few deciduous trees, and evergreen species may predominate. So damaged is the semi-deciduous forest at Mefou that it is difficult to say to what extent it is mixed with evergreen forest, apart from the *Gilbertiodendron dewevrei* forest that occurs along the major watercourses and is dealt with as a separate vegetation type. Mixed semi-deciduous and evergreen forest is mapped as unit 170 (totally degraded) by Letouzey (1985) in the area north of Mefou, immediately around Yaoundé, and as unit 171 (degradation pronounced) to just to the south of the Park, around Mbalmayo.

The understorey of semi-deciduous forest

Letouzey (1968), in characterising semi-deciduous forest, was mainly focussed on canopy trees and not the understorey. However it is clear that substantial differences exist between the understorey taxa of semi-deciduous forest at Mefou and those of evergreen forest elsewhere in Cameroon.

Diversity and frequency of *Combretum* species. *Combretum* species are rare in wet evergreen forest, yet nine specimens amounting to at least three species (*Combretum comosum*, *C. mucronatum* and *C. zenkeri*) were recorded at Mefou, and more taxa are probably present.

Presence of *Grewia* sensu stricto. *Grewia* species are absent from wet evergreen forest in Cameroon (excluding *Microcos* species which are often incorrectly included in *Grewia*). *Grewia pubescens* was recorded several times in semi-deciduous forest at Mefou.

Presence of deciduous shrubs. Typically these are notably absent from evergreen forest yet at Mefou, probably the most frequent understorey shrub is the deciduous *Chazaliella obovoidea* subsp. *villosistipula*.

Higher species diversity of herbaceous climbers such as Cucurbitaceae and Vitaceae.

The geographical range of most semi-deciduous forest species seen at Mefou and in Cameroon is extensive. The canopy tree species such as *Triplochiton scleroxylon*, *Terminalia superba*, *Celtis adolfi-friderici* and *Celtis zenkeri* extend with this vegetation type, in a band between the subsahelian

Guinean grassland and woodland to the N and evergreen forest to the S (where conditions are wet enough). This belt extends in the W to Guinea (Conakry) and Ivory Coast, and in the E to the farthest extremity of Congo (Kinshasa) with some elements reaching Uganda and Tanzania.

Thus most forest in W Africa is or was semi-deciduous. Yet, this vegetation type has been most profoundly affected by man especially when compared to evergreen forest. This is due firstly to the high proportion of timber species, secondly, the richer, or at least less well-leached soils and thirdly, the greater ease of clearing them for agricultural land. More can be found in the chapter on 'Human Impacts' regarding threats to this vegetation type at Mefou.

Conservation priority species (those which are rare or threatened) are rather few in number in this vegetation type apart from widespread timber species which have been ascribed VU status (see Red Data chapter, MELIACEAE). Moreover, it is often difficult to be sure that those conservation species that there are, are not better classified as evergreen — semi-deciduous transition species since so much disturbance has occurred in this vegetation type. Two species that appear clearly to fall into the semi-deciduous category (both being deciduous and occurring on high ground) are:

Phyllanthus kidna ined. CR

Chazaliella obovoidea subsp. *villosistipula* VU

These two species indicate an unexpected local element in this vegetation type at Mefou. The first being unique, so far, to the proposed National Park and the second being known only from Central Region Cameroon, to a little further E in CAR and Congo (Brazzaville).

MAPPING UNIT 2. EVERGREEN *GILBERTIODENDRON DEWEVREI* FOREST (600–700 m alt.)

In contrast to the adjoining semi-deciduous forest, the great bands of evergreen forest that flank the major water-courses in belts 100–200 m wide at Mefou are completely intact (see discussion in 'Human Impacts'). The thick canopy, 30–40 m above the ground, casts dense shade on the forest floor below and produces a thick leaf-litter and a sparse herb and shrub layer, making it easy to move around in the understorey. Generally evergreen forest is far more diverse in terms of species per unit area than is semi-deciduous forest but two species in this vegetation type at Mefou damp-down this diversity. Firstly the canopy trees are dominated by one species, *Gilbertiodendron dewevrei* and secondly the understorey shrubs and trees by *Scaphopetalum zenkeri*. *Gilbertiodendron dewevrei* is a Congolian species at the western extremity of its range here at Mefou. Remarkably 65–90% of all medium and large trees in this vegetation type at Mefou belong to this species. A member of the Leguminosae-Caesalpinioideae, this is the classical example in African forest of a species that forms monotypic stands. Several other Caesalpinioideae species also form monotypic stands in Cameroon (see for example Cheek *et al.* 2004: 28) but they have far smaller ranges and are less ecologically specialised.

Inundated forest at Mefou

Gilbertiodendron dewevrei is restricted to riverine habitats. It appears to be adapted to seasonal inundation. During October the rivers overspill their beds and flood the surrounding forest for as much as 100 or 200 m on each bank, to a depth of up to about 2 m and perhaps more. Over weeks and months the water-level drops and by the end of the dry season, in March, water-flow in the rivers has ceased and large parts of their beds are exposed allowing easy access for botanists. Consequently these river beds are well surveyed at Mefou. Although hundreds of kilometres from the sea at this point, the Mefou River, before it enters the Nyong, is sinuous and meandering, indicating only a low rate of fall. As a result, there appears to be no rheophytes. Moreover there is no exposed rock, nor rapids in the rivers which might support Podostemaceae. This family were actively searched for in 2002 by the writer, Drs Ameka and Rutishauser, without success. The slopes down to the water-courses at Mefou are never steep, and are often very gradual, with no well-defined river-bank, or with only a very low one.

Tree and shrub species at Mefou which appear to be specific to inundated riverine and swamp forest (as indicated by habitat information on other specimens of those species) are as follows:

Trees

Dactyladenia cf. *floribunda* (Chrysob.)

Cleistanthus sp. nov. (Euph.)

Pseudagrostistachys africana subsp. *africana* (Euph.)

Homalium africanum (Flac.)

Garcinia ovalifolia (Guttif.)

Symphonia globulifera (Guttif.)

Ficus vogeliana (Mor.)

Hallea stipulosa (Rub.)

Nauclea diderrichii (Rub.)

Raphia sp. of Mefou (Palm.)

Shrubs & lianas

Cordia dewevrei (Borag.)

Macaranga angolensis (Euph.)

Macaranga saccifera (Euph.)

Necepsia sp. 1 of Mefou (Euph.)

Euphorbia teke (Euph.) New to Cameroon

Octoknema genovefae (Olac.)

Pandanus sp. (Pandan.)

Lasiodiscus marmoratus (Rhamn.)

Pauridiantha liebrechtsiana (Rub.) New to Cameroon

Pauridiantha pyramidata (Rub.)

Urera thonneri (Urtic.) New to Cameroon

Several of these woody species are otherwise, like *Gilbertiodendron dewevrei*, best known from the Congo basin proper, from the same habitat, inundated forest. In fact, several of these species appear to be newly recorded here for Cameroon. The implications of this are discussed in the chapter 'The Nyong-Congo Hypothesis'. Others appear to be new to science (e.g. *Necepsia* sp. 1 of Mefou and *Cleistanthus* sp. nov.) indicating how much more research is needed into this vegetation type in Cameroon. It is unknown to what extent, for example, fish or water is important for seed-dispersal in this vegetation type, as it is known to be important in similar vegetation in the Amazon basin. *Octoknema genovefae* is a rare Cameroonian species restricted to this habitat type. The *Pseudagrostistachys* is newly recorded in Cameroon for lowland inundated forest, being previously known otherwise as a submontane species. An earlier lowland swamp record of this taxon is *Bos* 3747 from tidal flats near Kribi. In contrast, species such as *Homalium africanum, Symphonia globulifera, Ficus vogeliana, Raphia* sp. of Mefou, *Pandanus* sp. of Mefou, are widely spread in both West and Congolian Africa in riverine and swamp forest so it is to be expected that they would be found in inundated forest at Mefou.

Herbaceous species of open habitats along the river bed and swamp grassland are treated as the vegetation type 3 (see below).

It would be a mistake to assume that the specialised species above dominate the evergreen *Gilbertiodendron dewevrei* forest. This is not so. Most of these aquatic specialists only occur along the very edge of the river beds, or in them, or in low-lying swampy stretches. Otherwise the species that occur in this vegetation are standard evergreen forest species that appear to be able to withstand wet conditions, and much of this forest is not even seasonally submerged. The single 25 × 25 m Hall & Swaine plot placed in this forest type illustrates these points:

Plot MF1 3°37'N 11°35'E 13th March 2004

Lowland evergreen forest near slow-flowing, swamp-edged watercourse. Plot slope c. 15° to N and the stream. Alt. 700 m, 5 cm deep leaf-litter of *Gilbertiodendron dewevrei*. Canopy height 30–40 m. Voucher range MF1–MF124.

Trees above 10 cm diam. were 12, eight of which were *Gilbertiodendron dewevrei* (including the largest, at 60 cm, 70 cm, 70 cm and 120 cm diam.). The remaining four trees being:

Staudtia kamerunensis var. *gabonensis* 15 cm diam. (Myrist.)

Cola pachycarpa 12 cm diam. (Sterc.)

Strombosiopsis tetrandra 10 cm diam. (Olac.)

Xylopia acutiflora 15 cm diam. (Annon.)

The main understorey tree/shrub which dominates this layer, as *Gilbertiodendron dewevrei* dominates the canopy, is *Scaphopetalum zenkeri* which commonly makes up 50–80% of the individuals in the 1.5–8 m tall layer. Species of *Scaphopetalum* have also been reported as dominant in the understorey, suppressing species diversity, elsewhere in Cameroon (at Korup) and in Ghana. Nevertheless the understorey is still fairly diverse, and in plot MF2 indicator groups of good-quality evergreen forest, such as the ebonies, *Diospyros* spp., show respectable diversity, with four species, and *Cola* and *Strychnos* also, with three species each. This vegetation type has the greatest number of Red Data species. In plot MF2 the following threatened taxa occurred, all being understorey shrubs or small trees:

Tricalysia amplexicaulis (Rub.) VU

Poecilocalyx setiflorus (Rub.) VU

Cola letouzeyana (Sterc.) VU

These threatened species and others restricted at Mefou to this vegetation type (see introduction to Red Data chapter), such as *Aulotandra kamerunensis*, are restricted geographically to the Bipindi-Ébolowa-Eséka area and see their northernmost extension at Mefou, unexpectedly close to Yaoundé, and at probably their highest altitudes yet recorded.

A very high proportion of the remaining species show the same phytogeographic pattern, being also species of the evergreen forest of the coastal plain, extending from Campo-Ma'an in the E to Korup in the W, and sometimes further afield, even Guineo-Congolian wide, such as *Schefflera barteri*, *Myrianthus arboreus*, *Agelaea paradoxa*.

MAPPING UNIT 3. OPEN SWAMP GRASSLAND AND RIVER EDGE COMMUNITIES
(600–700 m alt.)

These two communities share so many species that they are treated together here. Both are made up largely of sun-demanding aquatic herbs, or at least those restricted to at least seasonally saturated soils. Although these communities abut inundated forest, indeed, are usually enclosed by them, they do not overlap since forest shade is too dense to allow these species to survive. Since the swamp grassland that we studied contains several species of floating herb, it is possible that when the forest is inundated, the swamp grassland areas are resupplied with aquatic species that float-in from the river system.

Periodically Inundated Grassland was mapped as unit 181 by Letouzey (1985). He depicted it at four points between Yaoundé in the N and a line between Akono and Mbalmayo to the S, although he did not show it at the proposed National Park itself.

The origin of this vegetation type is not clear. The absence of mature trees may well be due an iron-pan below the surface which might impede drainage and root penetration, but this is conjecture. At all seasons the ground underfoot is wet, and in October submerged to a depth of 10 cm or more — at least in our study area.

We studied this vegetation type along a 200 m stretch of path between Ndanan 1 and 2, close to the latter on the most direct path between the two. A single 25 × 25 m Hall and Swaine plot was placed here.

Plot RP1 16th March 2004

Raphia swamp-open canopy, soil: black dense wet humus —due to anaerobic conditions. Plot slope flat. Alt. c. 700 m. *Raphia* height 10–15 m, herbaceous layer diverse, to 1.5 m tall. Voucher range RP1–92, fertile specimen range Cheek 11744–11786.

The 134 records amounted to c. 125 species due to some duplication. The herbaceous layer had a surface cover of c. 90% where there were no fallen dead *Raphia* palm fronds. *Cyclosorus striatus* was especially common. There were a few open watery-muddy patches.

Specialist aquatic-swamp species:

Cyclosorus striatus (Pterid.)

Ceratopteris thalictroides (Pterid.)

Torenia thouarsii (Scroph.)

Bacopa crenata (Scroph.)

Struchium sparganophora (Comp.)

Thalia welwitschii (Marant.)

Isachne kiyalensis (Gram.)

Acroceras amplectans (Gram.)

Nephrolepis biserrata (Pterid.)

Pteris similis (Pterid.)

Raphia sp. of Mefou (Palm.)

Cyperus dilatatus (Cyp.)

Cyperus mannii (Cyp.)

Scleria verrucosa (Cyp.)

Aneilema disperma (Commel.)

Floscopa africana subsp. *petrophila* (Commel.)

Ludwigia africana (Onagr.)

Tristemma leiocalyx (Melast.)

Pentodon pentandrus (Rub.)

Oldenlandia lancifolia (Rub.)

Eichhornia natans (Ponted.)

Forest pioneer species

A second, large tranche of species were not aquatic specialists but shrubs & tree seedlings from the surrounding forest, mostly pioneer species attempting to colonise the open area. Since few exceeded 1–2 m, it seems likely that the permanently wet conditions kill them before they can develop further. These occurred thinly scattered, mostly as single individuals within the plot. They are listed here together with their height in the plot.

Leea guineensis (Lee.) 1 m

Boehmeria macrophylla (Urtic.) 1 m

Sterculia tragacantha (Sterc.) 1 m

Lannea welwitschii (Anac.) 2 m

Bridelia micrantha (Euph.) 1 m

Antidesma laciniatum (Euph.) 1 m

Alchornea cordifolia (Euph.) 1 m

Allophylus spp. indet of Mefou (Sapind.) 0.3 m

Dialium cf. *bipindense* (Leg.) 1 m

Zanthoxylum thomense (Rut.) 0.6 m

Alstonia boonei (apoc.) 0.6 m

Macaranga monandra (Euph.) 1 m

Macaranga spinosa (Euph.) 0.4 m

Sterculia oblonga (Sterc.) 0.5 m

Markhamia lutea (Bignon.) 1.5 m

Alangium chinense (Alang.) 3 m

Xylopia rubescens (Annon.) 3 m

Elaeis guineensis (Palm.) 0.5 m

Harungana madagascariensis (Guttif.) 3 m

Ceiba pentandra (Bombac.)

Numerous forest liana and other climbing species also occurred in the plot, as did some general herbs of waysides. These are not documented here for reasons of space.

None of the species in the plot or elsewhere in the swamp was endemic to Cameroon or of global conservation importance, most being found throughout tropical Africa, or at least being Guineo-Congolian in distribution.

Species of open river bed and associated swamp

Trekking along the river beds in March 2004, many of the above mentioned aquatic species, thinly scattered, were recorded again on an opportunistic basis. No plots or transects were executed. Most of the surface area was bare, having been submerged in opaque water until only a few months previously, so most of these herbaceous species would have stayed dormant, as seeds, over the wet season probably being unable to photosynthesise under water, although this should be verified. One long-standing hypothesis is that many of the World's annual tropical weeds originated in this niche where high seed production and a short life-cycle, with fast growth in full sun must be selected for strongly. In contrast herbs of forest shade are all perennials.

Additional perennial swamp species:

Lasiomorpha senegalensis (Arac.)

Pneumatopteris afra (Pterid.)

Crinum jagus (Amaryllid.)

Additional short-lived perennial and annual swamp and river-edge species:

Impatiens sp. aff. *mannii* (Balsam.)

Ludwigia abyssinica (Onagr.)

Polygonum setulosa (Polygon.)

Rhynchospora corymbosa (Cyp.)

Cyperus cylindrostachys (Cyp.)

Additional partly submerged or floating aquatic species:

Nymphaea lotus (Nymph.)

Caldesia reniformis (Alismat.)

Additional annual river bed or swamp species:

Ranalisma humile (Alismat.)

Pogostemon micangensis (Labiat.) New to Cameroon. CR

Lobelia gilletii (Camp.)   New to Cameroon. EN

These species are also widespread in W and/or Central Africa wherever aquatic conditions obtain. However the last two species are exceptions to this observation, both being extremely rarely recorded. *Pogostemon micangensis* was previously only known from the S of the Congo basin in Northern Angola, while *Lobelia gilletii* a rarity of Congo (Kinshasa) has only recently been discovered in Gabon. Both are prostrate herbs, apparently annual.

Inundated habitats and aquatic communities are still incompletely explored at Mefou.

Aquatic absences at Mefou

It is worth noting that *Utricularia*, *Genlisea*, *Xyris*, *Eriocaulon* and *Drosera* were not found in wet habitats at Mefou. It may be that seasonal water-level fluctuations are too high, the vegetation in inundated grassland/*Raphia* swamp is too dense or that nutrient levels are too rich for this group of small herbs. Free-floating aquatics were also not found, possibly because no areas of perennially still water are present, but, they may yet be discovered. The absence of Podostemaceae and other rheophytes has already been discussed.

The Submontane Element at Mefou

Unexpectedly, the following species usually found in submontane forest (800–2000 m alt.) in the Cameroon Highlands were encountered at Mefou (600–700 m alt.) in evergreen *Gilbertiodendron dewevrei* forest.

Polyscias fulva (Aral.)

Myrianthus preussii subsp. *preussii* (Cecrop.)

Heckeldora ledermannii (Meliac.)

Santiria trimera (Burser.)

In swamp or inundated forest:

Pseudagrostistachys africana subsp. *africana* (Euphorb.)

Symphonia globulifera (Guttif.)

Carapa sp. of Mefou (Meliac.)

Alangium chinense (Alang.)

Zenkerella citrina (Leg.- Caes.)

Several other records of *Alangium chinense* occur elsewhere in the South Cameroon Plateau. Otherwise this species is absent further east, throughout the Congo basin, until the Albertine Rift is reached. *Heckeldora ledermannii* appears to be newly recorded here for the S Cameroon Plateau being otherwise restricted to the Cameroon Highlands. The *Zenkerella*, *Polyscias* and *Myrianthus* listed appear provide supporting evidence for a weak but definite submontane element in the S Cameroon Plateau.

A curious phenomenom is that a few submontane forest species while absent from normal lowland forest, are not exclusively submontane, occurring as they do also in lowland swamp forest, *Symphonia globulifera* is such an example and is present in swamp at Mefou. New additions to this guild, discovered at Mefou are *Pseudagrostistachys africana* subsp. *africana* — found at Mefou along a river bed with *Raphia* and *Zenkerella citrina* found on a river bank and in *Gilbertiodendron* forest. The explanation for such a disjunction may be a dependence on continuously saturated soils.

THE NYONG-CONGO HYPOTHESIS

Martin Cheek

Herbarium, Royal Botanic Gardens, Kew, Richmond, Surrey, TW9 3AE, UK

Today the Congo basin drains to the Atlantic through the lower Congo River that flows west of the cities of Brazzaville and Kinshasa through the band of coastal hills and low mountains to the sea. For over a hundred kilometres, this river section comprises the most violent and continuous set of rapids imaginable- an indication that this outlet is geologically very recent. Previously another outlet must have existed. In an earlier paper it was conjectured that this might have been the Sanaga since this would help explain why it is such an important boundary today for animal species (Cheek, M. (2008), Nigerian Field 72 (2): 93–100).

Had the Congo basin drained through the Sanaga then that river would have been much broader, and more of a barrier to animals than it is today. Cheek (2008) addressed the Sanaga as a barrier to plant species. In that paper evidence from Onana pointed at the Nyong being more important as a barrier or boundary for his specialist group of plants, the Burseraceae. At that time we did not speculate on the possibility that the Congo Basin had drained through the Nyong instead of the Sanaga but this is the hypothesis considered here, prompted by new evidence uncovered in researching the plant species of Mefou proposed National Park. Several plant species recorded at Mefou have proven to be new records for Cameroon being previously only known from inundated forest or river beds in the Congo basin, specifically Congo (Kinshasa) itself, the northern part of Congo (Brazzaville), the southern extension of CAR and the northern part of Angola (see vegetation chapter). These are:

Euphorbia teke (Euphorb.)

Pauridiantha liebrechtsiana (Rub.)

Urera thonneri (Urtic.)

Pogostemon micangensis (Labiat.)

In addition, several other species have been found in Mefou along the river which are species primarily of inundated forest in the Congo basin but which are already known in Cameroon in the southern part of the S Cameroon Plateau, along the river Nyong and its tributaries and along the Dja and its tributaries. Examples are:

Cordia dewevrei (Borag.)

Tetracera rosiflora (Dillen.)

Macaranga angolensis (Euphorb.)

Macaranga saccifera (Euphorb.)

Pauridiantha pyramidata (Rub.)

Lasiodiscus marmoratus (Rhamn.)

Are these evidence that the Nyong was once connected to the Congo basin drainage and acted as a conduit for species from the extensive inundated forests of the Congo basin? We have seen in the chapter on Geology, Geomorphology, Soils and Climate how the rivers in Mefou, empty and mostly dry in the dry season, fill 'backwards' from the Nyong, carrying fish (Hence Mefou fish traps point downstream, to catch fish swimming upstream). It is possible that propagules of plants of riverine communities are transported upstream in this manner, either with the 'rising tide' or within the spawning fish.

Letouzey's work (1968 loc. cit.) on the mapping of *Gilbertiodendron dewevrei* provides further support for the Nyong-Congo hypothesis. As we have seen in the vegetation chapter this tree is specific to riverine forest where it forms monotypic stands and appears to be favoured by seasonal inundation. Fig. 4 reproduces Letouzey's distribution map showing the distribution in Cameroon of this largely Congo basin species. Areas of *Gilbertiodendron* forest are shown in black and their relationships to rivers seem clear enough. It is also clear that the headwaters of the Nyong and those of the Congo drainage system interdigitate repeatedly in the southern part of the S Cameroon Plateau. Fig. 4 shows that at several points, such as E and S of Abong Mbang or W of Sangmélima, and in the case of the Campo river (also a credible alternative), at Mintom, only a few kilometres separate the drainage systems. With the higher water table that would have obtained in the Congo Basin before the present Congo River had cut through the coastal ridge, it is possible that the Congo drainage did drain through Cameroon via the Nyong and possibly the Campo also. The distribution of *Gilbertiodendron dewevrei* shown by Fig. 4 supports this since it crosses from the Congolian drainage of the Dja to the upper system of the Nyong without any alteration in density or frequency of the mapped populations. At first glance one would assume that the drainages were connected from the continuity of this river dependent species. Letouzey's map (our Fig. 4) further illustrates the dependence of *Gilbertiodendron dewevrei* on riverine transport for colonisation. The species is absent from the Sanaga, Cameroon's largest and most extensive river, although there is no doubt that there is suitable habitat within its drainage and that, if introduced it would thrive.

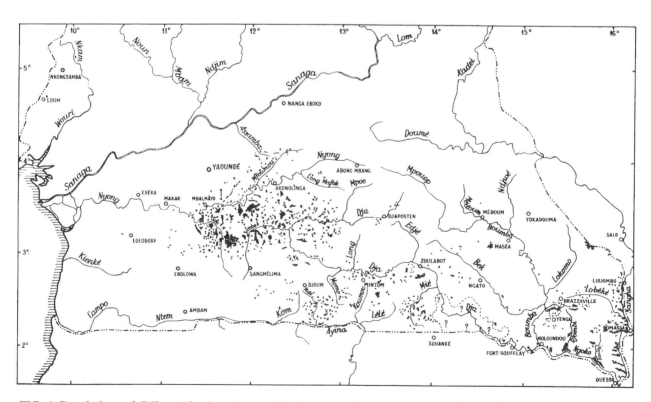

FIG. 4. Populations of *Gilbertiodendron dewevrei* in Cameroon (from Letouzey, 1968 carte 28, p. 251).

HUMAN IMPACTS AT MEFOU PROPOSED NATIONAL PARK

Martin Cheek[1] & M. Essama[2]

[1]Herbarium, Royal Botanic Gardens, Kew, Richmond, Surrey, TW9 3AE, UK
[2]Ministry of Environment and Forests, Yaoundé, Cameroon.

In common with other areas around Yaoundé, the forest in Mefou proposed National Park has been impacted by logging for timber. Many of the existing roads would have been constructed to allow access to timber trees. Some of these roads are overgrown and are reduced to tracks. Once the forest has been opened up, large trees removed, and access created, conditions for developing agriculture are improved, whether slash and burn or plantation, and this has certainly occurred at Mefou, although not with the same intensity as in some neighbouring areas.

TIMBER EXTRACTION

Logging within the confines of the proposed Park is reported to have ceased when the National Park project began in 1999. Certainly, we saw no evidence of current logging in 2002–2006. Some areas remain unlogged, such as that between Ndanan 2 and the Mefou River. The high value of the semi-deciduous forest (dominant at Mefou) in providing timber can be seen in the diversity of species (16) reported as having been logged. These species are listed below with their local/trade names. The value of these species varies greatly. ENEN (*Diospyros crassiflora* or ebony), while extremely valuable, is slow-growing and rarely found as a useful timber tree, while *Pycnanthus angolensis*, a common and fast-growing pioneer species, is much less valuable. Several species, such as *Ceiba pentandra*, *Celtis* spp. and *Distemonanthus benthamianus* were reported as being logged 20–25 years ago but as being no longer cut. One or two timber species that we recorded from specimens, such as *Pterocarpus soyauxii* (Padouk) are not among the 16 logged species noted, either by oversight or because they rarely occurred at Mefou as loggable individuals. Several of the 16 species (8, 9, 14, and 16 below) were not picked up in our specimen-based survey so do not always feature in our list although some have been inserted as sight records.

Timber species formerly logged at Mefou Proposed National Park.

1.	*Sterculia rhinopetala*	NKANA	(Sterc.)
2.	*Milicia excelsa*	ABANG	(Morac.)
3.	*Terminalia superba*	AKOM	(Combret.)
4.	*Triplochiton scleroxylon*	AYOUS	(Sterc.)
5.	*Baillonella toxisperma*	MOABI or ADZAP	(Sapot.)
6.	*Lophira alata*	AZOBÉ	(Ochn.)
7.	*Pycnanthus angolensis*		(Myristic.)

8. *Khaya ivorensis*	NGOLONG	(Meliac.)
9. *Entandophragma* sp.	ATOMASSI	(Meliac.)
10. *Lovoa trichilioides*	BIBOL	(Meliac.)
11. *Canarium schweinfurthii*	ABEU	(Burs.)
12. *Guarea cedrata*	EBEGBEMUR	(Meliac.)
13. *Guibourtia demeusei*	ESINGHAM	(Leg.-Caes.)
14. *Guibourtia tessmannii*	OVON	(Leg.-Caes.)
15. *Diospyros crassiflora*		(Eben.)
16. *Nauclea diderrichii*	AKONDOK	(Rub.)

It is important to note that the evergreen riverine forest that we surveyed at Mefou, has not been impacted by logging.

AGRICULTURE AND HORTICULTURE

During 2002–2006, small-holder agriculture (cassava, sweet potato, cacao and maize) continued in some areas, while some extensive plots were cleared for agriculture in former semi-deciduous forest along the road bisecting the reserve from Ndanan 1 southeastwards, and in 2006 areas of *Raphia* swamp were found to have been recently burnt, presumably to clear for planting. While these activities present a threat to the survival of some rare plant species, our impression was that the level of such activities was static or declining within the Park area.

The activities of man, in clearing and cultivating forest has allowed the invasion of a number of mainly non-native weed species, most of which occur in similar situations throughout the world. The main single source of these species is Central and S America. Should, as is expected, the native forest recover and recolonise in damaged areas at Mefou, these weed species will be gradually shaded out (since all are light-demanding), and become less common than today, and may even be lost to the Park.

Mefou weed species of cultivation, habitation and waysides.

Asystasia gangetica	(Acanth.)	*Ipomoea tenuirostris*	(Conv.)
Aerva lanata	(Amaranth.)	*Acalypha paniculata*	(Euph.)
Amaranthus dubius	(Amaranth.)	*Euphorbia hirta*	(Euph.)
Drymaria cordata	(Caryoph.)	*Euphorbia thymifolia*	(Euph.)
Ageratum conyzoides	(Comp.)	*Tragia volubilis*	(Euph.)
Chromolaena odorata	(Comp.)	*Chamaecrista mimosoides*	(Leg.-Caes.)
Conyza bonariensis	(Comp.)	*Senna obtusifolia*	(Leg.-Caes.)
Melanthera scandens	(Comp.)	*Mimosa pudica*	(Leg.-Mim.)
Tithonia diversifolia	(Comp.)	*Calopogonium mucunoides*	(Leg.-Pap.)
Ipomoea alba	(Conv.)	*Centrosema pubescens*	(Leg.-Pap.)
Ipomoea mauritiana	(Conv.)	*Urena lobata*	(Malv.)

Tristemma mauritianum	(Melast.)	*Paspalum conjugatum*	(Gram.)
Oxalis barrelieri	(Oxalid.)	*Paspalum paniculatum*	(Gram.)
Passiflora foetida	(Passifl.)	*Sporobolus tenuissimus*	(Gram.)
Eleusine indica	(Gram.)	*Stachytarpheta cayennensis*	(Verb.)

Non-crop species deliberately introduced by man at Mefou are as listed below. This list is far from exhaustive since our survey was focussed not on cultivated, but on wild species. Several of these species were found persisting in forest at the site of an abandoned village in the eastern parcel of the reserve (e.g. *Bauhinia* cf. *monandra*, and *Euphorbia cotinifolia*). In time it is expected that they will die out as they are out-competed by native forest species. Without the intervention of human settlement it is thought that in a central African forest context, none of these species is likely to be invasive. Certainly, no evidence of species invading intact forest was seen during our 2002–2006 surveys.

Introduced, non-crop, Mefou species and their uses

Crescentia cujete	(Bignon.)	Shade tree/Fruit
Terminalia mantaly	(Combret.)	Shade tree
Euphorbia cotinifolia	(Euph.)	Ornamental/Hedge
Codiaeum variegatum	(Euph.)	Ornamental/Hedge
Ocimum gratissimum subsp. *gratissimum* var. *gratissimum*	(Lab.)	Flavouring food
Bauhinia cf. *monandra*	(Leg.-Caes.)	Ornamental
Senna alata	(Leg.-Caes.)	Medicinal/Ornamental
Talinum fruticosum	(Port.)	Food
Brugmansia suaveolens	(Solan.)	Ornamental/Hedge
Capsicum annuum	(Solan.)	Flavouring food
Nicotiana tabacum	(Solan.)	Tobacco
Mangifera indica	(Anac.)	Fruit

CONTRIBUTION DE LA LANGUE EWONDO A LA CONNAISSANCE DES PLANTES DU SUD-CAMEROUN

Jean Michel Onana & Paul Mezili

Herbier national du Cameroun
BP. 1601 Yaoundé

Au Cameroun, coexistent près de 250 ethnies que l'on distingue par leur langue et cultures. Géographiquement, le sud-Cameroun est situé au sud du plateau de l'Adamaoua et à l'est de la ligne montagneuse de l'ouest. C'est un plateau couvert par la forêt dense humide semi-caducifoliée. Les habitants sont connus comme étant les peuples de la forêt. La forêt de Mfou fait partie de ce grand ensemble.

ETHNOGRAPHIE

Les peuples de la forêt sont des Bantous que l'on trouve en Afrique centrale. Selon Breton (1979) les peuples de Yaoundé et alentours sont du groupe Beti-Fang et dont les langues véhiculaires sont l'ewondo et le bulu. Au sud de Yaoundé, dans la zone de Mfou, les tribus sont les ewondo et les bané. Le mode de vie et l'économie sont basés sur l'usage et l'exploitation des ressources de la forêt, en particulier les plantes. En effet, traditionnellement les populations vivent essentiellement de ramassage des fruits, récolte des feuilles et écorces pour se nourrir ou se soigner. Le bois a longtemps servit à la fabrication d'objet courant tel que les pirogues, les cuillers, les meubles et la construction des cases. Les cases traditionnelles sont construites avec des piquets dont le bois particulièrement résistant est des espèces ewome (*Coula edulis*) ou mvanda (*Hylodendron gabunensis*) reliés entre eux par des rachis de feuilles de palmier et attaché par des rotins (*Calamus* sp., *Laccosperma* sp., *Eremospatha* spp.), les espaces sont alors comblées par de la terre humide pétrie par piétinement, le toit dont la charpente est faite les rachis de feuilles de raphia appelé ici "bambou" étant recouvert de limbes des folioles de raphia savamment tissés pour être imperméables: ce sont les cases en "poto-poto" caractéristiques dans les villages du sud Cameroun. De plus en plus les populations sédentaires pratiquent l'agriculture aussi bien de subsistance que de rente (cacaoyer). La chasse au gibier ou la pêche traditionnelle sont réduits à leur plus simple expression, les animaux n'étant plus représenté que par les rongeurs, souvent l'aulacaude (appelé hérisson) et le rat palmiste, les rivières étant peu poissonneuses. Il y a un petit élevage traditionnel fait essentiellement de poules, chèvres ou porcs.

L'empreinte humaine des peuples beti à la forêt est liée au mode de vie traditionnel des populations. La pratique de l'agriculture itinérante sur brûlis est fréquente, ce qui favorise la destruction du couvert végétal, la mise à nu puis l'érosion des sols qui s'indurent et deviennent peu propices aux cultures et plus tard l'installation des pestes végétales tel que l'"afan bikorgo" (*Chromolaena odorata*) qui signifie littéralement la plante des jachères de forêt. De même une pression est parfois exercée dans les milieux humides avec la destruction des raphiales pour la culture du riz. Dans l'ensemble cependant l'empreinte humaine sur le milieu naturel reste faible, si bien que le mode de vie traditionnel ne constitue pas un danger véritable pour l'environnement.

ETHNOBOTANIQUE

Au quotidien, les plantes jouent un rôle important dans la vie culturelle des populations, au point où certaines plantes ont des valeurs symboliques ou réservés à certains usages traditionnels. Dejà parmi les attributs du pouvoir traditionnel, il y a entre autre l'akwpahak (chasse -mouche) fait de pétiolules des folioles de palmier à huile et la canne souvent fait de tombo (*Carpolobia alba*).

Le patrimoine immatériel est important. Des arbres ont une valeur mystico religieuse reconnue, cas de 'l'essigang ou oveng (*Guibourtia tessmannii*) qui est reconnu disparaître en forêt lorsqu'on l'approche. Aussi la possession de l'écorce ou des meubles en bois est supposée protéger, éloigner les mauvais esprits de la maison ou les mauvais sorts, si bien qu'un rite particulier est recommandé avant de l'abattre. Parfois seul les initiés doivent le faire. *Codiaeum variegatum* appelé communément croton ou "fleur des tombes" est souvent planté à l'endroit des cimetières pour embellir les tombes et rendre plus gai la demeure des ancêtres; *Eremomastax speciosa* est réputé chassé les sorciers; les tests pour confondre les menteurs et les voleurs ont souvent été faits par la boisson d'un cadi à base de l'écorce de l'élôn (tali, *Erythrophleum suaveolens*) qui entraîne la mort si la personne est coupable. Le palmier à huile (*Elaeis guineensis*) dans toutes ses parties a un rôle culturel important: les feuilles placées le long des chemins sont un signe pour réserver un accueil chaleureux à des hôtes importants, elles sont placées devant les maisons dont les occupants sont en fête ou pour indiquer les lieux de vente des vins de palme ou de raphia, le vin de palme étant la boisson traditionnelle par excellence qui doit être consommée dans toutes les cérémonies, même de la vie moderne, de préférence dans des calebasses dont certaines ont des formes de bouteille et alors appelés ndeg (*Lagenaria siceraria*), les noix produisent une huile et un jus pour les plats traditionnels et la cuisson est faite au feu de bois. Il est d'ailleurs établi que le goût des plats traditionnels est authentique seulement lorsqu'on a utilisé le foyer de bois plutôt que le foyer moderne au gaz domestique.

Les mets (gâteaux traditionnels) d'arachide ou de concombre (egusi) ou de feuilles de manioc sont emballés dans des feuilles de bananier ou de nken (*Thalia welwitschii*) ou de okoé (*Megaphrynium macrostachyum*), ceux-ci étant irremplaçables pour une bonne cuisson et un goût agréable.

Du côté esthétique et les cérémonies festives, des colliers traditionnels appelés "missanga" sont fabriqués avec les graine de mvenduu (*Coix lacryma-jobi*) et on se badigeonne le corps avec le "ba'a" qui est une poudre fait de la pulvérisation du bois de cœur de mbel (*Pterocarpus soyauxii*). Les instruments de musique sont le mvet (guitarre) et le balafon (xylophone) fabriqués à base de bois et les caisses de résonance sont les calebasses (*Lagenaria siceraria*) Les enfants de leur côté utilisent à la moelle des bambous (en fait la moelle des rachis de *Raphia* spp.) pour fabriquer divers objets de jeux tel que des voiturettes.

Ce sont les femmes âgées à la recherche de plantes alimentaires et les soins des bébés et surtout les guérisseurs ou tradipratherapeutes qui ont développés un savoir local en botanique et détiennent le savoir local.

CONTRIBUTION DE LA LANGUE EWONDO A LA BOTANIQUE

Comme le souligne Mbida (2006), les Fang-Betis ont développé un riche patrimoine à travers la connaissance des plantes et de leurs vertus thérapeutiques.

Morphologiquement, une distinction nette est faite des types biologiques: ele = arbre, alen = port de palmier, ndig = liane ou nkol = corde, elog = herbe, zeng = fougère.

Du point de vue taxonomique, c'est à travers les différents usages que les populations ont appris à connaitre et à nommer les plantes. La nomenclature n'est pas codifiée, mais en général, elle repose sur deux principes:

1) les noms ont sont en fait de rang génériques. On trouve ainsi le même nom pour les espèces du même genre, cas de "mevini" (bois noir) qui désigne toutes espèces de *Diospyros* et donc les Ebenacées, abam qui désigne *Chrysophyllum* spp. et autres Sapotaceae, ou alors la plupart des espèces d'une famille: "awonog" désigne ainsi la plupart des Sapindaceae (à cause de ceci, "awonog" est considéré par les prospecteurs qui identifient les arbres avec les noms en langue ewondo comme l'espèce d'arbre la plus répandue dans les forêts camerounaises);

2) les appellations varient selon les usages ou les ressemblances morphologiques ou d'usage. En effet la nomenclature en ewondo est liée à l'usage de la plante et celle sans usage sont souvent sans dénomination. Quelques exemples:

• la même plante porte différents noms selon l'usage de l'organe. Cas du manioc ou le tubercule (racine) est appelé:"mbong" alors que les feuilles sont appelés du nom du plat qu'elles compose "kye kpwem"; de même "otu ele" qui signifie l'arbre à encens (otu) est l'autre nom de "abel" qui désigne le bois du même arbre tous deux désigant *Canarium schweinfurthii*.

• pour les ressemblances: "eteng" désigne *Pycnanthus angolensis* et " nom eteng" désigne *Coelocaryon preussii* car ils se ressemblent ou alors *C. preussii* est le mâle (nom = mari ou qui ressemble en ewondo) de *P. angolensis*. Ainsi le préfixe "nom" est placé pour noter l'affinité entre deux espèces.

• par similarité d'usage: les "Ekop" qui désigne tous les arbres dont l'écorce est succeptible de s'arracher en large lanière comme des tissus végétaux. Il désigne alors environ 19 espèces différentes appartenant à près de 9 genres des Legumineues-Cesalpinoidées (Letouzey & Mouranche, 1952).

A fil de l'histoire, la fang-beti ont ainsi nommé la plupart des arbres de la forêt de leur environnement au point où aujourd'hui, il existe une liste de correspondance de noms ewondo-bulu avec les noms scientifiques qui compte près de 800 noms. Cette liste non éditée est cependant la référence pour les inventaires forestiers, la plupart des prospecteurs indigènes connaissant les arbres souvent uniquement en ewondo ou bulu et les botanistes s'y référent pour connaître les noms scientifiques. Comme les noms en langue pigmée (Letouzey, 1976) la langue ewondo apporte une contribution de grande valeur aux études taxononomiques et écologiques de la forêt camerounaise. Ce document précieux devrait être publié afin de valoriser ce savoir traditionnel et de le pérenniser à travers le temps. Ce sont à la fois les populations et la communauté scientifique qui en profiteront.

REFERENCES

Breton, R. (1979) Ethnies et langues: 31–33. In Laclavère, G. & Loung J F.(eds) *Atlas de République Unie du Cameroun*. Editions J.A. Paris.

Letouzey, R. & Mouranche, R. (1952). *Ekop du Cameroun*. Publication n°14 du Centre technique forestier tropical. Centre technique forestier tropical. Nogent Sur-Marne. France

Letouzey, R, (1976) *Contribution de la botanique au problème d'une éventuelle langue pigmée*. Selaf. Paris

Mdida, M.C. (2006) Patrimoine culturel: 70. In Ben Yamed, D. (ed.) *Atlas de l'Afrique*. Cameroun. Les Editions J.A. Paris.

THE EVOLUTION OF THIS CHECKLIST

Martin Cheek

Herbarium, Royal Botanic Gardens, Kew, Richmond, Surrey, TW9 3AE, UK

The impetus for this book came from Naah Sylvestre Ondoa, then Minister for the Environment and Forestry for the Govt. of Cameroon. During his official visit to the Herbarium, RBG, Kew, in Feb. 2001, having been presented with the newly published book "The Plants of Mount Oku and the Ijim Ridge, A Conservation Checklist", he requested RBG, Kew's assistance in producing something similar for the proposed Mefou National Park, for which he had been the champion (see Frontispiece of this book, and also, Kew Scientist 19: 3 April 2001: "Cameroonian Minister Visit".). Later that year, in Nov. 2001, RBG, Kew's Benedict Pollard was sent to Mefou proposed National Park to execute a reconnaissance for possible future survey missions, supported by Chris Mitchell, then Director of CWAF. Recently CWAF has been renamed as AAA (Ape Action Africa) as explained in the Preface. As the project succeeded, more space was needed. And so, thanks to the then Minister of MINEF, the proposed Mefou National Park had come into being. Supported by Bristol Zoo, CWAF, in partnership with MINEF, had, in 2000–2001 established large enclosures in the area in which to rehabilitate rescued primates.

Following Benedict Pollard's reconnaissance to the proposed Park in late 2001, application was made to the Earthwatch Institute, Oxford, to include the Park in the series of Kew – National Herbarium expeditions in Cameroon, which at that time, Earthwatch had supported for many years. This application being successful, we ran our first two week, Earthwatch supported, survey in October 2002. By this time Chris Mitchell had left CWAF and Bryan Carrol at Bristol Zoo was our main contact with the project. In March 2004 we redoubled our efforts with two consecutive teams spending a total of four weeks surveying until early April. Having completed in late 2004 our "Plant Diversity of Western Cameroon" project, we applied in 2005 for support for a further project with the National Herbarium of Cameroon, entitled the "Cameroon Red Data Book Plants" project within which there was scope to excecute three more Conservation Checklists for areas in Cameroon. It was decided that one of these Checklist projects should be focussed on the proposed Mefou National Park.

Specimen identifications were spear-headed by Dr Jean Michel Onana (see biography in 'History of Botanical Collection', this book), head of the National Herbarium Cameroon, who visited Kew for this purpose in July and August 2005 under Bentham-Moxon sponsorship. Identifications were continued in October 2005 by Charlotte Couch and Emma Fenton. In June-July 2006 under Darwin Initiative sponsorship, Dr Onana continued. Apart from Kew staff, from around the globe, numerous World specialists in different African plant groups were involved in identifications. Eberhard Fischer (Scrophulariaceae), Mats Thulin (Campanulaceae), Carel Jongkind (Combretaceae), Geoffrey Mwachala (Dracaenaceae), Frans Breteler (Dichapetalaceae), David Harris (Zingiberaceae), Paul & Hiltje Maas (Costaceae) are examples. These are credited as coauthors inside the Checklist under the different family accounts. Identifications were not conducted continuously, full-time, but as resources and the needs of other projects allowed. During 2006 Emma Fenton did most of the databasing of the identifications. During the field surveys of 2002 and 2004, we had the impression that very few Red Data (conservation priority) species were present at Mefou, and there was no impression of species new to science being found, such was the degraded state of the semi-deciduous forest areas which in any case was a forest type not renowned for its conservation significance. The evergreen forest, while intact and more species-rich, was clearly also much less diverse than that in the coastal forests or in the submontane forest of the Cameroon Highlands. Despite these initial conclusions based on field observations, new species to science began to emerge as the identifications progressed. Most spectacular was the discovery of what we have come to regard as the 'flagship species' of the proposed National Park, kidna, *Phyllanthus kidna*. This tree is not only new to science, so far unique to Mefou, but also has a great interest for science, having a strange combination of characters and being shown by molecular phylogenetics to be basal within its genus with its closest living relative in India, not Africa. Discoveries such as these prompted a return to Mefou for a few days in March 2006 in order to take photographs to illustrate this book and to collect more data on *Phyllanthus kidna*. However, a steady trickle of other new or very rare species continued to emerge as identifications progressed.

By 2009 most of the identifications were complete. However, it was realised that two major gaps remained.

Among the top set (most complete) the survey specimens that had been extracted at Yaoundé for the National Herbarium before the remainder were sent for identification at Kew, were a significant number that were unicates and so were only to be found at the National Herbarium. Dr Onana drew attention to this and arranged to send these specimens on loan to Kew for identification.

The 335 vouchers from the three 25 × 25 m plots that had been executed at Mefou had been set to one side so had remained unidentified. In 2009 Xander van der Burgt was tasked with integrating these into the separate family bundles so that they could be identified. Several families additional to the Mefou were then revealed, such as Alangiaceae and Oleaceace. A draft checklist was then outputted from George Gosline's Access Cameroon database (in which the Kew team keep all specimen and species data for checklist and mapping purposes) by Yvette (Tivvy) Harvey so that gaps could be gauged for rectification.

Finally, in 2010, gaps in the checklist were filled family by family, final checks were made on identifications and the introductory chapters were written, the manuscript sent for scientific review before being sent for publishing.

THE HISTORY OF BOTANICAL INVENTORY AT THE MEFOU PROPOSED NATIONAL PARK AND IN THE YAOUNDÉ AREA

Martin Cheek[1], Iain Darbyshire[1], Jean Michel Onana[2] & Yvette Harvey[1]

[1]Herbarium, Royal Botanic Gardens, Kew, Richmond, Surrey, TW9 3AE, UK
[2]Herbier National du Cameroun, BP. 1601 Yaoundé

Apart from our own botanical exploration of the proposed Mefou National Park, in 2002–2006, no other has taken place, so far is known. However, before we document our surveys, we include a history of botanical inventory of the greater Yaoundé area, since the proposed National Park falls within this, broadly speaking, and some of the unique species of the Yaoundé area, such as *Leptonychia subtomentosa* K. Schum., are found at Mefou.

The information that we present here on the historical collectors of the Yaoundé area is taken from:

1. Letouzey's Les Botanistes au Cameroun (Flore du Cameroun 7, 1968).
2. Letouzey's Les Botanistes Hollandais au Cameroun (pp 245–252 in Miscellaneous Papers 19 (1980) Landbouwhogeschool, Wageningen, The Netherlands.
3. Analysis of specimen records held on the 'Cameroon Database' developed by George Gosline at RBG, Kew for the Wet Tropics Africa team.

HERBARIUM COLLECTORS OF THE YAOUNDÉ AREA

GEORG AUGUST ZENKER (1855–1920)

Having been Head Gardener at the Naples Botanic Garden, he accompanied an Italian mission to Gabon in 1886 where he stayed, working for the Woermann agency on the plantation at Sibange, which is now within the confines of the capital Libreville. From here he visited Cameroon, on account of a pharmaceutical company of Hamburg in order to collect and cultivate medicinal plants. He was then recruited by Captain Kund, the leader of the second of a series of German military expeditions which over 15 years, established German rule in the interior of what was then German Kamerun. These German military expeditions usually included biologists to record the organisms and vegetation in the areas being controlled. Zenker replaced the botanist Johannes Braun who had died in 1893, but who, while traversing Mbalmayo and Yaoundé as early as 1888, had made very few collections. Only 200 in total were listed for Braun's stay in Cameroon, most of which seem to have been live plants sent to Berlin and so were not permanent specimen records. Arriving in Cameroon at Kribi on 5 November 1889, Zenker spent three years as deputy to the Morgen expedition. According to Letouzey (1968 op. cit.) he then spent two years, 1893–1895 collecting specimens in the Yaoundé area with A. Staudt (see below).

These collections can be regarded as the foundations of our botanical knowledge in Yaoundé. The two began a numbering series with printed labels recording the specimens as 'Zenker & Staudt' with locations recorded as 'Kamerun Jaunde' at 'Yaunde-station', 1893–1895. So far, our work at Kew has uncovered that this number range appears to have extended from 1–762 (highest number found to date) of which we found 20 numbers in the literature or at Kew. Since at that time collectors for the German authorities habitually seem to have made only a single specimen of each number (a 'unicate') that was destined for Berlin alone, and later, in the main, destroyed there in the allied bombing of March 1943, it is likely that this was normally the case with this number series also. While the number of 20 is likely to increase with further research, it is probable that all 762 (or more) of the numbers, will never be recovered. As was usual at the time, the specimens that survive bear very little data indeed on their label. Apart from 'Yaoundé', no more precise locality is given, and no information on habitat, or habit is found; often even the date of collection, and sometime 'Yaoundé' is omitted.

Analysis of our specimen database at Kew shows that early in his career as a specimen collector of plants, and before he began the better-known partnership with Staudt, Zenker was collecting under his own numbers in Yaoundé. This appears to have escaped previous attention. *Zenker* 363 is the earliest traced so far as labelled Yaoundé, but the series *Zenker* 483–675 that survive at Kew are nearly all also given as Yaoundé — also, with the dates mainly given as 1890: years before the Staudt partnership.

At the end of 1895 Zenker left Cameroon, returning in 1896 in a private capacity to set up his own plantations of cocoa, coffee and rubber at Bipindi, while also continuing to collect plant specimens, until his death in 1922, although after 1913 few, if any, specimens were collected. While most of these specimens were from Bipindi and areas nearby, e.g. Mimfia, some numbers were labelled as Yaoundé, e.g. *Zenker* 756 in 1896 and *Zenker* 1412 in 1897. The last number that we have of his is *Zenker* 8471 in 1913. Fortunately for science, from 1896 onwards when in a private capacity, Zenker was a commercial collector, earning money from the sale of specimens to subscribers in Europe as was common at that time. Consequently he made many duplicates of each number and so, although the Berlin set were mainly destroyed, many other sets remain and so species based on these have a reliable evidential basis and are not a subject of guesswork, confusion and mystery, as is the case with so many other collections that arose during the period of German administration in Cameroon. Today Zenker ranks as pioneer of our knowledge of the plants of Cameroon, due to the sheer number of specimens that he collected, and which survive.

ALOIS STAUDT (fl. 1893–1897)

A gardener, he appears to have begun his plant collecting in partnership with Zenker, in the Yaoundé area in 1893–1895 (see above). He then continued collecting alone, in Lolodorf, 1895–96, after Zenker's departure, later continuing in Kumba in 1896–1897. Our highest number thus far recorded at Kew for Staudt is 935. We have only 19 records so far recorded from this range (see explanation under the entry for Zenker, above). Little other information is known concerning Staudt, although many species are named for him from his collections.

In the 20th century, few botanists have concentrated on collecting in Yaoundé itself, even if they have been based there. Mildbraed (numerous missions in the first three decades of the century in Cameroon) concentrated on expeditions in remote areas in the provinces, as did that giant of botany in Cameroon, Réné Letouzey, who founded the National Herbarium at Yaoundé in 1950.

W.J.J.O. DE WILDE (1936–present) and B.E.E. DE WILDE-DUYFJES

Based in N'kolbison Yaoundé, this couple spent late 1963–late 1964 collecting around Cameroon, but they also made many collections in the Yaoundé area, especially N'kolbison. Willem de Wilde focussed on *Adenia* (Passifloraceae) which he monographed, but he and Birgitte also made numerous specimens of other plant groups. They were joined, and followed in the mid 1960s by other botanical collectors from Wageningen University, such as Frans Breteler and Toon Leeuwenberg who augmented their records from the Yaoundé area. The penultimate is commemorated by *Talbotiella breteleri*, a tree endemic to the Yaoundé area. Many of the species unique to Yaoundé were uncovered by these hardworking and meticulous scientists. Their contribution is recorded by Letouzey in "Les Botanistes Hollandais au Cameroun" (op. cit. 1980)

STEPHEN D. MANNING (d.o.b. unknown – present)

A doctoral student gathering material for his research on *Pavetta* subgenus *Baconia* (Annals Miss. Bot. Garden 83: 87–150 (1996)), Manning was initially based with Duncan Thomas at his house in Kumba, but in May-July 1987 and probably for longer outside that period, he was based in Yaoundé and collected several hundred numbers within its confines, e.g. at Mt Febe and at Etoug-Ebe. He collected many forest species, especially of Rubiaceae (other habitats less rich in Rubiaceae such as inselbergs were not so well investigated) obtaining records e.g. of the rare *Psychotria sycophylla* (K. Schum.) Petit). The first records we have for *Vitex* sp. nov. of Mefou were made by him in Yaoundé. He was famous for his spartan existence, and a diet of rice and beans. His duplicates were distributed by Missouri Botanical Garden. Fittingly he is to be commemorated by the name *Chassalia manningii* O. Lachenaud, which also happens to occur at Mefou.

HERBARIUM COLLECTORS OF MEFOU PROPOSED NATIONAL PARK

DR JEAN MICHEL ONANA (1961– present)

He is currently head of IRAD-National Herbarium of Cameroon (YA) and Editor-in-charge of the Flore Du Cameroun programme, and also Head of Biodiversity Programmes of IRAD-Ministry of Research and Scientific Innovation.

After high school, where he obtained a Baccalaureate in Mathematics and Natural Sciences, his university studies were conducted at the Faculty of Sciences of the University of Yaoundé. His research in Plant Systematics and Phytogeography were sanctioned for a doctorate in Botany in 1998. He was also awarded an HDR doctorate (*Habilitation à Diriger des Recherches*) in Botany and Conservation of Diversity in 2010.

His professional career began as a researcher at the National Herbarium of Cameroon. He has conducted field research in many regions of Cameroon and led collection teams, making 150 specimens during the survey at Mefou (see details in the edited expedition reports in this chapter). Jean Michel has collected nearly 3000 specimens in total, some of which are nomenclatural types. In addition, two species — *Ledermaniella onanae* Cheek (Podostemonaceae) and *Diospyros onanae* Gosline (Ebenaceae) — have been named in his honour, the *Ledermaniella* having been a species first collected by him.

His major projects include the Burseraceae, for future publication in the Flora of Cameroon series, and the Cameroon Red Data book for plants (with Martin Cheek), and rare and endemic species in Cameroon for the conservation of ecosystems of Cameroon. Dr Onana is also completing a Checklist of accepted names of vascular plants of Cameroon.

He has published several papers, mainly on systematic research on the family Burseraceae, which included a revision of the genus *Dacryodes* for Africa, but also on a range of other taxa, including several new species, e.g. *Allexis zygomorpha*, Violaceae (with Achoundong), *Indigofera patula* subsp. *okuensis*, Leguminosae, with Schrire; *Bertiera heterophylla*, Rubiaceae, with Sonke; *Dacryodes trapnellii* (with Cheek), *D. camerunensis* and *D.villiersiana* (Burseraceae, sole author).

Dr Onana has also co-authored several books with RBG, Kew staff in the Cameroon Conservation Checklist series, documenting the species present, vegetation and Red data species of different areas in Cameroon: Mt Oku (2000), Kupe, Mwanenguba & Bakossi Mountains (2004), Bali Ngemba (2004), Dom (2010) and Mefou (2010).

These studies have helped reveal that Cameroon's forests are floristically among the richest in tropical Africa and contain many endangered species that need to be protected. He is married with two children and in his spare time, apart from writing, he is a keen horticulturalist.

M. FULBERT TADJOUTEU (April 28, 1965 in Dschang – present)

A Cameroonian with a Masters degree in Plant Biology, Fulbert has worked for many years at IRAD-National Herbarium of Cameroon as a technician in botanical research. He began under the GEF-Cameroon biodiversity programme (September 1996–2000); continuing in the Project RIHA (Network Computing Herbaria of Africa) funded by " Fonds Francophone des Info-route" from 2000 to 2005; then in the API (African Plant Initiative) of the Mellon Foundation of the U.S. in 2006; the Gluing Phase of RIHA in 2007, and in the SEP Project (Sud Expert Plante) from the French Ministry of Foreign Affairs from April 2008 to the present.

At intervals during his 14 years at the National Herbarium, he has participated as a collector in botanical surveys and forestry inventories in: the Taita Hills in Kenya (Earthwatch sponsorship) and in Cameroon in Oku-Laikom, Bali Ngemba in NW Region; Bakossi, Mt Kupe, Konye in the SW Region; Mts. Bamboutos in the W Region; Makak, Reserve Mefou, Mbaminkom and Mt. Elounden in the Central Region; Edéa Reserve in the Littoral Region; The Mt Elephant, Bipindi and the Dja Reserve in Southern Region.

At Mefou he led collection teams (see details in the edited expedition reports in this chapter) and collected 74 numbers.

During the survey at Mefou, Fulbert collections helped revealed the existence of some of the most notable rarities: *Aulotandra kamerunensis* (Zingiberaceae) and *Phyllanthus kidna* (Euphorbiaceae Phyllanthaceae)

MME FELICITÉ TONJ EPSE BLISS NANA (7 May 1959, Yaoundé – present)

Mme Nana was assigned from IRAD to the National Herbarium on 12 October 1998, as part of the administrative staff but quickly became interested in doing fieldwork and joined numerous collecting expeditions where her dynamism and skills in local negotiations made her an especially valued team member. She also supported several expatriate botanists during their collecting expeditions around Cameroon, e.g. Drs Alex Wortley and Anna Saltmarsh. During her years at the National Herbarium she attended many botanical training workshops.

51 of her specimens were collected from Mefou, among the 167 specimens recorded on our database that she collected in the period 2003–2007. Mme Nana's first experience as a botanical collector was in Kenya with the expeditions led by Dr Mwachala in the Mbale area in 2000 (Earthwatch sponsored). She also joined joint Kew–IRAD HNC expeditions in NW Region at Bali Ngemba, Kilum-Ijim (Mt Oku & Laikom) and also in SouthWest Region (Mt Kupe area). Felicité is married with seven children.

NATIONAL HERBARIUM CAMEROON – RBG, KEW BOTANICAL INVENTORIES OF THE PROPOSED MEFOU NATIONAL PARK, YAOUNDÉ 2002–2006

In October 2002, March 2004 and March 2006, RBG Kew in collaboration with the National Herbarium of Cameroon and the support of CWAF mounted three botanical surveys of the proposed Mefou National Park. All were based at the village of Ndanan 1. The first two of these periods, the most important in terms of volume of data collected, were supported by funding and volunteers from the Earthwatch Europe, based in Oxford UK.

The first survey, conducted over two weeks, in October 2002 was conducted when river levels were high during the wet season, inundating the forest nearby. At this time we were based inside Ndanan 1 itself, renting rooms from the villagers as well as camping outside their houses. Access from Yaoundé was then through a tortuous maze of small roads so vulnerable to rain on slopes that at one or two points they became impassable in wet weather.

The second survey was conducted in the dry season, over four weeks in March, and early April 2004 when water levels were low and it was possible to access the riverine forest by walking along the river beds. At this point we were asked by CWAF to camp in an area that had been cleared just outside the village.

A third survey in March 2006 amounted to just a few days when Emma Jane Fenton was based at Ndanan 1 to fill in collecting gaps having been introduced to the area by a visiting group preparing for the AETFAT Congress due in Yaoundé the following year, led by Professor Simon Owens of RBG, Kew and accompanied by Martin Cheek. The following details are extracted from the relevant expedition reports. Edits are indicated in [square brackets]. The reports vary greatly in their information content and style, depending on their authors, but have been only minimally edited so as to capture the maximum amount of the data that they contain.

MEFOU SURVEY 1

EXPEDITION REPORT

The Proposed Mefou National Park, Central Province, Cameroon
October 11th – 24th 2002

Martin Cheek & Yvette Harvey
Herbarium, Royal Botanic Gardens, Kew, Richmond, Surrey, TW9 3AE

Summary. This report documents an expedition to Central Province, Cameroon to conduct a botanical inventory of the proposed Mefou National Park, to assist future conservation management and to aid interpretation by future visiting students and ecotourists. The expedition was organised by RBG Kew (funded by Earthwatch) in collaboration with the Herbier National Camerounais (HNC–IRAD), Cameroon at the request of CWAF (Cameroon Wildlife Action Fund).

The proposed Mefou National Park lies near the capital city, Yaoundé. It is largely lowland forest below 600 m alt. with several hills and rivers. Parts in the south around the village of Ndanan 1, where CWAF's base camp is placed, are used to rehabilitate gorillas, chimpanzees and other primates captured from poachers by officers of the Ministry of the Environment & Forests (MINEF). No previous botanical inventory work had been conducted in the area, so far as is known.

The objectives of the expedition were:

• To begin to discover all vascular plant species present by collecting specimens for future identification, to record local names and use where available.

• To investigate which of these if any are threatened with extinction (Red Data Species), to assess their conservation status and to document them in detail so as to facilitate their mapping; and their identification for protection by those managing the park in future.

• To characterise the main vegetation types present within the park boundary so as to aid interpretation for visitors and to assess their value for conservation.

The inventory teams split into two or three teams each day. In addition, Lou Fay forayed semi-independently for ferns. Using the satellite derived map provided by CWAF, different routes through the proposed protected area were taken, specimens being collected for further identification of all fertile plant species seen.

Three semi-autonomous research groups joined the expedition for part of the two week duration, departing for other areas after the mid expedition break at Kribi. These were:-

1. Podostemaceae — Dr Rolf Rutishauser, Dr Gabriel Ameka and Jean-Paul Ghogue

2. *Thomandersia* (Acanthaceae) —— Alex Wortley & Felicité Nana

3. Leguminosae and their nodulation — Barbara Mackinder & Hermine Kiam

4. Celastraceae, Asparagaceae & Labiatae — Sebsebe Demissew, University of Addis Abeba

FIELDWORKERS

Herbier National Camerounais:

Jean Michel Onana, Herbier National Cameroun (researcher, leader of HNC team, Burseraceae specialist)

Victor Nana, Herbier National Cameroun (technician)

Madame Nana, Herbier National Cameroun (technician)

Jean-Paul Ghogue, Herbier National Cameroun (technician)

University of Yaoundé I:

Hermine Angele Kiam, University of Yaoundé I, Cameroon (legume nodule specialist)

RBG, Kew:

Martin Cheek, RBG, Kew, UK (generalist, leader of the Kew team)

Barbara Mackinder, RBG, Kew, UK (legume specialist)

George Gosline, RBG, Kew, UK (specialist in Ebenaceae, Annonaceae & databases)

Tivvy Harvey, RBG, Kew, UK (Sapotaceae specialist, deputy leader of Kew team)

University of Oxford:

Alex Wortley, University of Oxford, UK (specialist in *Thomandersia* (Acanthaceae))

Other participating botanists:

Sebsebe Demissew, University of Addis Abeba (specialist in Asparagaceae, Celastraceae and Labiatae)

Gabriel Ameka, University of Ghana (specialist in Podostemaceae)

Rolf Rutishauser, University of Zurich, Switzerland (specialist in Podostemaceae)

Observer from MINEF: M. Essama

Earthwatch volunteers (& sponsors if applicable):

Bronwyn Baker, Yellowknife, NT, Canada (RT)

Francis Condon, London, UK

Audrey Demidov, Russia (BAT)

Jimmy Drekore, Port Moresby, Papua New Guinea (RT)

Louis Fay, Milledgeville, GA, USA (Fern collector)

Bethanie George, Bondi Beach, NSW, Australia (RT)

Vaneshrie Govender, Pietersburg, South Africa (RT)

David Kruger, Cheshire, UK

Ziaur Rahman, Dhaka, Bangladesh (BAT)

Ken Siminji, LAE, Morobe Province, Papua New Guinea (BAT)

James Wandayi, Bungoma, Kenya (BAT)

Shigeo Yasuda, Hodogaya, Yokohama, Japan

Earthwatch funded Cameroonian staff: Camp organisers Martin Mbong (camp manager) & Edmondo Njume (equipment organiser & botanical assistant); Japhet Wain (driver)

Earthwatch funded local workers from Ndanan 1: Virginie Ngomo (cook); Samedi Amougou (cook); Florence Ebeudeu (cleaner); Janvier Brice (water carrier); Joseph Ndi (wood collector); Bernard Ndi (guide).

RESULTS

DOCUMENTATION OF PLANT SPECIES PRESENT

726 numbers* of mostly fertile specimens in sets of up to 5 duplicates were gathered by the inventory teams. All these specimens were pressed, dried, filed, databased, labels printed, and the top set pulled for HNC under supervision of Jean Michel Onana. Both specimens for HNC

(awaiting incorporation) and Kew (awaiting export permits; usually takes 6–12 months) are being kept at HNC. 212 plot voucher specimens (unicates) were made – they will be sent as on loan from HNC to Kew for identification. Local names and uses were provided for many specimens by Belinga of Ndanan 2. About 50% of the specimens were determined to species in the field, i.e. provisionally. Confirmation of these names, and identification of the remainder will not be possible until the specimens arrive at Kew.

*Collectors and numbers at Mefou:

Botanist	Number sequence	Total
Martin Cheek	10999–11267	268
Yvette Harvey	70–215	145
George Gosline	400–445	45
Jean Michel Onana	2186–2204	18
Louis Fay	4700–4737	38
25 × 25 m plot	1–212	212
Total		726

RED DATA SPECIES

Two rare, and very distinctive species were identified; both are potential Red Data Species and therefore are potential priorities for conservation. Neither of these species had been seen alive by any of the botanists concerned, causing considerable interest:

Cola letouzeyana Nkongm. (Sterculiaceae). This small tree is distinguished among all other *Cola* species in all the leaves being subsessile, oblong – cordate, with filiform stipules over 1 cm long and with thickly brown pilose stems and midribs. It is known from only c. 6 sites in the Yaoundé – Edéa – Kribi – Eséka polygon with an outlier at Mt Cameroon. Only two plants were found at Mefou despite much searching. Evidently even where it occurs it is rare. Both plants were threatened: the first was nearly shaded out by overgrowth of *Scaphopetalum*. The second had been broken badly by a nearby tree fall.

Tricalysia amplexicaulis Robbr. (Rubiaceae). Initially this was suspected of being a new species to science. Follow up work at the National Herbarium of Cameroon soon showed it to have been named 15 years ago. The material seen was erect, c. 60 cm tall, unbranched, with oblanceolate– oblong, subsessile, cordate leaves angled at c. 45° above the horizontal. The 4–5 petalled white flowers, c. 3 mm diameter, are held in cymes of 1–3 and twisted into the leaf axils. Known also from c. 6 sites in the same polygon as *C. letouzeyana*, *T. amplexicaulis* is also a Red Data Candidate.

No other Red Data Species (or candidate) or new species were identified in the course of the inventory, but these may yet be uncovered once the material reaches Kew.

VEGETATION TYPES AND THEIR CONSERVATION VALUE

A. Farmland/Farmbush. Most of the proposed park area that was seen comprises of this vegetation type. These areas are believed to have been logged and then to have been converted to cocoa and cassava farms which are still very much apparent. They have an extremely low value for conservation. However, if the park is to go ahead they might be useful for interpreting succession for visitors for education purposes. If further animal enclosures are planned, this vegetation type is the obvious place for them since they would cause little or no negative conservation impact here.

B. Open *Raphia* Swamp and Cyperceae meadows. This habitat is easily reached on the route due south of Ndanan I. During our visit a large area was flooded at least c. 10 cm deep in tannin–rich water which *Raphia ×africana,* and *Thalia welwitschii* were common, with an adjoining sward of Cyperceae and Pteridophytes forming an extensive meadow. This habitat contains species not found elsewhere in the park and has value for education purposes and to a lesser extent for conservation. More investigations needed.

C. Inundated fresh water swamp forest. Found south of Ndanan 2 in 200 m wide strips along the Mefou river, and in Eastern sector, Northern side. Dominated by *Gilbertiodendron dewevrei, Nauclea, Hallea, Uapaca ×vanhouttei* and *Ficus vogeliana*. This interesting forest type has high value for education purposes and may have conservation value.

D. Semi-Deciduous Forest. Rich in timber species, this forest type may have been dominant in the park before logging. The example studied, immediately North of the village is on a higher, well–drained location that may favour development of this type. The main canopy indicators are *Triplochiton scleroxylon, Celtis* spp., *Terminalia superba*. Here we mounted our 25 × 25 m 'intensive' (all species) plot. With a species count (v. approx) of c. 230 taxa – i.e. moderately diverse. Valuable for education purposes, possibly of some conservation value. Moderately disturbed.

E. Evergreen forest. Found in lower lying areas, yet above the swamp forest, this habitat type is probably the most species diverse in the park. Four parcels were identified:

Track south east of Ndanan 1,

a) parcel East of road, midway to bridge. This site was dominated by a diversity of Caesalpinioideae canopy species and with an understorey dominated by *Scaphopetalum* sp. Both of the potential Red Data species were found here. This site is of importance for conservation and would be valuable for education. *Omphalocarpum procerum* located here by Yvette Harvey.

b) parcel West of road, past monkey cages. This site has numerous additional Caesalpinioideae taxa.

c) track East of road immediately after bridge. Understorey again dominated by *Scaphopetalum*. Several spectacular species located here including *Pararistolochia* sp. and *Oncoba flagelliflora*. The first is a cauliflorous liane with foul-scented bizarre flowers c. 15 cm long, the second a tree 4 m tall, with string-like inflorescences produced from the trunk, bearing flowers up to 4 m from the trunk, from the ground.

d) parcel south of Ndanan 2, close to inundated forest. This area also appears to be undisturbed 'primary' forest, rich in Caesalpinioideae and Annonaceae, and also of conservation value.

SUMMARY OF INITIAL RESULTS

The initial results of our survey are that although possibly most of the area of the proposed park has been degraded, significant fairly intact areas remain of a surprisingly varied range of vegetation types. These areas are easily accessible on foot from the base camp. If the area is formally gazetted as a National Park, they would form an excellent basis for interpreting forest-zone habitats to the general public, especially school children and tourists.

Two Red Data candidate species were identified during the survey. More may yet be uncovered after specimens arrive in London and are identified. Not surprisingly however, it looks unlikely that the Park is a major repository of Red Data species. No new species to science have yet come to light from our survey.

ITINERARY

Saturday 5th Oct

Harvey, Gosline & Cheek arrive in Yaoundé in eve (AF 840) Overnight at L'Independence.

Sunday 6th Oct

Cheek to Bamenda by Vatican Express. Meeting with DeMarco.

Monday 7th Oct

Gosline and YH to HNC — research permit applications and discussions with staff. Harvey & Gosline to zoo, meet CWAF reps. Gosline offered seat on CWAF shuttle to Ndanan 1 and makes recce. YH to BAT to discuss BAT volunteer visa anomalies. Cheek picks up car and driver at BHFP and travels to Nyasoso, visits CRES.

Tuesday 8th Oct

EW car loaded with first part of equipment from store, MC travels with Edmondo Njume and Martin Mbong to Mfou town, c. 5–6 pm at MINEF, then proceeds to Ndanan 1. Unloads.

Wednesday 9th Oct

EW car (with Bernard as company) to Nyasoso for second load. MC drops at airport, to Meumi. At Yaoundé airport YH, GG & MC meet first arrivals on AF flight (Drekore, Siminji, Yasuda, Ziaur & George)

Thursday 10th Oct

YH to airport to pick up Congdon (am). EW Car delivers second load to Ndanan 1. Mbong, Njume & Wain to set up camp. MC, GG & YH to airport to meet Sabena flight (Wortley (Sonke), Baker, Kruger).

Friday 11th Oct

First day of expedition.

Main expedition move from Nsimalen airport to Ndanan 1. In afternoon MC leads group botanising SE of village. GG to airport in pm to pick up Govender & Wandayi — spend night (with Fay) in Ideal Hotel, Yaoundé. MC, EN & MM explain protocols and pressing techniques

Saturday 12th Oct

MC/GG collect in semi deciduous forest N of Ndanan 1

YH group SE track to bridge

Sunday 13th Oct

YH group collect route to gorilla (SW of Ndanan 1)

GG & MC am to church Metet. In afternoon reconnaissance S of Ndanan 1 across *Raphia* swamp, to Ndanan 2, to Mefou R, return to Ndanan 2 then via western route to Ndanan 1

Monday 14th Oct

YH group further along route past gorilla enclosure.

GG & MC via westerly route to forest before Ndanan 2.

Tuesday 15th Oct

YH group track S towards *Raphia* swamp

GG & MC as 14th Oct extending to bridge on edge of the inundated Mefou R.

Wednesday 16th Oct

YH group track SSE towards (and past) water supply

MC group returns to edge of inundated Mefou R

GG group in landrover recce eastern sector of park. Traverse park from Ndanan 1 on SE track and return (fallen tree), take route north towards Mfou town, then E and S to boundary. Return after heavy rain. Park roads impassable.

Thursday 17th Oct

All to Kribi

Friday 18th Oct

Return from Kribi

Saturday 19th Oct

YH chimpanzee forest area

GG & MC forest off track se Ndanan 1 Discovery of *Cola letouzeyana* & *Tricalysia* ? sp nov. by ace volunteer collector, Jimmy Drekore.

Sunday 20th Oct

GG leads group to church then sets up specimen db operation

MC and YH reorganise

No field groups

Shigeo's Midge bites cause notable swelling on face

Evening "Health & Safety forum"

Monday 21st Oct

GG leads data entry operation with volunteers

MC/YH group conducts intensive 25 × 25 m plot

c. 212 vascular species. Car to clinic in Yaoundé for Govender, tonsilitis diagnosed.

Tuesday 22nd Oct

GG leads data entry operation

YH group on SE track, then SW past Monkey enclosures pm SW to chimpanzee enclosures

MC/JMO group SE track to bridge then east into E sector of park

Evening 'end of team' party; Bethanie George is ill briefly

Wednesday 23rd Oct

First group depart for Yaoundé airport

Remainder continue data entry specimen sorting

JMO supervises extraction of top specimen set for Nat. Herb. Cameroon

Thursday 24th Oct

First group depart for Yaoundé airport

Conclusion of data entry, specimen sorting etc

Packing of equipment

Second group departs for Yaoundé airport with specimens for Nat. Herb.

Friday 25th Oct

Car departs with all equipment to Nyasoso for storage (Mbong, Njume, Wain)

Saturday 26th Oct

Ndanan 1 vacated by EW expedition

Sunday 27th Oct

All rest

Monday 28th Oct

Car Nyasoso to Bamenda (where based)

First day of 2 week BAT sponsored International Herbarium Techniques Course at British Council, Yaoundé (taught by YH, BM, GG & colleagues from Kew)

Expedition report drafted and passed to CWAF acting director Francis Kameni

MC returns to London

RBG, Kew & Herbier National du Cameroun expedition for botanical inventory and conservation management in Mefou Proposed National Park and Bali Ngemba Forest Reserve, March–April 2004 by Iain Darbyshire, Martin Cheek & J.-M. Onana, 10th June 2004

Team I: Mefou Proposed National Park, Central Province.
7th – 20th March 2004.

Team II: Mefou Proposed National Park, Central Province.
21st March – 3rd April 2004.

Executed by RBG, Kew (funded in part by the Earthwatch Institute) in collaboration with the Herbier National du Cameroun (HNC) – IRAD, Cameroon (funded in part through a grant from the Overseas Fieldwork Committee, RBG Kew, backed by British & American Tobacco) with the aid and co-operation of local conservation organisations in Central Province (CWAF).

Abbreviations used within this document:

ANCO	Apicultural and Nature Conservation Organisation (formerly North-West Beekeeping Association – NOWEBA)
BHFP	Bamenda Highlands Forest Project
BNFR	Bali Ngemba Forest Reserve
CRES	Center for the Reproduction of Endangered Species
CWAF	Cameroon Wildlife Aid Fund
ERuDeF	Environment and Rural Development Foundation
EW	Earthwatch
HNC	Herbier National du Cameroun
K	Herbarium of the Royal Botanic Gardens, Kew
LBG	Limbe Botanic Garden
MINEF	Ministry of the Environment and Forests
SCA	Limbe Botanic Garden herbarium
YA	Herbarium of HNC

TEAMS I AND II: MEFOU PROPOSED NATIONAL PARK

The first two teams of this season's fieldwork followed on from botanical inventory work conducted in October 2002 (see expedition report) in the Mefou Proposed National Park in Central Province, Cameroon. Work was hosted by CWA, an NGO based at the park, after discussion with their primary sponsors, Bristol Zoo (UK).

Mefou Proposed National Park lies approximately 30 km SSE of the capital city, Yaoundé and is an area of remnant forest patches in the otherwise largely denuded landscape of Central Province in the vicinity of Yaoundé. The park is largely below 600 m alt., with rolling terrain. Along the rivers, well developed and intact evergreen rainforest occurs of the *Gilbertiodendron dewevrei*-type, while the better-drained, higher areas supporting semi-deciduous forest with a *Triplochiton*-Ulmaceae association. Several *Raphia* swamps occur where drainage is impaired. Most of the semi-deciduous forest is secondary in nature, with large tracts of open forest and farmbush; active arable farming is still practised by residents from the several villages located within the park's boundaries. Sections in the central park, around the village of Ndanan I where CWAF's camp is located, are used to rehabilitate gorillas, chimpanzees and other primates captured from poachers by MINEF officers.

The objectives of the two expeditions were:

• To further the inventory work in this area, unexplored botanically prior to the October 2002 expedition, by adding to the 726 collections made on that trip. It was aimed to collect all plant taxa flowering and/or fruiting, together with identifiable sterile collections, for future identification at K, HNC and other herbaria with west African plant specialists.

• To investigate which of these if any are threatened with extinction (Red Data Species), to assess their conservation status and to document them in detail so as to facilitate their mapping; and their identification for protection by those managing the park in future.

• To continue in the characterisation of the main vegetation types present within the park boundary and to assess their value for conservation.

• To provide interpretation of the vegetation communities in the park, thus aiding CWAF in their future environmental education programme on forest conservation. With its close proximity to Yaoundé, Mefou is recognised as a possible site for promoting environmental sustainability and conservation to the Cameroonian public and to foreign visitors.

It is aimed that through the combination of the collections and botanical and ecological data gathered during the 2002 and 2004 expeditions, an understanding of the plant diversity and its importance for conservation sufficient to produce a "Conservation Checklist of the Plants of Mefou Proposed National Park" will have been acquired. Work on preparation of the checklist will begin in late 2004. This document is seen as a potential major contribution to the future gazettement and conservation management of this site as National Park.

Team I fieldworkers and initials:

RBG, Kew: Martin Cheek (MC — Project Leader, UK).

HNC: Nana Felicité (NF), Nana Victor (NV) (Technicians).

University of Coimbra, Portugal: Celia Cabral (CC — PhD student, joining the expedition to gather specimens and field data of the genus *Vitex* (Labiatae)).

MINEF: Essama M. (EM — principal guide and ethnobotanical informant).

Earthwatch volunteers and corporate sponsor (where appropriate):

Jennifer Bury (JB — USA, private volunteer).

Louis Fay (LF — Milledgeville GA, USA, private volunteer).

David Gallagher (DG — Manchester, UK, HSBC).

Elizabeth Greene (EG — Seattle WA, USA, private volunteer).

Marianne Holding (MH — Sheffield, UK, HSBC).

Earthwatch-funded Cameroonian field staff:

Ngolle Ngolle Hoffmann (NH — camp manager)

Edmondo Njume (EN — equipment organiser and botanical assistant).

Martin Etuge (ME — botanist and research assistant).

Gilbert Tanyi (GT — driver of EW vehicle).

Earthwatch-funded local Cameroonian camp assistants:

Jeane Mbiah (cleaner).

Rosalie (week 1), Ngah Bernadette (from week 2), Ngwano Margeritte (cooks).

TEAM I PROGRAMME

Monday 1st March

MC departs London Heathrow, meeting CC at Charles de Gaulle, Paris, on flight to Cameroon, arriving at Nsimalen, Yaoundé, c. 7.30pm. Met by Dr. G. Achoundong and J.-M. Onana (HNC), transferred to Hotel Noka, dining there with Dr Achoundong. Research permit applications handed to Dr Achoundong together with reimbursement for last specimen consignment to K.

Tuesday 2nd March

MC and CC deposit luggage at HNC and take bus to Bamenda. Meet Dr Vabi (head), Chrysanthus, and Grace Timah at BHFP headquarters (project in the process of closing). Overnight at Skyline Hotel, Bamenda, dinner with Rita Ngolan, formerly of Kilum-Ijim Forest Project, prospective counterpart for Anna Saltmarsh (University of Montpellier) who will be joining the expedition at the onset of team II in order to study *Leonardoxa* (Leguminosae: Caesalpinioideae).

Wednesday 3rd March

MC and CC breakfast with Dr Vabi at his home, visit official offices in order to gain clearance for forthcoming fieldwork in the Bamenda Highlands during team III (but officers absent) then to BHFP HQ. Discussion on premature closure of BHFP and possible succession to it as institutional partner in NW Province for RBG, Kew. MC and CC then visit ANCO in EW vehicle with new driver, GT. Vehicle in excellent order apart from lack of logbook. Depart for Nyasoso at noon, arriving c. 5.30pm. Meet with Bethan Morgan, deputy head, CRES, Nyasoso, and show latest draft of the "Conservation Checklist of the Plants of Mt Kupe, Mwanenguba and the Bakossi Mts". Discuss hiring of ME for the 3 EW teams and learn that Martin Mbong (usual camp manager) indisposed but NH and EN available. Check EW equipment store. Arrange to pack first load early next morning.

Thursday 4th March

MC visits herbarium and orchid garden at CRES HQ at request of B. Morgan. Loading delayed by dramatic thunderstorm. Eventually depart Tombel after further delays in collecting belongings of EN and NH at noon. MC misdirects car through Douala, then delayed 1 hr by Gendarmes as Carte de Grise has not been extended. Phone Dr Vabi at BHFP re this problem; eventually pay 10,000 fcfa, ME negotiating. GT drives slowly to Yaoundé. Further Gendarme – Carte de Grise delay near Nsimalen. Arrive Mefou town c. 9pm, spend night at hotels, no electricity.

Friday 5th March

All to Ndanan 1 village, Mefou Proposed NP, view campsite laid out by CWAF and unload first load, leaving NH and EN to set up; vehicle returns to Nyasoso for second load, dropping MC and CC at Garanti Bamenda office, Yaoundé, where GT picks up newly extended Carte de Grise sent express from BHFP. MC and CC to HNC. Dr Achoundong arranges meeting for following morning for CC with Maximilianne of Univ. Yaoundé Phytochemistry lab, which specialises in extracting plant oils by steam distillation. Night at Motel Sikam, Auberge opposite HNC.

Saturday 6th March

MC and J.-M. Onana give data on invasive alien plant species to IRAD officer representing Cameroon at an international meeting on the subject in Ghana. Maximilianne visits HNC and J.-M. Onana, MC and CC go with her to the lab, passing *Vitex* tree on the way!

Second tranche of permit applications submitted. Tadjouteu Fulbert (HNC) buys old newspapers for expedition. MC to airport to meet EW team I volunteers, all arriving on AF 940 at 7.30pm. All transferred in 2 taxis to Hotel Ideal and met for dinner. MC overnight in Motel Sikam.

Sunday 7th March

Expedition team I begins. MC and CC depart Motel Sikam passing by Hotel Noka en route to Hotel Ideal. Following difficulties in delivering letters, arrive at Ideal at noon. Three taxis with all to Mvog-Mbi, Gare Routiere Mefou. Hire Hi-ace to Mefou, transfer there to EW vehicle, thence to Ndanan 1. Tents set up. Lecture on Health & Safety by MC. Walk along road with volunteers for initial introduction to plant characters.

Monday 8th March

Demonstration of plant collecting and pressing techniques by MC.

MC, EN, JB, DG, MH, EG, CC to forest left of road to Ndangan 1, before bridge, on downwards slope; cross stream to the right and follow its course. Return to village at 4.00pm to begin drying specimens. LF executes programme studying ferns and continues to end of stay.

NF, NV and EM arrive in evening with EW vehicle. GT reports rattling sound in vehicle engine, suggesting engine needs attention otherwise may seize.

EN demonstrates preparations for drying specimens. Evening: MC delivers presentation on history of Kew-EW programme in Cameroon.

Tuesday 9th March

Expedition divides into 2 groups. Main group: MC (leader) NV, EM, JB, EG, DG continue on route to Ndangan 1, exploring routes right of the road into the forest immediately next to Ndanan 1, down to abandoned chimp cages at bottom of slope before bridge.

Vitex group: CC, NF, EN, MH search for *Vitex* after whole group collects *V. grandifolia* and *Dialium* (both between camp and village along road), demonstrating use of poles.

Wednesday 10th March

Main group: MC (leader) NV, EM, JB, EG, DG continue on route to Ndangan 1, beginning on right just before, at, and after the bridge. Unusual and possibly new *Impatiens* at river, *Lasiodiscus* sp., a shrubby *Pseudosabicea* sp. are collected. Returned to camp at noon, DG going for water for camp with Divine (CWAF) and EG relieving LF as camp guard. Main group continue on route after bridge, left, to good forest. An unusual *Uvariopsis* sp. in fruit and *Scaphopetalum* sp. in flower are notable. Return at 5pm.

Vitex group: CC, NF, EM MH return at 3pm after finding three more trees of *V. grandifolia* (EM) and general collecting (NF).

EN and GT depart c. 8.30am to Bamenda for EW vehicle engine inspection; NH with them to Mefou for shopping, returns by motorbike 1.30pm.

NF gives evening presentation.

Thursday 11th March

Main group: MC, NV, EG, MH, joined by 4 CWAF volunteers for the day (G.Perry, A.Nelson, K.Crystal and S.MacDonald), continue at *Uvariopsis* site and go further E, all in good forest. No more fertile *Uvariopsis* found but about 9 more *Cola letouzeyana* in addition to the three at the start of this path, which grow with three *Tricalysia amplexicaule*, both Red Data candidates. Interesting material included *Dictyophleba* sp. and 2 *Drypetes*, all in flower. On return, explore *Impatiens* habitat further.

Vitex group: CC, NF, ME, DG spend morning collecting leaves for steam distillation on Friday.

JB guards camp and fetches water. EG gives evening presentation. ME arrives from Nyasoso by bike.

Friday 12th March

Vitex group: CC and NF depart at 7.30am for Univ. Yaoundé I in CWAF car for distillation of leaves; accompanied by NH for camp shopping.

Group 2: ME (leader), JB, EG, ME collect along W side of park towards Ndanan II

Group 3: MC, NV, DG continue along E side along road to Ndangan I, getting to within 1 km of boundary. About 10 plants of *Tricalysia amplexicaulis* seen in evergreen forest on left before river, near bank. More areas being actively farmed; apart from cocoa plots being cleared of regrowth, new areas or at least incompletely cleared areas are being prepared for planting. One plot being burnt. Cocoa areas mostly with canopy trees so will not show on satellite images.

MH guards camp and fetches water with Divine.

ME gives presentation in evening. Israeli CWAF volunteer applies to join expedition next week. EW vehicle with GT and EN returns at 8.00pm.

Saturday 13th March

MC discusses Hall & Swaine plot methodology with volunteers and field staff. Meeting with Teketay Demel, Africa Regional Director of Forest Stewardship Council, visiting the Park that morning before returning to Ethiopia; will move to Yaoundé with his family in July.

MC, ME, EN, ME, NV, DG, JB, MH go to site of *Uvariopsis* (see 10th & 11th March) to do 25 × 25 m intensive plot.

EG guards camp and fetches water with GT.

Sunday 14th March

Rest day. All volunteers go with MC to Greek Orthodox church at Metet, afterwards meeting Chief of Metet, then viewing the new road bridge under construction over the Mefou River, collecting roadside plants on return, including *Rytigynia* sp. and *Strophanthus* sp.

Monday 15th March

Group 1: ME, EM, JB collect towards Ndanan 2 by western route, in forest towards Mefou River. About 15 wild, tan-coloured, monkeys seen at 9.30am.

Group 2: MC, MH, EN, Alalouf (CWAF) do survey of forest in the Baboon enclosure now under construction. Three plants of *Tricalysia amplexicaulis* found. Then to forest where new Chimp enclosure planned. This proves to be the best preserved area of intact semi-deciduous forest yet encountered in the National Park! Lots of *Triplochiton* and *Celtis*, sparse understorey, large trees and lianas recorded. Finally visited *Raphia* swamp; no surface water now present (unlike October 2002, when 10 cm deep) but soil still damp under foot. Collection of fertile swamp taxa.

Later discussed with CWAF's Sarah MacDonald moving the above-mentoned *T. amplexicaule* plants to the interpretation area for visitors near the main Chimp enclosure.

NH, GT and EG to Yaoundé for shopping, to deliver NV to HNC and to pick up Dr Achoundong and Nicole Ngueje; the latter not able to join the expedition owing to last minute changes. Newspapers collected. CC visited in clinic where she is suffering from suspected malaria. Rainstorm 4.30–6.00pm.

Evening presentation by JB.

Tuesday 16th March

Group 1: ME, EM, MH, EG reach Mefou River by western route and collect aquatics (e.g. *Nymphaea* and *Ceratopteris*) and in excellent, prolific riverside forest.

Group 2: MC, JB, DG, EN do 25 × 25 m plot in *Raphia* swamp and collect further fertile swamp specimens including *Impatiens irvingii* and *Raphia* sp.

Evening presentation by MH.

Wednesday 17th March

Group 1: ME, EM, DG, JB continue at Mefou River beyond Ndanan 2, this time entering by going W of the gorilla pen and walking down river, collecting for 3 hours until Ndanan II reached. One shotgun and more than 15 fence traps seen. Many *Raphia* cut down for mimbo.

Group 2: MC, EN, EG collect towards southern end of western block along road to Ndangan 2, reaching T-junction with water pump. Many plants infested with ants. Notable species: a *Rinorea* with brown-pilose fruits, very common in this area, a 4 m *Stereospermum*, two *Dichapetalum* spp. and two Asclepiadaceae.

Shopping trip to Yaoundé: NH, MH, GT. JB guards camp.

Evening presentation by DG.

Thursday 18th March

Group 1: ME, EN, MH continue along Mefou River.

Group 2: MC, EM, EG, JB travel to eastern side of park, going W from Ndangan I. Understorey dominated by *Scaphopetalum*. No *Triplochiton* or *Gilbertiodendron* leaves seen. Possibly an intermediate forest type? Access by current road to sand extraction sites and by old logging roads. No signs of hunting or farming here. Several species not found in W of park: *Irvingia robur*, *Coleotrype laurentii*, *Warneckea* sp., *Aframomum* sp. with a white stem.

DG has aches and pains, guards camp and sorts specimens.

Evening presentation by EN.

CC has returned to Portugal and is recovering; communication via satellite phone.

Friday 19ᵗʰ March

Group 1: ME, EN, JB, EG continue at Mefou River.

Group 2: MC, EM, MH return to Ndangan I; EM leads through new logging by footpath through *Triplochiton* forest and abandoned village to a new river site with *Gilbertiodendron-Scaphopetalum* forest. Notable discoveries: *Thonningia* (Balanophoraceae), first record for the park collected by MH, several Caesalpiniod legumes and another *Strophanthus* along the river.

DG sorts specimens and guards the camp.

Evening presentations by LF and GT.

Saturday 20ᵗʰ March

All volunteers pack bags ready for departure and assemble the new 8-person tent for the next team; ashes put in toilet to reduce smell; EG and EN dig holes at start of nature trail for *Tricalysia amplexicaulis* plants to be transplanted from Baboon enclosure.

MC, ME, EM and DG walk proposed CWAF nature trail and identify notable plants along route for signposting and interpretation as part of CWAF's future environmental education programme on forest conservation. DG leaves ahead of the group and is wounded between the eyes by a stone thrown by a gorilla. Thunder at 3pm results in early departure for airport in case the roads become dangerous in the rain. EN dropped at Mefou to travel onwards to Douala in order to meet Charles Pigott, a volunteer on team II, at the airport there early next morning.

MC. JB, EG, MH, DG dropped at airport after 3hr journey delayed by road blockage (personal representative of Chief of State of Cameroonian government and entourage leaving airport).

Team II fieldworkers and initials:

RBG, Kew: Martin Cheek (MC — Project Leader, UK); Iain Darbyshire (ID — EW Principal Investigator); David Roberts (DR — Orchidaceae specialist); Nina Rønsted (NR — PhD. Student, working on the genus *Ficus* (Moraceae), sponsored by Carlsberg).

HNC: Jean Michel Onana (JMO, senior researcher and EW Principal Investigator), Tadjouteu Fulbert (TF – technician).

University of Yaoundé I: Dorisse Jiofak (DJ – student under Dr L. Zapfack, joining the expedition to gain experience in plant inventory work and to get ideas for her MSc. project).

Earthwatch volunteers and corporate sponsor (where appropriate):

Catherine Blanc (CB – Geneva, Switzerland, HSBC).

Richard Brinklow (RB — Kirkinch, UK, private volunteer).

Ali Butt (AB — Islamabad, Pakistan, British & American Tobacco).

Louis Fay (LF — Milledgeville, USA, private volunteer).

Gillian Kirk (GK — Cheshunt, UK, HSBC).

Anura Peiris (AP — Colombo, Sri Lanka, British & American Tobacco).

Charles Pigott (CP — Glasgow, Uk, private volunteer).

Gulnara Sabirova (GS — Moscow, Russia, British & American Tobacco).

Earthwatch funded Cameroonian staff:

As previous team with the addition of:

Bernard Nde (BN – guide from Ndanan I village, from 21st April).

TEAM II PROGRAMME

Saturday 20th March

ID meets DR, NR and GK at Heathrow and check in equipment including computer for databasing, EW vehicle chains and cardboard for specimen bundles; flight to Paris Charles de Gaulle, where meet Anna Saltmarsh, student at University of Montpellier who is to meet Rita Ngolan in Cameroon in order to study plant-ant interactions in the genus *Leonardoxa*, collecting at previous known sites for the genus in Cameroon, e.g. Korup National Park and Kribi. RB and CB join the group in Paris, take flight AF 940 to Cameroon, delayed by 2 hours on the runway. Visit of EW volunteer Wajed Khan of Bangladesh cancelled due to visa problems.

MC meets Rita Ngolan at Nsimalen Airport, Yaoundé in preparation for meeting the incoming team. Transfer to Ideal and Meumi Hotels. BAT fellows AB, AP and GS arrive on different flights and are met by BAT representatives; meet at the Ideal Hotel in the evening.

Sunday 21st March

Expedition begins; official rendezvous at Ideal Hotel 9.30am. Nana Felicité (HNC), Prof. B. Sonké and Dr. L. Zapfack (both Univ. Yaoundé I) and John De Marco (former head of BHFP) visit; De Marco agrees to meet up with DR prior to team III in order to look over his orchid collections and to provide ID with possible sites for short botanical trips during team III. EN and CP arrive from Douala at 11.00am; proposed candidate to join the expedition from LBG does not arrive. Depart by 4 taxis to Mefou town at 11.30am, transferring in 3 loads by EW vehicle to camp at Ndanan 1.

After settling into tents and taking lunch, MC leads short reconnaissance into forest to introduce volunteers to plant characters and vegetation types. Instruction in plant pressing by MC and ID, using specimen of *Trema orientalis* (ID 180), followed by lecture on Health & Safety (MC and ID).

Monday 22nd March

Demonstration of specimen databasing (ID) and preparing specimens for drying (EN and ID). Divide into 2 groups. Group 1: ME, ID (leaders), EN, CB, GK, CP, NR continue in the *Raphia* swamp and riverine forest near the Mefou River; ME explains to ID and NR which areas and taxa have already been collected. NR begins fig herbarium sheet and silica gel collections by collecting *Ficus sur*. Fertile material and silica gel material of *Oncoba welwitschii* collected for Sue Zmarzty (K).

Group 2: MC, JMO, TF, EM, AB, GS, DR proceed S on footpath to Ndanan 2, doing survey of area of new Chimp enclosure; continue to lower part of Mbeme River, before confluence with Ndza'a (small river) and Mefou River, at point SE of Ndanan II.

LF and RB collect ferns and lichens N of camp-village road.

A. Saltmarsh and R. Ngolan leave to begin the *Leonardoxa* collecting expedition.

Evening presentation by ID on the aim of producing a Conservation Checklist of the plants of Mefou Proposed National Park and how the current work will contribute to this.

Tuesday 23rd March

Group 1: ME, ID, EN, NR, RB, GK return ot Mefou River and adjacent *Raphia* swamp NW of Ndanan II. Notable finds include *Leonardoxa* × *africana* in flower, *Rhabdophyllum affine* subsp. *myrioneurum* and several sterile plants of *Ancistrocladus* sp. in dense riverbank forest.

Group 2: MC, JMO, TF, AB, AP, GS, DR return to Mbene River near Ndangan 1, proceeding further down river from the same site as 19th March. Notable finds include *Neostenanthera* sp. in flower and fruit, several new Caesalpiniod legume fruit records, three new *Ficus* sp. and several sterile orchid records, new to the park, identified by DR.

CB sorts specimens, guards camp and fetches water with CWAF car. CP begins databasing of Team I specimens.

Evening presentation by NR on *Ficus* in Africa.

Wednesday 24th March

Group 1: ME, ID, NR, EN, CP, AP collect between Ndanan 2 and Metet, reaching as far as the new bridge over the Mefou River and returning via a track from Metet village, collecting in semi-deciduous forest. Collections include several fertile *Ficus* spp., *Celtis* cf. *gomphopylla* in fruit, a woody *Dracaena* sp. in bud, and a sterile Loranthaceae sp. A large area within the park boundary between Ndanan 2 and the bridge under construction was noted to have been recently burnt and subsequently colonised by *Pteridium aquilinum* with some pineapple plantation.

Group 2: JMO, EM, DR, AB, GS to water source near village, then onwards to swamp edge. Interesting finds include a new site for *Pararistolochia* (the third location) and *Tricalysia amplexicaule*, and a bizarre *Pyrenacantha* sp.

CB accompanies NH, GT and TF to Yaoundé for supplies including gas for the specimen driers. TF collects permits from HNC. DJ is collected from Yaoundé. RB continues with specimen data entry, GK sorts specimens, guards camp and collects water with the CWAF car. MC remains at the village to go over the accounts, reports and collecting books.

Rain 3–4pm, the first for several days. Evening presentation by LF on his history of fern collection.

Thursday 25th March

Group 1: ME, ID, EN, DR, CP, CB, GK collect in the forest to the left of the path between Ndanan 1 and Ndanan 2, which includes areas of steep terrain. Notable discoveries include *Begonia* sp., a further *Aframomum* sp. and *Auxopus macranthus* (Orchidaceae), the first saprophyte record for Mefou. This brings the orchid total for the park now to 21 taxa (DR). A further site for *Cola letouzeyana* is found (EN). However, few taxa in flower in this part of the forest.

Group 2: MC, JMO, TF, EM, AP, RB, NR, DJ return to Mbeme River forest as 19th and 23rd March, trekking along river bed back to Ndanan 1 via the bridge. According to EM, river bed leg only takes 25–30 minutes if non-stop. Notable discoveries include an additional pink-flowered *Aframomum* sp. with long ligules, *Calanthe sylvestris* (brought in by AP), several more *Ficus*, bringing the total now to c. 15 taxa, a further site for *Leonardoxa* × *africana* (JMO) and a new site for 5 plants of *Tricalysia amplexicaulis*.

AB and GS remain in camp to continue clearing data entry backlog and specimen sorting. GT guards camp. Shower facilities built in evening by ME and GT.

Evening presentation by JMO on Cameroon, HNC and himself. Presentation by DJ. Songs by CP, AB, GS.

Friday 26th March

Group 1: ME, ID, EN, NR, GK, CP return to the Mefou River and further explore the *Raphia* swamp. Notable discoveries include an unusual *Pancovia/Placodiscus* in fruit (ID) and *Octoknema* sp. in flower and fruit (GK). Sterile plants of *Pandanus* sp. are observed on the opposite bank of the river but are not accessible.

Group 2: JMO, TF, EM, CB, GS, AB collected along the River N of Didoumou; collections included *Sorindeia* sp., *Mostuea* sp. and *Lindackeria* (*Oncoba*) *dentata*.

RB supervises DJ (specimen sorting) and AP (databasing) whilst processing lichen specimens. MC does accounts and reports. Staff from BAT Cameroon arrive in late afternoon to drop off AP's luggage and to look around the field site.

Saturday 27th March

Group 1: ID, NR, RB, GK, EN, BN take circular route from Ndanan 1 to good forest and *Raphia* swamp to left of path to Ndanan 2, returning via Baboon enclosure. Notable finds include 1–2 *Cyathea* spp., a *Begonia* sp. with peltate leaves and winged fruits, and fertile material of *Mapania* sp.

Group 2: JMO, TF, EM, AP, AB continued collecting in secondary forest along the trail from Ndanan 1 towards Ndanan 2.

DR remains in camp to identify and process orchid specimens from previous days and to observe chimpanzee feeding behaviour, including their reaction to the *Pancovia/Placodiscus* fruits collected by ID on the previous day, which the chimpanzees consumed readily. DR interviews MC for his radio broadcast.

GS accompanies NH to market for purchase of provisions. CB and CP remain in camp for specimen sorting and data entry. ID begins entering his data onto the database.

MC is taken by GT to Nsimalen Airport at 5.00pm for return flight to UK. ME returns to Nyasoso to see his family, agreeing to return for the final stage of team II and for team III.

Evening presentations by CP, CB and NR.

Sunday 28th March

Rest day. JMO and TF take CB, GK, AB, AP and GS to Yaoundé for sightseeing and shopping; the trip is, however, cut short by the onset of heavy rains in the mid-afternoon. ID, NR, DR, EN, CP, RB, DJ and LF remain in camp / in the village. DR completes orchid identifications. ID, DJ and EN begin sorting of all specimens to date. RB continues with lichen collection and processing of specimens, collected under ID numbers.

Monday 29th March

Early am: JMO, CB and EM visit the school director in Metet to arrange for the volunteers to visit the classes on 1st April. JMO collects several specimens en route.

Group 1: TF, DR, GK, RB, BN collect in forest W of the Chimpanzee enclosure; collections include 2 *Memecylon* spp.

Group 2: JMO, EM, CB, CP, collect in forest adjacent to Metet village following the visit to the school. Collections included a different *Pararistolochia* sp.

ID and NR remain in camp to complete the sorting of the specimens and to continue databasing and begin label production; AB and GS begin label insertion, guard camp and fetch water with GT.

Evening presentations by AP and AB.

Tuesday 30th March

Group 1: ID, AB, DJ, CP, BN return to the *Raphia* swamp near Ndanan 2 and forest to the E of the Ndanan 1 – Ndanan 2 trail. More taxa have come into flower and fruit following the rains of the 28th.

Group 2: JMO, TF, GS, GK, EM, NR took the EW vehicle to collect near the village of Biyeme in the NE sector of the park where a semi-deciduous forest and farmbush mosaic was found. The area was productive; notable finds included *Corymborkis minima*, identified by DR, only the second record of this Red Data candidate, the first being from Korup NP. Other collections include *Baillonella toxisperma*.

EN accompanies LF to the site of the *Cyathea* spp. collected by ID on 27th March; LF makes supplementary collections of these and other ferns at this site.

DR remains in camp to process new orchid specimens under Cheek numbers. RB collects lichens locally and continues to process these specimens. CB and AP remain in the village to process and database specimens. Late pm: following difficulties with the generator, a power surge damages the laptop transformer, rendering it unusable.

Evening presentations by GK, DR and ID.

Wednesday 31st March

Group 1: ID, EN, GS, CP, RB, BN take a route through the *Raphia* swamp then along the Mefou River for approximately 2 km until meeting the bridge under construction near Metet; return via trail from Metet. Several orchid taxa recorded, together with *Ficus cyathistipula* (EN), *Ficus* sp., a tree to 30 m with copious figs on short spurs throughout the trunk and main branches, and first records of *Scadoxus* and *Polygonum* for the park. GS suffers from dehydration but recovers.

Group 2: JMO, NR, DR, DJ, CB return to the NE sector of the park but find large-scale encroachment of farms and cocoa plantations. Few plants collected, but a second site for *Thonningia sanguinea* is recorded. Return 6.00pm after walking the return journey.

TF, AB, NH to Yaoundé for camp supplies and to try to purchase a replacement laptop transformer; these cost 99,000 fcfa and are therefore not purchased – it is decided that we will wait for a spare transformer for NR's old laptop to be sent out with one of the volunteers on team III. GK guards camp. AP is feeling feverish so remains in camp.

Evening: NR prepares *Ficus* specimens with aid of volunteers.

Thursday 1st April

Am: all volunteers except LF and RB visit the school at Metet with JMO. JMO and CB give presentations on our research in Cameroon and the involvement of volunteers from around the world. The plight of conservation of the Mefou Proposed National Park is promoted. The children sing the Cameroonian National Anthem.

RB and LF continue with processing of the lichen and fern specimens respectively. TF collects in forest around Ndanan 1 with DJ. ID and NR guard camp and continue fig identification. DR processes the remaining orchid specimens from 31st March.

Pm: Group 1: ID, CB, GK, AB, EN, DJ collect ruderals and grass taxa around Ndanan 2 and by the Mefou River.

Group 2: TF, RB, CP, EM collect additional taxa around Metet and the Mbeme River.

AP and GS remain in camp to recover from illness. JMO works on manuscripts in camp and prepares his collecting books.

Early evening: ID and GT travel to Mefou to collect ME, expected at 6.00pm. Heavy downpours throughout the journey. ME does not arrive.

Evening presentations by ID on the role of plant taxonomists and use of herbarium specimens, and by DR on Kew's role in conserving tropical biodiversity. General discussion on the importance of conservation, led by ID and DR. Late pm, ME arrives from delayed journey (2 buses broken down).

Friday 2nd April

Whole group involved in processing specimens for sending to YA – YA and K sets separated, separation of CITES material, bundles labelled and bagged. Majority of specimens sent with GT in EW vehicle in pm, accompanied by TF and RB. RB visits HNC and is shown around by TF.

Pm: ID begins payments for facilities and staff. On the request of J. Dransfield (K), DR and EM collect silica gel material of the *Raphia* sp. from the fertile tree from which an earlier herbarium collection had been made by MC.

Evening presentation by Cathy (CWAF volunteer, Canada) on CWAF's role in primate rehabilitation, the fight against the bushmeat trade and the future plans for Mefou Proposed National Park. Invites volunteers to visit the primate nursery the following morning. JMO talks on the Cameroonian conservation system. Père of Ndanan 1 village gives farewell speech, provides palm wine and hosts small gathering in his house, attended by AP, CP, EN and ID.

ME falls ill, suspected malaria.

Saturday 3rd April

Immediately after breakfast, GT takes JMO, TF, CP, GK, AB, AP to Yaoundé; JMO and TF dropped off at HNC together with remaining specimens. CP dropped off for onwards journey to Douala at midday where he will meet a birdwatching guide for trip to Mt Kupe (guide arranged via NH). The remainder to Meumi Hotel.

CB and GS visit CWAF nursery. ID, DR, NR, ME, EN begin preparations for first load transfer to Mantum, nr. Bamenda for team III. ME still suffering from fever, but taking medication. ID finalises payments of staff at Ndanan I. NH falls ill, suspected malaria.

3pm: ID and NR accompany LF, RB, CB and GS to Nsimalen Airport for evening flight home, then transfer by taxi to Meumi Hotel, Yaoundé. GT returns to Ndanan I where DR, EN, NH and ME are continuing preparations for transfer to Bamenda.

RESULTS OF TEAMS I AND II

1. SUMMARY OF COLLECTIONS MADE

A total of 1560 collections were made from Mefou Proposed National Park over the four week period 7th March – 3rd April 2004 (Table 2). Of these, 216 were voucher specimens taken from two 25 × 25 m plot surveys; these specimens are unicates and will be sent to K as a loan from YA, for identification and subsequent return. The remaining specimens represent largely fertile material, collecting in sets of approximately 5 duplicates, the first set being separated out and now awaiting incorporation at YA, the remainder to be sent to K following the receipt of export permits from the Cameroonian government and, in the case of taxa covered by CITES (Orchidaceae, Cyatheaceae and Cactaceae taxa), the completion of all CITES documentation. Only preliminary determinations were given to the specimens in the field; verification and determination of as-yet un-named specimens will take place largely at K, with assistance from YA and from specialists in other herbaria.

The lichen collections made by RB, under the collecting permit of ID, represent the first exploration of the lichens of Mefou Proposed Nationa Park and will thus provide a significant further contribution of the Kew-Earthwatch expedition to the documentation of the biodiversity of this site. The sets sent to K will be forwarded as a loan to RB for study and identification.

Specimen data entry onto the Cameroon Specimens Database, and subsequent label generation, was incomplete due to the problems with the laptop transformer. K do not currently hold the data for the collections made by C. Cabral.

TABLE 1. Summary of collections made at Mefou Proposed NP March 7th – April 3rd 2004.

Collector	Opening number	Closing number	Total collections
Brinklow R. in Darbyshire I.	38	155	118
Cabral C.	101	111	11
Cheek M.	11481	12017	537
Darbyshire I.	180	335	156
Etuge M.	5099	5307	209
Fay L. in Darbyshire I.	4799	4838	40
Nana F.	24	74	51
Onana J.-M.	2800	2928	129
Rønsted N.	200	221	22
Tadjouteu F.	566	636	71
Forest Plot	1	124	124
Raphia swamp plot	1	92	92
TOTAL			**1560**

2. VEGETATION AND HABITAT TYPES AND THEIR SIGNIFICANCE FOR CONSERVATION

By more complete coverage of the park, with a greater number of botanists collecting, aided by the input of the expertise of Essama M. (MINEF) and with our trip coinciding with a period of widespread flowering and fruiting of plant taxa, a much greater understanding of the habitat types and their importance has been gained than on the October 2002 expedition. The park has proved much more diverse than at first suspected and with a greater number of taxa potentially important for promoting conservation.

a. Evergreen forest

This forest type, located patchily within the park in low-lying areas in strips along rivers which are not seasonally inundated, is largely dominated by *Gilbertiodendron dewevrei* (Leguminosae: Caesalpinioideae) together with a diverse understorey assemblage often dominated by *Scaphopetalum* (Sterculiaceae). It is this habitat that contains the greatest number of rare and potential Red Data species, including *Cola letouzeyana* (Sterculiaceae) and *Tricalysia amplexicaule* (Rubiaceae), together with interesting taxa such as *Uvariopsis* sp. (Annonaceae), *Octoknema* sp., *Ancistrocladus* sp. and several Leguminosae: Caesalpiniodeae and *Ficus* taxa. Several Apocynaceae taxa were in abundant flower during the current expedition.

Collection of specimens from a 25 × 25 m plot in representative evergreen forest between Ndanan 1 and Ndanan 2 villages should further elucidate the characteristic species of this vegetation type at Mefou.

b. Semi-deciduous forest

This forest type is located in the better-drained areas of the park, dominating some of the undisturbed areas. This vegetation type is dominated by *Triplochiton scleroxylon* (Sterculiaceae) and Ulmaceae taxa: *Celtis* and *Holoptelea*. The presence of *Lophira alata* (Ochnaceae), one of several timber species, in this habitat is of conservation significance (see section 3).

Good semi-deciduous forest was recorded at the proposed site of the new chimpanzee enclosure between Ndanan 1 and Ndangan 1 villages.

c. *Raphia* swamp forest

Located in seasonally-inundated low lying areas of the park between Ndanan 1 and Ndanan 2 and particularly adjacent to the Mefou River near Ndanan 2 village. Unlike in October 2002, standing water was largely absent in March-April 2004, though the soil remained wet, with pools persisting in depressions. It is dominated by two *Raphia* spp. (Palmae), but proved surprisingly diverse, with associated taxa including several Apocynaceae spp. abundantly in flower, *Drypetes* sp. (Euphorbiaceae) and herbaceous taxa including Cyperaceae spp., Marantaceae spp. and, along dried water channels, *Lasiomorpha* (Araceae). Taxa potentially important for conservation include *Leonardoxa* × *africana* (Leguminosae: Caesalpiniodeae) and an unusual *Pancovia/Placodiscus* sp. (Sapindaceae).

A 25 × 25 m plot conducted in representative *Raphia* swamp, will further elucidate the components of this vegetation community.

d. Riverine herbaceous communities and aquatics

Herbaceous communities were recorded along the Mefou River and along the river between Ndanan 1 and Ndangan 1. Dominated by Cyperaceae and Poaceae, interesting taxa include an *Impatiens* sp. (Balsaminaceae) recorded under *I. mannii* in Fl. Cameroun, but differing significantly in habitat, being restricted to lowland river courses with seasonal inundation, and in having pink, slightly hairy, flowers, and thus potentially a separate species, restricted to Cameroon. Other taxa include at least 2 *Ludwigia* spp. (Onagraceae).

The slower-moving sections of these rivers support interesting aquatics, including *Nymphaea* sp. (Nymphaceae), a potential *Nymphoides* sp. (Menyanthaceae) and *Ceratopteris* sp. (Parkeriaceae).

e. Farmbush and farmland

Large areas of the park are dominated by this vegetation type. Active clearance of understorey vegetation, and burning of secondary growth, were observed during the current field period. The main cultivation types are:

Cocoa plantation — the canopy trees are largely left intact in the cultivation type, but the understorey is largely cleared.

Manioc farming — involving complete clearance of natural vegetation, with only weedy species persisting.

Other crops cultivated include yams, pineapples, maize and mangos.

The communities developing in abandoned farmland were largely of weedy species such as *Pteridium aquilinum*, *Sida* spp. and grasses, together with shrubs such as *Harungana madagascariensis*.

As stated in the expedition report for October 2002, this vegetation type is of little value to conservation but could be used in public education to demonstrate forest regeneration following agricultural abandonment. It is also unfortunate that these degraded habitats are not being preferred by CWAF as sites for new primate enclosures, instead of using prime forest sites which will inevitably become degraded following release of animals into the enclosures.

3. RED DATA TAXA

Our knowledge of the status in the parkof the two taxa identified in the October 2002 expedition report as Red Data candidates has been improved significantly:

Tricalysia amplexicaulis Robbr. (Rubiaceae)

Several new sites were discovered for this species (see day by day accounts). In addition, the moving of 3 individuals from the site designated as a new Chimp enclosure to the nature trail should benefit this species through alerting its importance to visitors and through allowing CWAF staff to familiarise themselves with this taxon and thus to discover and protect new plants in other areas of the park.

Cola letouzeyana Nkongm. (Sterculiaceae)

In October 2002, only 2 trees of this species were found, both in precarious sites. The current fieldwork revealed several extra sites in undisturbed forest, with flowering individuals. Its status in the park therefore appears more stable than previously thought.

A third Red Data candidate was recorded on this expedition:

Corymborkis minima P.J.Cribb

Previously known only from the Korup National Park, collected by D.W. Thomas, this species was recorded at one site close to the park boundary near Metet in undergrowth on a stream bank. If the determination is verified, it is likely to qualify as EN under criterion B of the IUCN Red Data guidelines, as its future at Mefou appears highly uncertain.

In addition, one taxon on the Red Data list (www.redlist.org) was collected:

Lophira alata Banks ex Gaertn.f.

VU A1cd

This species is recognised as VU because of its quality timber, resulting in widespread over exploitation of mature trees. In Mefou Proposed National Park, it is local in semi-deciduous forest, being recorded between Ndanan 1 and Ndangan 1 (MC) and between Ndanan 1 and Ndanan 2 (ID, sight record).

Further Red Data species or candidates are likely to be found on identification of the specimens at K. The most likely additions are *Impatiens* sp. "*mannii*", *Leonardoxa* × *africana*, *Uvariopsis* sp. (Annonaceae), *Octoknema* sp. (Olacaceae) and *Neostenanthera* sp. (Annonaceae).

PRELIMINARY CONCLUSIONS

The large number of collections made on the current trip, supplemented by those collected in October 2002, together with the wide coverage of the park and the detailed knowledge gained of habitat types, are sufficient to produce a Conservation Checklist of the Plants of Mefou Proposed National Park once the specimens are received at K. This document will contribute to the growing understanding of the biodiversity of this important remnant forest location and will hopefully contribute to the awarding of full National Park status to this site in the near future.

Lichen data derived from the specimens collected by RB should provide a significant contribution to our knowledge of Cameroonian lichens in general and the only known information on this group at Mefou. Data on bird species collected by CP and ID may also contribute to biodiversity information in the park.

The continued spread of agriculture, despite the proposition of National Park status and the resultant presence of MINEF officials and the CWAF project, is of concern. The ongoing delay to compensate villagers within the park for the change in status of the land is one likely cause of the current failure to halt loss of natural habitat here; it is suspected that for this reason also, trapping and hunting continues within the park boundary. Recent incidents of sabotage of CWAF property and enclosure fencing is likely to have been carried out by disgruntled local residents. However, the resultant placing of a permanent army presence in Ndanan I seems to have brought this particular problem to a halt. On a more positive note, the village of Ndanan 1 appears appreciative of the presence of CWAF and the opportunities that have arisen from the project, and are hopeful that the future gazettement of the site as a National Park will bring with it further opportunities through tourism and land management.

The diversity of plant habitats and associated high species diversity, together with the recognition of 4 clear-cut Red Data candidates and several more likely candidates, including potential new species, should create a strong botanical case for speedy gazettement.

ENDEMIC, NEAR ENDEMIC & NEW TAXA AT MEFOU

Martin Cheek

Herbarium, Royal Botanic Gardens, Kew, Richmond, Surrey, TW9 3AE, UK

ENDEMIC TAXA

Endemic taxa are those unique to an area. Elsewhere we have treated near-endemic taxa as those that occur only in the study area and at one other location. Mefou proposed National Park, topologically and ecologically, is not a unique or isolated location. Consequently it should be no surprise that numbers of endemic and near endemic species are not high. In fact one would not expect any. Five of the apparently new species to science are so far unique, and so apparently endemic, to Mefou. But we believe that this is an artefact of our relatively intensive sampling at Mefou which contrasts with the very low sampling elsewhere in the general Yaoundé-Mbalmayo area. Further surveys will find them elsewhere in the area, so long as relatively intact vegetation can be found in which to sample — or so we believe.

Apparently endemic species at Mefou proposed National Park:

Cleistanthus sp. nov. (Euphorbiaceae)
Necepsia sp. 1 of Mefou (Euphorbiaceae)
Phyllanthus kidna sp. nov. (Euphorbiaceae)
Rhapiostylis sp. 1 of Mefou (Icacinaceae)
Pavetta subg. *Baconia* sp. 1 of Mefou (Rubiaceae)

NEAR ENDEMIC TAXON

Pogostemon micangensis (Labiatae) also known from River Micongo, Angola.

NEW TAXA

To date 12 new taxa have been found at Mefou. In addition to the five listed above as endemics there are:

Vitex sp. of Mefou (Labiatae)
Chassalia manningii Lachenaud ined. (Rubiaceae)
Morinda mefou Cheek (Rubiaceae)
Psychotria longicalyx Lachenaud ined. (Rubiaceae)
Psychotria sp. nov. aff. *fernandopoensis* Petit (Rubiaceae)
Cissus sp. 2 of Mefou (Vitaceae)
Rhapidophora bogneri Boyce ined. (Araceae)

The significant number of new species to science discovered at Mefou was not expected at the outset of our field surveys. Several of the species were already known to be new from specimens collected elsewhere, having been analysed by botanists in the cause of other studies. They simply have not yet been published, although all are globally rare. Other new species such as the *Vitex*, were only discovered to be new to science in the course of identifying the Mefou specimens, when they were matched with specimens collected in Yaoundé. In many cases these new species of Mefou require additional material before they can be published as such. Caution must also be used since although we sincerely believe these to be new to science additional data may become available that disproves them as such.

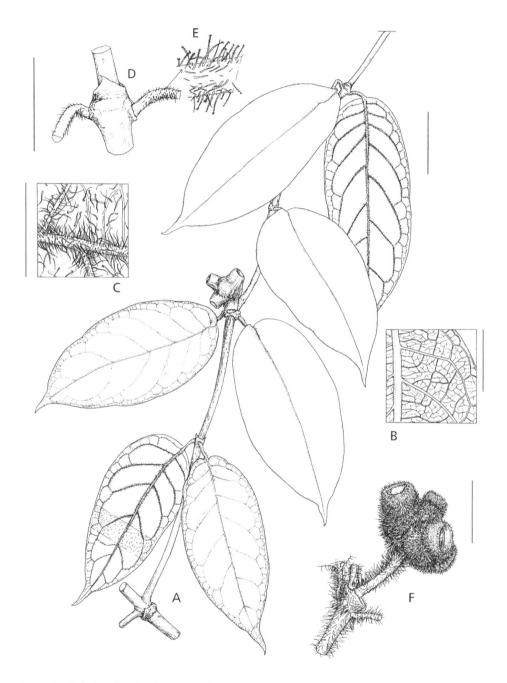

FIG. 5 *Morinda mefou* **A** habit, fruiting branch; **B** lower surface of leaf showing vein detail (hairs not shown); **C** lower surface of leaf showing detail of hairs; **D** node, showing stipule and petiolar insertion; **E** petiolar hairs, including dendritic hairs (artefacts of fungal growth); **F** immature syncarp. **A–E** Drawn from *Etuge* 5214; **F** from *Breteler* 6400. Scale bars: **A** = 3 cm; **B,D** & **E** = 1 cm; **C** & **E** = 5 mm. ALL DRAWN BY JULIET WILLIAMSON.

MORINDA MEFOU (RUBIACEAE),
A NEW FOREST CLIMBER

Martin Cheek[1], Seren Thomas[1] & Emile Kami[2]

[1]Herbarium, Royal Botanic Gardens, Kew, Richmond, Surrey, TW9 3AE, U.K.
[2]CERVE-Herbier National Congolais, Brazzaville, B.P. 1249, Congo-Brazzaville

Abstract: a new species, *Morinda mefou* Cheek discovered in Mefou proposed National Park, Central Province, Cameroon is described as only the third climbing species in that genus for continental Africa, differing from the other two species in being densely hairy, not glabrous, and in having few- fruited (3–4) syncarps, not 6–12-fruited. Its range extends from Cameroon, through Gabon, to Congo (Brazzaville). It is assessed as Vulnerable according to IUCN 2001.

During the identification at RBG Kew of specimens collected during the botanical surveys at the proposed National Park of Mefou 2002–2006 one specimen, *Etuge* 5214 caused interest. A climber, it was identified as a *Morinda* L. since it had fleshy syncarps from which emerged the upper surface of the ovary, bearing the distinctive, fleshy, cylindrical-truncate calyx. The U-shaped seeds and the sheathing stipule with an entire, triangular apex seen in this specimen are also consistent with the genus *Morinda*.

Morinda is a pantropical genus of about 130 species of climbers and trees which is most species-diverse in SE Asia. In tropical continental Africa, only about eight species occur. The genus has never been revised globally, nor for Africa, but the species of E Africa have been treated (Verdcourt 1976: 145–148) as have those for W Africa (Hepper 1963) and for Congolian Africa (Petit 1962). No account has covered the Cameroon, Gabon, Congo (Brazzaville) area (lower Guinea) in which our new species appears.

In tropical Africa only two species of climbing *Morinda* are currently recognised. *Etuge* 5214 differs greatly from these two species in having very hairy leaves, not being glabrous (apart from domatia) and in having very few fruits (3–4) in the syncarps (not 6–12). It also differs in not drying black, but usually an unusual shade of yellow-green, perhaps signifying metal-accumulation of some sort. At Kew the anomalous specimen from Mefou was found to match several other specimens from Gabon and Cameroon, all showing the characters described above. They had previously been unidentified to species or misidentified as *Morinda morindioides* (Baker) Milne-Redh. One specimen, collected about 100 years ago in Bipindi, Cameroon, *Zenker* 2864, the oldest record of the taxon known so far, is annotated in a script from decades ago as "*Morinda* spec. nov." so evidently this paper is not the first hypothesis that this taxon is a separate, undescribed species. It is here formally described as *Morinda mefou* despite the absence of flowers. The species referred to above are separable using the key below, drawn partly from Hepper (1963).

Key to the climbing, scrambling *Morinda* species

1. Leaves usually drying bright light yellow-green, long-hairy on lower surface; syncarps with 3–4 fruits. Cameroon & Gabon . . . . ***M. mefou***
1. Leaves drying black or dark brown, glabrous on lower surface apart from the hairy domatia; Syncarps with 6–12 fruits. Guinea to eastern Congo (Kinshasa) 2
2. Peduncles in pairs; corolla tube long and slender (3–) 4–8 cm, pubescent in throat; calyx undulate; fruit 2–3 cm diam. not lobed, horned by persistent appendices . . . . . . ***M. longiflora***
2. Peduncles solitary; corolla tube short and stout, to 3 cm; glabrous in throat; calyx truncate; fruit lobulate, 4 cm diam. . . . . ***M. morindioides***

Morinda mefou Cheek sp nov, a *M. morindioides* (Bak.) Milne-Redhead et *M. longiflora* G. Don syncarpio 3–4 fructibus tantum composito (non 6–12), lamina foliorum siccitate flavovirenti tota parietate inferiore longe pilosis (non siccitate atrobrunnea glabra domatiis exceptis) differt. Typus; Cameroon, Central Province, S of Yaoundé, Mefou proposed National Park 3°37'N, 11°34'E, fr. 18 March 2004, *Etuge* 5214 (holotypus K; isotypi MO, WAG, YA).

Liana, scrambling on trees to 6 m tall. Stem terete or drying with a shallow groove on each side, green, brown or blackish green, internodes (3–)4–6.5(–8) cm long, 3–4(–5) mm diam., glabrous or sparsely to densely hairy with two hair types; longer hairs 0.75–2 mm, multicellular uniseriate, with 3–6 cells, erect, translucent or slightly yellow, often crinkled or curved, apex entire or bifid; sometimes appearing dendritic due to fungal growth (Fig. 1E), short hairs 0.1–0.2 mm, erect, straight, unicellular, inconspicuous; leaves opposite, equal; blades drying bright yellow-green, less usually brown, elliptic or oblong- elliptic, rarely ovate (5.7–)7–10(–11.2) × (2.8–)3.2–5.8(–6) cm, acumen 0.9–1.2 cm long, gradually tapering, base rounded or truncate, rarely obtuse, midrib slightly sunken above, lateral nerves on each side of the midrib 5–9, parting at 45° from the midrib, later arching upwards and uniting with the nerves above to form a looping marginal nerve, domatia long hairy; quaternary nerves reticulate, raised on the lower surface; indumentum hairy, as the stem, dense on the lower surface midrib, moderately sparsely hairy on the remainder of the lower surface; upper surface glabrous or moderately hairy, the hairs shorter and straighter than on the lower surface; petiole plano-convex in section, 5–8(–9) mm long, 0.75–1 mm wide, glabrous or hairy (see stem); stipule sheath (1–)2–3 × 4–6(–10) mm the basal half swollen, thickened, the apical half submembranous with conspicuous white rod-like raphides visible at the surface; limb broadly triangular, 1–2(–3) mm long, apex acute. Inflorescence and flowers not known. Infructescence a syncarp comprising 3–4 united fruits, subglobose-lobed, green, ripening yellow or orange, hard, 2.2–3.1 cm diam., subglabrous with only inconspicuous short hairs (*Etuge* 5214) or densely long hairy (*Louis et al.* 705); individual fruits subovoid, the free part volcano shaped, c. 15 × 15 mm, the upper 6 mm comprising the subcylindric, thickened, truncate calyx tube which is confluent with the ovary; calyx throat aperture c. 4 mm; disc persisting, annular, flat, 3 mm diam. Seed, bony, grey, lumpy, 10 × 8 × 5 mm, U-shaped, folded transversely, comprising mainly embryo, with the cotyledons and radical discernible from the exterior.

DISTRIBUTION. Cameroon, Gabon and Congo (Brazzaville).

HABITAT. Lowland evergreen semi-deciduous to submontane forest; either forest edge or secondary; 350–600 m alt.

LOCAL NAME. Moeya (Mitsogho language, *Wilks* 1693)

ETYMOLOGY. Named (noun in apposition) for the proposed Mefou National Park on the river Mefou, where the species first came to our attention and which is currently the location for this species which is most likely to be protected.

SPECIMENS. CAMEROON. S Region, Bipindi, st. 1904, *Zenker* 2864 (K); ibid, st. 1908, *Zenker* 3551 (K); Station du Cacaoyer de N'Koemvone, 2°49'N, 10°08'E, fr. 25 March 1975, *de Wilde* 8113 (K, WAG n.v., YA n.v.); C Region, Mefou proposed National Park, 3°37'18"N, 11°34'9"E, fr. 18 March 2004, *Etuge* 5214 (K, MO, WAG, YA n.v.). GABON. Route Owendo- Ikoi Komo, 0°21'N, 9°32'E, fr. 23 Sep. 1987, *Wilks* 1639 (K, MO n.v.); Moanda-Franceville, 1°33'S, 13°15'E, fr. 10 Sep. 1970, *Breteler* 6400 (K, WAG n.v.); Région des Abeilles, 0°13'S, 11°51'E, fr. 15 Nov. 1983, *Louis et al* 705 (K, WAG n.v.). CONGO (BRAZZAVILLE). Environs de Brazzaville, tronçon de route entre N'Tonkama et Moutampa, st. 1967, *Sita* 1935 (IEC)

CONSERVATION. Here we assess *Morinda mefou* as Vulnerable VU B2ab(iii) according to the criteria of IUCN (2001) since it is only known from six locations (see specimens cited above) equating to an area of occupancy of 24 km^2 using the recommended 4 km^2 cells, and since the species is threatened by forest clearance for agriculture, for example at Mefou itself. It is to be hoped that this and other Red Data species at Mefou proposed National Park identified in this volume will be incorporated into a future management plan for the Park and protected for the future. The first step, in this instance, is to refind the original plant, survey for additional individuals in the area, establish actual and possible threats to those individuals so that they may be addressed, evaluate actual regeneration levels in the wild, begin seed collection for propagation in a nursery and future planting out in suitable safe sites to bolster the wild population.

VARIATION. The specimens cited vary widely in certain aspects and particularly in the distribution of the species-characteristic long, multicellular hairs. While present on the lower surface of the leaf-blade in all specimens cited, in some specimens they are also found on the upper surface of the leaf-blade and/or on the petiole, and/or on the stem. See Table 1 below.

OCCURRENCE IN CONGO (BRAZZAVILLE). The discovery of this species in Congo (Brazzaville) was made by the third author (EK) who uncovered the specimen cited (*Sita* 1935) among the specimens at IEC and who sent photos to MC at K. This specimen is stated to have white hairs on the lower surface of the leaf-blades, clearly placing it as *Morinda mefou*.

TABLE 1. Variation in the distribution of long-haired indumentum among specimens of *Morinda mefou*.
✓ = present; ✗ = absent

	Stem	Petiole	Upper leaf blade	Lower leaf blade
Louis *et al.* 705	✓	✓	✓	✓
de Wilde 8113	✗	✓	✗	✓
Zenker 2864	✓	✓	✓	✓
Wilks 1639	✗	✗	✗	✓
Breteler 6400	✓	✓	✓	✓
Etuge 5214	✗	✓	✗	✓

ACKNOWLEDGEMENTS

We thank Jean Michel Onana (IRAD-National Herbarium of Cameroon) for facilitating our research and the Darwin Initiative of the UK Government for part funding the fieldwork and research which led to this paper. Earthwatch Europe supported the fieldwork including Martin Etuge's costs in March 2004 during which the type specimen was collected. Mark Coode translated the Latin. Diane Bridson and Aaron Davis reviewed an earlier version of this manuscript. The first author thanks CWAF and the people of Ndanan for hosting him and his plant surveys in 2002 and 2004. Thanks to Xander van der Burgt for assisting studies of Morinda at IEC.

REFERENCES

Hepper, F.N. (1963) *Morinda*, pp. 188–190 in Hepper (Ed.) *Flora of West Tropical Africa* ed. 2,2. Crown agents, London

IUCN (2001). *IUCN Red List Categories and Criteria: Version 3.1*. Prepared by the IUCN Species Survival Commission. IUCN, Gland, Switzerland and Cambridge, UK.

Petit, E, (1962) Rubiaceae Africanae IX. pp. 173–198 *Bull. Jard. Bot. État. Bruxelles* 32.

Verdcourt, B.V. (1976). Rubiaceae (Part 1). *Flora of Tropical East Africa*. Crowns Agents, London.

This paper is an abbreviated version of that in press with the scientific journal Systematic Botany (reproduced here with permission); publication is expected in 2011. This does not constitute formal publication of the epithet in accordance with the Code.

PHYLOGENETICALLY DISTINCT AND CRITICALLY ENDANGERED NEW TREE SPECIES OF *PHYLLANTHUS* FROM CAMEROON (PHYLLANTHACEAE, EUPHORBIACEAE S.L.)

Gill Challen[1,3], Maria S.Vorontsova[1,2], Harald Schneider[2], Martin Cheek[1]

[1]The Herbarium, Royal Botanic Gardens, Kew, Richmond Surrey TW9 3AE, UK.
[2]Department of Botany, The Natural History Museum, Cromwell Road, London SW7 5BD, UK.
[3]Author for correspondence (g.challen@kew.org)

Abstract: *Phyllanthus kidna*, the only tree species of *Phyllanthus* from Cameroon, is described and analysed. Its bilocular berries and two connate stamens indicate no affiliation with any known species or infrageneric group within *Phyllanthus*. Phylogenetic analyses place it as sister to the Indian and E African species *Phyllanthus pinnatus* and is also related to the New World species of *Phyllanthus acidus* and *Phyllanthus chacoensis*. Only five individuals are currently known and the species is assessed as Critically Endangered. The Proposed Mefou National Park near Yaoundé could save this species from extinction.

Keywords: Cameroon, conservation, Critically Endangered, *Phyllanthus*, Phyllanthaceae, tree.

The Phyllanthaceae are a poorly known and morphologically diverse pantropical plant family of c. 60 genera and 2000 species, segregated from Euphorbiaceae sensu lato following molecular phylogenetic research (Savolainen 2000, APG 2003, 2009). *Phyllanthus* L. is by far the largest genus in the family with 833 species sensu Govaerts *et al.* (2000). It is widespread in tropical and subtropical regions of the world and contains annual and perennial herbs, shrubs, small trees, and even a floating aquatic. Most species of *Phyllanthus* are recognisable by their characteristic phyllanthoid branching, where terminal branches resemble a legume compound leaf: leaves on lateral (plagiotropic) axes develop normally while leaves on the main axes are reduced to scales called cataphylls (Webster 1956). The modern infrageneric classification of *Phyllanthus* includes 10 subgenera and over 30 sections delimited by Webster (1956, 1957, 1958, 2002) based on floral characters, vegetative architecture, and pollen morphology; further contributions to the infrageneric classification have been made by Brunel (1987) for African taxa and Ralimanana & Hoffmann (in press) for groups in Madagascar. The only molecular phylogenetic analyses to date to encompass the whole genus (Kathriarachchi *et al.* 2006) has found several existing subgenera and sections to be polyphyletic and paraphyletic; any new phylogenetic classification will require further research. *Phyllanthus* remains understudied and poorly understood due to its confusing species richness and taxonomically important characters hidden inside small caducous flowers.

African *Phyllanthus* is one of the better-known parts of the genus, studied by Brunel (1987). W Africa is home to more than 100 herbaceous and shrubby species and 33 taxa have been recorded in Cameroon. Vegetation surveys in Cameroon carried out by Martin Cheek and colleagues during the last 12 years have already revealed two new herbaceous species of *Phyllanthus* (Hoffmann and Cheek 2003). The species described here is a tree collected during botanical surveys of the proposed Mefou National Park made by the staff of IRD-Herbier National, Cameroon and the Royal Botanic Gardens, Kew between 2002–2006. We describe the new species, place it in phylogenetic context within *Phyllanthus*, and discuss implications for conservation planning in Cameroon.

MATERIALS AND METHODS

Taxonomic work was carried out by Gill Challen and Petra Hoffmann using collections at the Herbarium, Royal Botanic Gardens, Kew. DNA amplification from *Phyllanthus kidna* and preliminary phylogenetic analysis was carried out by Maria Vorontsova at the Jodrell Laboratory, Royal Botanic Gardens, Kew. DNA sequences for this analysis are from Kathriarachchi *et al.* (2006), DNA extraction, amplification, and sequencing of *Phyllanthus kidna* was carried out using the methods described in Vorontsova *et al.* (2007b), and the matrix used was assembled as part of the work towards Vorontsova *et al.* (2007a). The matrix alignment was amended and final phylogenetic analysis was carried out by Harald Schneider at the Natural History Museum, London.

RESULTS

Maximum Likelihood analysis places *Phyllanthus kidna* in Kathriarachchi *et al.*'s (2006) clade O, a heterogenous assemblage including the Asian *P. acidus* (L.) Skeels from Webster's subgenus *Cicca* (L.) Webster, the New World *P. chacoensis* Morong from section *Aporosella*, and the African/Indian *P. pinnatus* (Wight) G.L.Webster from section *Chorisandra. Phyllanthus kidna* is sister to *P. pinnatus* (Fig. 6).

DISCUSSION

Phyllanthus kidna has no obvious affiliation with any known species or infrageneric group within *Phyllanthus*. In Webster's (1956) classification a tree with phyllanthoid branching and baccate fruit would be placed in the Neotropical subgenus *Cicca. Phyllanthus kidna*, however, does not have the (3–4)5–6 free stamens normally associated with subgenus *Cicca*, but has two connate stamens and smooth seeds. Taxa with connate stamens and smooth seeds are rare in Phyllanthaceae, otherwise only found in the mainly SE Asian subgenus *Eriococcus* (Hassk.) Croiz.& Metc. (usually with lacerate calyx lobes and capsular fruits), *P. warnockii* Webster (previously recognised as *Reverchonia* A.Gray), and N American *P. abnormis* Baill. The genera *Antidesma* L. and *Aporosa* Blume may also have two stamens and may show superficial similarities in their inflorescence and fruit structure (Webster 1956).

Phyllanthus kidna is unique among W African *Phyllanthus* in its combination of tree habit, two connate stamens, and baccate bilocular fruits. The morphologically closest W African species are *P. reticulatus* Poir., *P. letouzeyanus* Jean F.Brunel, *P. muellerianus* (Kuntze) Exell, and *P. raynallii* Jean F.Brunel & J.P.Roux, all of which are large shrubs or small trees. *Phyllanthus muellerianus* and *P. reticulatus* are also similar in their baccate fruits but both have 5–(6) stamens, of which a central group of 2–3 are connate.

The sister group relationship with the E African/Indian *P. pinnatus* is unexpected: the two species share smooth seeds hollowed out at the hilum and bifid styles, but *P. pinnatus* has six free stamens and six calyx lobes. Comparison between these species and *P. acidus* in the same clade is provided in Table 1. *Phyllanthus pinnatus* is another morphological outlier placed in the segregate genus *Chorizonema* Jean F.Brunel (1987), a placement proved incorrect by Wurdack *et al.* (2004). It is tempting to speculate that the most recent common ancestor of *P. kidna* and *P. pinnatus* gave rise to the Old World clade containing these taxa.

The discovery of this morphologically aberrant species highlights the complex and seemingly counterintuitive evolutionary history of *Phyllanthus*. It is hoped that further molecular phylogenetic work across the genus will shed light on character evolution in this enigmatic group.

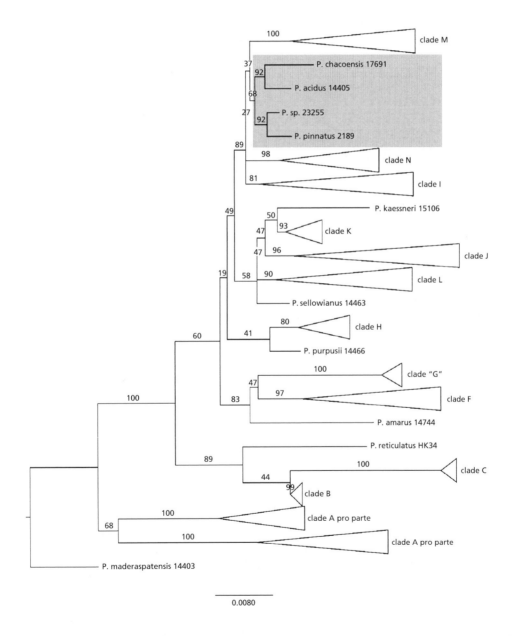

FIG. 6. Maximum Likelihood bootstrap analysis. Clades are defined after Kathirarachchi *et al.* (2006). Clade O has been expanded and is outlined to show the relationships of *Phyllanthus kidna* Challen & Petra Hoffm., sp.nov.

TABLE 1. Morphological characters of the three Old World species in Clade O: *Phyllanthus kidna, P. pinnatus* and *P. acidus*.

	P. kidna sp. nov.	*P. pinnatus*	*P. acidus*
Habit	tree	shrub	tree
Leaf lamina size	0.9–2.3 × 0.5–1.3 cm	1–3(4) × 1–2(5) cm	5–9 × 2.5–4 cm
Calyx lobes	4–5	(5)–6	4
Male disc	4-lobed	entire	4-lobed
Female disc	entire	entire	entire or 4-lobed
Stamens	2, connate	6, free	3–4, free
Staminodes	unknown (insufficient material)	absent	present
Carpel number	2	3–4	3–4
Fruit	baccate	dehiscent capsule	baccate/drupe
Seed	smooth	smooth	smooth
Distribution	Cameroon	E Africa and India	pantropical by introduction

TAXONOMIC TREATMENT

Phyllanthus kidna Challen & Petra Hoffm., **sp.nov.** TYPE: CAMEROON. Central Province: road to Ndangan 1, 03°37'28"N 11°34'53"E, 09 Mar 2004, *M. Cheek 11531* (holotype: K000338996!; isotypes: MO!, P!, US!, WAG!, YA!).

Tree, monoecious or dioecious, most likely deciduous, 10–15 m high. Trunk cylindrical, 15–20 cm diam. at 1.5 m height, dilated at ground level. Bark white-grey, in plaques 6–12 cm diam.; outer slash dark red, thin inner slash mottled white with longitudinal pink-red stripes, with a scent resembling green beans. Orthotropic branchlets terete, simple, with up to 29 leaves per shoot. Brachyblasts solitary or in fascicles of 2–3 arising from short woody stems, sometimes co-axillary with the plagiotropic leafy shoots. Cataphylls deciduous, 1.5 × 1 mm, triangular, dark brown, entire, indurate, glabrous. Cataphyllary stipules triangular, erose, indurate, 0.5 × 0.5 mm. Leaves distichous, glabrous, chartaceous, green tinged red when young; petioles 1.0–1.3 × 0.3 mm. Leaf blades oval to oblong (orbicular), 0.9–2.3 × 0.5–1.3 cm, entire, basally rounded to obtuse, apically rounded to obtuse(mucronate). Drying discolorous, dull olive adaxially, pale glaucous green abaxially. Venation brochidodromous, faint adaxially, barely visible abaxially, veins (5)7–9 pairs. Stipules triangular, 0.5 × 0.3 mm, erose, light brown. Inflorescence bracts 0.3 × 0.3 mm erose, light green to light brown, roughly triangular. Flowers solitary or fasciculate, the fascicles in the axils of lower leaves of brachyblasts, consisting of 1 pistillate and many staminate flowers, also some solitary pistillate flowers in the axils of leafy plagiotropic shoots. Staminate flower pedicels c. 1.3 mm long, capillary. Sepals 4, 0.7–1.3 × 0.5–0.8 mm, equal, suborbicular, erect, apically acute, glabrous, white, occasionally tinged red, margins hyaline. Petals 0. Disc segments 4, discrete, alternisepalous, 0.25 × 0.25 mm, irregularly globular, shortly stipitate. Stamens 2. Filaments 0.07–1.2 mm long, connate for c. $^{7}/_{8}$ of length. Anthers basifixed, 0.1 × 0.1 mm, extrorse, ellipsoid, dehiscing longitudinally. Pistillate flowers seen at early fruiting stage only. Sepals 4–5, 1.2–1.4 ×

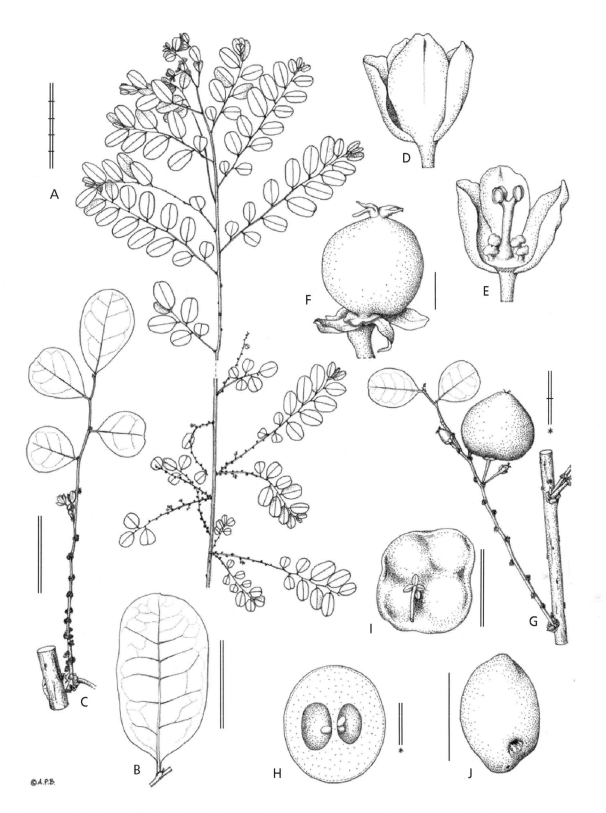

Fig. 7 *Phyllanthus kidna* Challen & Petra Hoffm., sp. nov. **A** Habit showing upper sterile shoots and lower fertile shoots; the dashed line represents a 10 cm section of the stem which has been omitted. **B** Leaf, abaxial surface. **C** Fertile shoot with a male inflorescence. **D** Male flower. **E** Male flower with one sepal removed. **F** Developing fruit. **G** Branch with a young infructescence. **H** Young fruit, transverse section. **I** mature dry fruit with pedicel and calyx. **J** Seed. A–F, I, J drawn from herbarium specimens: A–E from *Cheek 11531*; F, I from *Tadjouteu 611*; J from *Harvey 73*. G, H drawn from field photographs by M. Cheek. Scale bars: single bar = 1 mm; double bar = 1 cm; graduated double bar in A = 5 cm; graduated double bar in G = 2 cm. A star indicates that the scale bar is only approximate. Drawn by Andrew Brown.

0.8–0.9 mm, equal, oblong, apically rounded, erect when young, spreading at maturity. Petals 0. Disc annular, entire. Staminodes 0. Styles 2, free, bifid almost to base, style arms 0.5–0.6 mm, recurved, basally inflated, apically acute. Fruiting pedicels 4.5–5.2 × 0.3–0.4 mm, articulate just above base. Young fruit glabrous, sub-globose. Mature fruit 2 locular, with 1–2 ovules per locule, indehiscent, 1.3 × 1.1–1.3 cm, subglobose with a fleshy pericarp, drying mid-brown. Seeds 0.6–0.7 × 0.7–0.8 cm, ecarunculate, irregularly ovoid, dorsally convex, smooth, light brown when dried, hollowed out at the hilum. Fig. 7.

ETYMOLOGY AND USES. *Phyllanthus kidna* takes its specific epithet, as a noun in apposition, from the name for the species at the type locality. Elders in the village of Ndanan 1 inside the proposed National Park of Mefou know it as Kidna (Bety language) and know it to have an unspecified medicinal use according to Victor Ngueben who works at the Park (pers. comm. to Cheek, 2006).

DISTRIBUTION AND HABITAT. Cameroon, Central Region, Mefou Proposed National Park, near park entrance. Cut-over semi-deciduous forest planted with Cocoa (*Theobroma*). Other surviving indigenous tree species include *Triplochiton scleroxylon* K.Schum., *Celtis* L. spp., and *Baillonella toxisperma* Pierre. Fig. 8.

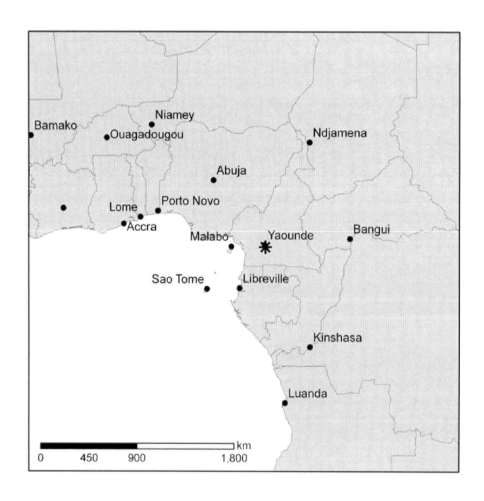

FIG. 8 The only known collection locality of *Phyllanthus kidna* in the proposed Mefou National Park near Yaoundé, Cameroon, is marked with a star.

84

CONSERVATION. *Phyllanthus kidna* is assessed using the criteria of IUCN (2001) as CR B2ab(iii) + D, i.e. Critically Endangered. This is because it is currently known from only a single location (AOO 4 km^2 using 4 km^2 cells) where it is threatened by agriculture. One of the two individuals (*Cheek* 13098) seen by Cheek in 2006 was regrowing from a stump of a tree that had been cut in the previous 1–2 years, possibly to facilitate cocoa cultivation or to furnish firewood or poles. Much of the area of the proposed National Park has previously been logged for timber for export and subsequently been cultivated for small-holder cacao or manioc which was ongoing in 2002–2006. It is estimated that, at the moment, only five individuals have been seen of the species during that time frame, when it was discovered during botanical surveys of the proposed park. It is to be hoped that the species will be found elsewhere in the Yaoundé area and that in future its threatened status will be reduced.

NOTES. Assignment to a section awaits clarification of infrageneric boundaries by further molecular phylogenetic studies.

SPECIMENS EXAMINED. CAMEROON. Central province: Ndanan 1, next to village, 03°37'33"N 11°34'53"E, 12 Oct 2002, *Y.B. Harvey 73* (K, YA); Ndanan 1, Mefou National Park, west of big chimpanzee enclosure, 03°37'00"N 11°35'00"E, 29 Mar 2004, *F. Tadjouteu 611* (K, MO, P, US, WAG); Ndanan 1, Mefou National Park, road from Ndanan 1, bisecting path, 03°37'20"N 11°35'01"E, 26 Feb 2006, *M. Cheek 13098* (K,YA); Ndanan 1, Mefou National Park, road from Ndanan 1, bisecting path, 03°37'20"N 11°35'01"E, 26 Feb 2006, *M. Cheek 13102* (K, WAG, YA).

ACKNOWLEDGEMENTS

The fieldwork and research which led to the discovery of this species at the proposed Mefou National Park was initiated in 2001 by an invitation from the then Minister for the Environment of Cameroon, Sylvestre Ondoa. We thank him and his successor, Elvis Ngollengolle and their staff for their support. This work was partly supported by Earthwatch Europe and by the Darwin initiative programme of the UK Government through the Red Data Plants, Cameroon project. MC thanks Dr Jean Michel Onana head of IRD-National Herbarium, for facilitating the involvement of his staff with the survey and obtaining permits, Cameroon Wildlife Action Fund (CWAF), particularly Rachel Hogan for hosting the survey teams in the proposed National Park, and Bryan Carroll of Bristol Zoo for discussion on the Park's status. We would also like to thank Andrew Brown for the beautiful illustration.

LITERATURE CITED

Angiosperm Phylogeny Group II (2003). An update of the Angiosperm Phylogeny Group classification for the orders and families of flowering plants: APG II. *Botanical Journal of the Linnean Society* 141: 399–436.

Angiosperm Phylogeny Group III (2009). An update of the Angiosperm Phylogeny Group classification for the orders and families of flowering plants: APG III. *Botanical Journal of the Linnean Society* 161: 105–121.

Brunel, J. F. (1987). Sur le genre *Phyllanthus* L. et quelques genres voisins de la tribu des Phyllantheae Dumort. (Euphorbiaceae, Phyllantheae) en Afrique intertropicale et à Madagascar. Thèse doctorale, Université Louis Pasteur, Strasbourg, France.

Govaerts, R., Frodin, D. G. and Radcliffe-Smith, A. (2000). *World checklist and bibliography of Euphorbiaceae (with Pandaceae)*, vols. 1–4. Royal Botanic Gardens, Kew, UK.

Hoffmann, P. and Cheek, M. (2003). Two new species of *Phyllanthus* (Euphorbiaceae) from south west Cameroon. *Kew Bulletin* 58: 437–446.

IUCN (2001). *IUCN Red List Categories and Criteria*: Version 3.1. Prepared by the IUCN Species Survival Commission. IUCN, Gland, Switzerland and Cambridge, UK.

Kathriarachchi, H., Samuel, R., Hoffmann, P., Mlinarec, J., Wurdack, K.J., Ralimanana, H., Stuessy, T.F. and Chase, M.W. (2006). Phylogeny of the tribe Phyllantheae (Phyllanthaceae) based on nrITS and plastid *matK* sequence data. *American Journal of Botany* 93: 637–655.

Ralimanana, H., and Hoffmann, P. Taxonomic revision of *Phyllanthus* (Phyllanthaceae) in Madagascar and the Comoro Islands I: Synopsis and Subgenera *Isocladus*, *Betsileani*, *Kirganelia* and *Tenellanthus*. *Kew Bulletin*, in press.

Savolainen, V., Fay, M.F., Albach, D.C., Backlund, A., van der Bank, M., Cameron, K.M., Johnson, S.A., Lledó, M.D., Pintaud, J.-C., Powell, M., Sheahan, M.C., Soltis, D.E., Soltis, P.S., Weston, P., Whitten, W.M., Wurdack, K.J. and Chase, M.W. (2000). Phylogeny of the eudicots: A nearly complete familial analysis based on *rbcL* gene sequences. *Kew Bulletin* 55: 257–309.

Vorontsova, M.S., Hoffmann, P., Kathriarachchi, H., Kolterman, D.A., and Chase, M.W. (2007a). *Andrachne cuneifolia* (Phyllanthaceae; Euphorbiaceae s.l.) is a *Phyllanthus*. *Botanical Journal of the Linnean Society* 155: 519–525.

Vorontsova, M.S., Hoffmann, P., Maurin, O. and Chase, M.W. (2007b). Molecular phylogenetics of tribe Poranthereae (Phyllanthaceae; Euphorbiaceae sensu lato). *American Journal of Botany* 94: 2026–2040.

Webster, G. L. (1956). A monographic study of the West Indian species of *Phyllanthus*. *Journal of the Arnold Arboretum* 37: 91–122, 217–268, 340–359.

Webster, G. L. (1957). A monographic study of the West Indian species of *Phyllanthus*. *Journal of the Arnold Arboretum* 38: 51–80, 170–198, 295–373.

Webster, G. L. (1958). A monographic study of the West Indian species of *Phyllanthus*. *Journal of the Arnold Arboretum* 39: 49–100, 111–212.

Webster, G. L. (2002). Three new sections and a new subgenus of *Phyllanthus* (Euphorbiaceae). *Novon* 12: 290–298.

Wurdack, K. J., Hoffmann, P., Samuel, R., de Bruijn, A., van der Bank, M. and Chase, M. W (2004). Molecular phylogenetic analysis of Phyllanthaceae (Phyllanthoideae pro parte, Euphorbiaceae sensu lato) using plastid *rbcL* DNA sequences. *American Journal of Botany* 91: 1882–1900.

RED DATA TAXA OF MEFOU

Martin Cheek

Herbarium, Royal Botanic Gardens, Kew, Richmond, Surrey, TW9 3AE, UK

INTRODUCTION

As in the previous conservation checklists in this series (Harvey *et al.* 2004, Cheek *et al.* 2004), **all** flowering plant taxa recorded in the checklist have been assessed on a family by family basis for their level of threat, i.e. as threatened (CR — Critically Endangered; EN — Endangered; VU — Vulnerable), Near Threatened (NT) or of Least Concern (LC).

The main part of this chapter consists of taxon treatments, giving detailed information on the 26 Red Data species known to be present in the checklist area. Several of these taxa were assessed for the first time in the process of writing this book while others had been assessed as threatened in other publications and are reassessed here. IUCN rules do not allow acceptance of assessments of taxa unless they are either published or on the brink of publication. Consequently, the several new species to science that are not yet at this point, some of which are only known from Mefou, have not been assessed and are therefore not mentioned in this chapter.

This, the introductory part of the chapter, details the methodology used in making the assessments, followed by a series of lists in which Red Data taxa are detailed by vegetation type.

ASSESSMENTS — METHODOLOGY

TAXA OF LEAST CONCERN (LC)

In the first place all taxa that were found to be fairly widespread e.g. Extent of Occurrence (IUCN 2001) greater than 20,000 km^2 (and/or from 20 or more localities) were listed as LC. These facts were established principally using F.W.T.A. as an indication of range and number of collections sites, supplemented by other published sources, such as Fl. Cameroun, or by research into specimens at the Herbarium, Royal Botanic Gardens, Kew (K), in cases of doubt. Taxa which, by these measures, are widespread and common do not qualify as threatened under Criterion B of IUCN (2001), the main criterion used in our previous checklists for assessing threatened species. Under Criterion A, widespread and common taxa may, in contrast, still be assessed as threatened, but only if their habitat, or some other indicator of their population size has been, or is projected to be, reduced by at least 30% in the space of three generations, so long as this does not exceed 100 years.

THREATENED AND NEAR THREATENED TAXA

Those taxa in the checklist that were not assessed as LC, were then checked for level of threat using IUCN (2001). Most of these taxa were at least fairly narrowly endemic (restricted) in their distributions, e.g. endemic to Cameroon, or to NW Province, Cameroon and Nigeria. Criterion C, which demands a knowledge of the number of individuals of a species, was not used, since these data were not usually available. Criterion E, which depends on quantitative analysis to calculate the probability of extinction over time, was also not used. Criterion D was used for the only two species *Phyllanthus kidna* and *Pogostemon micangensis.*

i) Using Criterion B

Almost all of the remaining assessments were made using Criterion B (usually B2ab), since the nature of the data available to us for our taxa lends itself to this criterion. Knowledge of the populations and distributions of most tropical plant species is mainly dependent on the existence of herbarium specimens. This is because there are so many taxa, most of which are poorly known and have never been illustrated. For this reason observations based only on sight-records are particularly unreliable and so undesirable in plant surveys of diverse tropical forest. Exceptions can be made when a family or genus specialist, working with a monograph at hand, or a proficient tree spotter identifying timber tree species. In contrast, surveys of birds and primates are not specimen dependent, since species diversity in these groups is comparatively low, and comprehensive, well-illustrated identification guides are available.

For the purpose of Criterion B we have almost always taken herbarium specimens to represent 'locations'. Deciding whether two specimens from one general area represent one or two locations is open to interpretation, unless they are from the same individual plant. Generally in the case of several specimens labelled as being from one town e.g. 'Bipindi', or one forest reserve e.g. 'Bafut Ngemba FR', these have been interpreted as one 'location'. Where a protected area has been divided into several geographical subunits, as at Mt Cameroon (see Cable & Cheek, 1998), and it is known that, say six specimens of a taxon occurring at Mt Cameroon fall into two such subunits, then this is usually treated as two locations. 'Area of occupancy' (AOO) and 'extent of occurrence' (EOO) have been measured by extrapolating from the number of locations at which a species is known. The grid cell size used for calculating area of occupancy is that currently advocated by IUCN, that is 4 km². Information on population declines in relation to Criterion B has been obtained from personal observations, sometimes supplemented by local observers.

ii) Using Criterion A

Very few assessments were made using Criterion A. This criterion is used when a population reduction of over 30% over three generations (but not exceeding 100 years) can be demonstrated or reasonably estimated. This has been the case for timber species with a range across W Africa due to forest disappearance in such countries as Ivory Coast, Ghana and Nigeria. Most members of the mahogany family (Meliaceae) included here had previously been assessed using this criterion.

iii) Using Criterion D

Assessments using Criterion D generally depend on a knowledge of the global numbers of mature individuals (D). This knowledge is generally not available for plant taxa in Cameroon except under VU D2. However, in the case of narrow endemics, where all known observations of a taxon have been made by us, the assessors, or can reasonably be inferred, this criterion can be invoked.

RED DATA SPECIES BY VEGETATION TYPE

The Red Data species of Mefou proposed National Park are presented below under each vegetation type in which they occur. The vegetation types follow the classification used in the vegetation chapter. No species occurs in more than one vegetation type.

In the Red Data taxon treatments that conclude this chapter the species are presented in the same order as in the main checklist, i.e. Dicotyledonae, Monocotyledonae, then alphabetically by family and within family first by genus, then by species. The author(s) of the family assessment is given.

The most important vegetation type for threatened taxa is the evergreen *Gilbertiodendron dewevrei* forest (15 Red Data taxa), with significant numbers also occurring in semi-deciduous vegetation (10 Red Data taxa) and with some in open inundated swamp and river bed habitats (2 Red Data taxa).

RED DATA TAXON TREATMENTS

Several of the following Red Data taxon treatments are modified and updated from accounts written earlier, both published (e.g. The Plants of Bali Ngemba, Harvey *et al.* 2004) and unpublished (e.g. Cameroon Plants Red Data book).

1. SEMI-DECIDUOUS FOREST MOSAIC (600–700 m alt.)

Fernandoa ferdinandi VU

Phyllanthus kidna CR

Entandrophragma angolense VU

Entandrophragma cylindricum VU

Entandrophragma utile VU

Khaya ivorensis VU

Lovoa trichilioides VU

Morinda mefou VU

Chazaliella obovoidea subsp. *villosistipula* VU

Cyphostemma camerounense EN

2. EVERGREEN *GILBERTIODENDRON DEWEVREI* FOREST (600–700 m alt.)

Dictyophleba setosa VU
Momordica camerounensis EN
Momordica enneaphylla VU
Guarea cedrata VU
Heckeldora ledermannii EN
Trichilia zewaldae EN
Platytinospora bucholzii var. *macrophylla* EN
Chassalia manningii EN

Hymenocoleus nervopilosus var. *orientalis* EN
Poecilocalyx setiflorus VU
Pseudosabicea sthenula VU
Tricalysia amplexicaulis VU
Cola letouzeyana VU
Leptonychia subtomentosa EN
Aulotandra kamerunensis EN

3. OPEN, PERIODICALLY INUNDATED HABITATS

Lobelia gilletii EN
Pogostemon micangensis CR

DICOTYLEDONAE

APOCYNACEAE (M. Cheek)

Dictyophleba setosa de Hoogh

VU B2ab(iii)

Range: Cameroon (S: Kribi-Ébolowa, km 16; Kribi-Mpolongwe, km 15; Kribi-Lolodorf, km 19; Campo, 35 km E, Muini; Kribi 40 km S, Ebodje; Campo-Kribi, 6 km. C: Mefou proposed National Park) Gabon (Ngounié at Fougamou).

Here *Dictyophleba setosa* is assessed as Vulnerable since eight locations are known (see range, above, AOO 32 km^2 with 4 km^2 cells, threats as below). EOO is calculated as 48839 km^2.

Habitat: coastal and riverine evergreen lowland forest.

Threats: apart from slash and burn agriculture, known threats are few at all sites apart from the Kribi area where extractive industry infrastructure represent a clear threat.

Management suggestions: rediscovery of this species at one or more of its known sites; surveys to provide basic populational data for future monitoring and intervention assessments; advising leaders of local communities nearby, to seek their support in protecting this species. Care should be taken not to confuse this species with others of the genus in Cameroon. Gazettement of the proposed Mefou National Park would help secure the protection of this species.

BIGNONIACEAE (M. Cheek)

Fernandoa ferdinandi (Welw.) K.Schum

VU B2ab(iii)

Range: Cameroon (S: 25 km S Djoum. E: 36 km NE Moloundou. C: Mefou proposed National Park), Gabon (Minkebe; Lastoursville-Poungou; ibid-Bounzoco; Moila-Mounabounou), Angola (Golungo Alto, type).

This tree, known from only the eight sites above (AOO 32 km^2 with 4 km^2 cells) and with the threats below is here assessed as Vulnerable. EOO is calculated as 372113 km^2.

Habitat: semi-deciduous and riverine-evergreen forest mosaic; c. 700 m; fl. April and Nov.

Threats: the Djoum and Moloundou sites are believed to be in logging areas and at the Mefou site tree clearance for agriculture continues (pers.obs. 2006).

Management suggestions: upgrading the Mefou proposed National Park to official status, and cessation of clearance for agriculture within its boundaries would provide a secure site for this rare tree. Clarification on the status of the E and S Region sites by field survey is advisable. Information on the threat status of *F. ferdinandi* in Gabon is needed.

CAMPANULACEAE (M. Cheek)

Lobelia gilletii De Wild.

EN B2ab(iii)

Range: Cameroon (C: Mefou proposed National Park), Gabon (Billagore [Mbilagoné]; Doudou Mts), Congo (Brazzaville), and Congo (Kinshasa) (Kasai, vallée de la Djuma).

Here *Lobelia gilletii* is assessed as Endangered since five locations (see range, above; AOO 20 km^2 with 4 km^2 cells) are known, with threats below. EOO is calculated as 588081 km^2. Three specimens are known from Congo (Kinshasa) but only one location. All material known has been identified by Mats Thulin (UPS) author of the Flore D'Afrique Centrale Lobeliaceae account (1985) from which the data cited is taken – apart from the Doudou Mts and Congo (Brazzaville) locations which derive from Olivier Lachenaud (G. Walters pers. comm to Cheek, v.2010).

Habitat: lowland swamps in forest areas; c. 700 m alt.

Threats: agriculture at Mefou – a small community, only 100 m distant from the swamp location, use the habitat for *Raphia* cultivation and extraction.

Management suggestions: the ecological requirements of this species should be researched to ensure it can be managed for survival. Some disturbance may be beneficial for the species. Basic population data should be gathered to form a baseline for future monitoring.

CUCURBITACEAE (M. Cheek)

Momordica camerounensis Kerauden
EN B2ab(iii)

Range: Cameroon endemic (C: N'kolbison; Mefou proposed National Park. S: Nkane, 27 km WNW Ambam).

Here *M. camerounensis* is assessed as Endangered since only three locations are known (above AOO 12 km^2 with 4 km^2 cells) and threats are as below.

Habitat: riverine forest edge; 600–700 m alt.

Threats: forest clearance for timber followed by agriculture.

Management suggestions: formal recognition of the proposed National Park would provide a protected site for this species. Comparison with the type material at Paris is needed to confirm identification of the Mefou material.

Momordica enneaphylla Cogn. (I. Darbyshire, updated by M. Cheek)
VU B2ab(iii)

Range: Cameroon (SW: Kupe-Bakossi (3 coll., 2 sites). C: Mefou proposed National Park; 20 km NW of Eséka (1 coll.). S: Bipindi (1 coll.)), Gabon (La Mvoum, Fougamou), Congo (Kinshasa) (Forestier Central: Yangambi (3 coll.); Watsi, Boende (1 coll.); Yabwesa, Isangi (1 coll.)) and Congo (Brazzaville) (Massif du Chaillu).

First assessed as VU B2ab(iii) by Darbyshire in Cheek *et al.* 2004, the assessment is maintained here since it is still the case that no more than 10 locations are known (AOO 40 km^2 with 4 km^2 cells) and threats are as below.

This little known liana was first collected in Gabon, being described in 1888. Collections from Congo (Kinshasa) are all from the central Congo Basin. In Cameroon, the discovery in 2004 of 2 sites in the Kupe-Bakossi area, on Mt Kupe above Kupe Village and at Ngomboku in the Bakossi Mts, extended this species' range northwestwards.

This is a highly distinctive member of the Cucurbitaceae family, being the only bipinnate species of west Africa, with 3 ternate leaflets. *Sidwell* 443 appears to be the first collection of the currently undescribed female flowers of this species; this specimen demonstrates how this taxon could be easily overlooked, as it is flowering prior to the development of leaves. A collection of a single *Momordica* flower on a trailing leafless stem at Nyandong, western Bakossi (c. 500 m alt.) by M. Etuge in March 2003 may prove to be this species on further investigation.

Habitat: primary forest, including riverine and swamp forest; occasionally in secondary forest growth; 450–1070 m alt.

Threats: on Mt Kupe, this species is recorded in forest down to 900 m, which is below the lower limit of effective protection on this mountain, thus this population is threatened by forest clearance. The population at Mefou proposed National Park is not yet secure.

Management suggestions: Formal protection of Mefou as a National Park would give this species a protected area in Cameroon. If the low levels of forest clearance continue in Gabon and Congo (Kinshasa), this taxon should remain extant there, though its rediscovery in these two countries is advisable in light of the age of the known collections.

EBENACEAE (M. Cheek)

Diospyros crassiflora Hiern

Often known as the 'ebonies', the genus *Diospyros* are famed above all for the hard black wood produced by a minority of the species. The main Central African ebony-producing species in *D. crassiflora* Hiern which occurs from Nigeria to Congo and is relatively common in the forests of Cameroon, although trees large enough to provide the commercially useful heartwood are rarely seen. Although listed by IUCN www.redlist.org (viewed Oct. 2010) as EN we disagree for the reasons stated and would instead list it as Near Threatened

EUPHORBIACEAE (M. Cheek)

Phyllanthus kidna Challen & Petra Hoffm. sp. nov.

CR B2ab(iii) + D

Range: Cameroon endemic (C: Mefou proposed National Park).

Since it has been estimated that, at the moment, only five individuals of *Phyllanthus kidna* are known from the single location so far discovered (AOO 4 km² using 4 km² cells) where it is threatened by agriculture, this species was assessed as Critically Endangered by Cheek in Challen *et al.* (Syst. Bot. in press).

Habitat: Cut-over semi-deciduous forest planted with Cocoa (*Theobroma*). Other surviving indigenous tree species include *Triplochiton scleroxylon* K.Schum., *Celtis* L. spp., and *Baillonella toxisperma* Pierre.

Threats: One of the two individuals (*Cheek* 13098) seen by the author in 2006 was regrowing from a stump of a tree that had been cut in the previous 1–2 years, possibly to facilitate cocoa cultivation or to furnish firewood or poles. Much of the area of the proposed National Park has previously been logged for timber for export and subsequently been cultivated for small-holder cocoa or manioc which was ongoing during fieldwork there in 2002–2006.

Management suggestions: It is to be hoped that the species will be found elsewhere in the Yaoundé area and that in future its threatened status will be reduced. Surveys to address this are suggested by asking in communities for this tree using the Ewondo name "kidna". The surviving trees in the Park should be flagged, signposted, and made the subject of an education campaign in the area to avoid their being cut down by mistake. Seed should be collected and both seed-banked and raised and replanted in the Park so that the wild population in the proposed National Park can be strengthened.

LABIATAE (M. Cheek)

Pogostemon micangensis G. Taylor

CR D

Range: Angola (Ganguellas, Cuiriri, River Micango) and Cameroon (C: Mefou proposed National Park).

Here *Pogostemon micangensis* is assessed as Critically Endangered on the basis that it is known from only an estimated 12 individuals seen by field botanists and with threats as below.

The Mefou specimens were confirmed in their identification by Dr A. Paton at Kew; data on the Angolan material comes from the species protologue. In view of the geographical disjunction perhaps the identity of the Mefou material should be revisited. Ours is the first record of the genus in W-C Africa.

Habitat: river marshes (Angola); drying sandy river beds; 600–710 m alt. (Cameroon).

Threats: sand extraction for construction of buildings and infrastructure.

Management suggestions: gazettement of Mefou proposed National Park would help secure resources to protect this species. A survey along river beds in Mefou and similar areas would yield a better conservation assessment, a baseline for future monitoring, and an evaluation of management needs.

MELIACEAE (M. Cheek)

Entandrophragma angolense (Welw.) C.DC.

Entandrophragma cylindricum (Sprague) Sprague

Entandrophragma utile (Dowe & Sprague) Sprague

Guarea cedrata (A. Chev.) Pellegr.

Khaya ivorensis A. Chev.

Lovoa trichilioides Harms

VU A1cd

The *Guarea*, *Khaya* and *Lovoa* species listed above all occur at Mefou. *Entandophragma* also occurs but has not been identified to species due to lack of an adequate specimen. All are internationally traded timber species of the mahogany family which were listed as Vulnerable by Hawthorne (Hawthorne 1997 in www.redlist.org) using the 1994 criteria of IUCN. They all have a wide range in Africa and, were they reassessed in this book, without reference to their use as timber, they would probably be downlisted. Hawthorne variously cites over-exploitation, poor levels of regeneration, fire damage, and slow growth to support his assessments of these species.

The assessment of VU under criterion A refers to the fact that more than 30% of the population has been lost in the last 100 years, mainly due to replacement of forest by agriculture and farm fallow in countries such as Ghana, Nigeria, Ivory Coast and Guinea which represent major parts of the range for these species.

On the other hand these species are all diligently raised from seed and planted in reserves or otherwise favoured by governmental forest departments in most countries within their range, so as to maintain national timber revenues, so their extinction is unlikely, although large trees are certainly becoming increasingly rare because of the high value of their timber.

Heckeldora ledermannii (Harms) J.J. de Wilde

EN B2ab(iii)

Range: Cameroon endemic (SW: Mts Ekomane (Kongoa); Fosimondi and Bakossi Mts. C: Mefou proposed National Park).

Here *H. ledermannii* is assessed as Endangered since three locations are known (AOO 12 km^2 with 4 km^2 cells) with threats as below. The type locality, Kongoa Mts, has not been visited by botanists in many decades. A fourth possible location has been identified from a specimen at the proposed Mefou National Park near Yaoundé, but better material is needed to confirm the identification. *Heckeldora* was revised recently by J.J.F.E. de Wilde and the Mefou specimen was identified as *H. ledermannii* by his student Erich.

Habitat: submontane evergreen forest; 900–1500 m alt.

Threats: forest clearance for agriculture and wood at lower altitudes only on Mt Kupe, but wholesale at the Fosimondi site where extremely threatened by farm expansion.

Management suggestions: *Heckeldora ledermannii* appears secure in Bakossi, especially at Mt Kupe where it occurs at high density (given the number of specimens made there), so long as submontane forest continues to be protected here, the species is likely to survive.

Trichilia zewaldae J.J. de Wilde

EN B2ab(iii)

Range: Cameroon endemic (S: Bipindi, 10 km NW Eséka. C: Mefou proposed National Park) Known from six specimens at two or three sites (AOO 8 km^2 with 4 km^2 cells) and threats as below, *T. zewaldae* is here assessed as Endangered.

Habitat: lowland evergreen forest; 200 m alt.

Threats: slash and burn agriculture in S Region is a low level threat.

Management suggestions: inclusion of a site in a protected area is advisable. The Mefou proposed National Park offers this potential if gazetted. Data on population density, range, regeneration levels and threats need to be gathered to inform management of this species.

MENISPERMACEAE (B.J. Pollard, updated by M. Cheek)

Platytinospora buchholzii (Engl.) Diels var. *macrophylla* Diels

EN B2ab(iii)

Range: Nigeria (1 coll.) and Cameroon (SW: Mt Kupe (2 coll.). Littoral: Douala-Edéa Reserve (1 coll.). S: Bipindi (1 coll.). C: Yangamok II near Bafia (1 coll.); Mefou proposed National Park (2 coll.)).

Zenker 3014a was cited by Diels (in Engler, Pflanzenr. 46, 1910: 168) as the type, collected at Bipindi, whence *Zenker* 4008, st., 1911 was also collected. From 1969, three collections were made within 10 years of each other: *Letouzey* 9610, fr. 26 Nov., at Yangafok II, 25 km ENE from Bafia, the fruiting Nigerian collection of *Wit & Daramola* in FHI 64887, from the Ago-Owu Forest Reserve, West State, 28 Dec. 1971, and *McKey* 260, st., 26 May 1979, fallow fields by the village of Cité-Lac Tissongo, Douala-Edéa Reserve. Two further collections were made from Bakossi, along Walter's trail on Mt. Kupe itself in October 1995 (*Cheek* 7535, st., 1000 m) and June 1996 (*Cable* 2898B, st., 1000 m). Additional material is desirable to enhance our knowledge of this monospecific genus.

This taxon was assessed as CR A2c + 3c by Pollard in Cheek *et al.* 2004 but is reassessed here as Endangered since the evidence of 80% population reduction is ambiguous, but it is only known from five locations (above; AOO 20 km² with 4 km² cells) with threats as below. EOO was calculated as 69545 km²

Habitat: lowland evergreen and semi-deciduous forest and submontane forest; 0–1000 m alt.

Threats: forest clearance, especially at Mefou proposed National Park and at lower altitudes at Mt Kupe, for agriculture. Even if the trees are left standing, but the undergrowth is cleared, vines can all be cut at ground level, as in the Loum FR understorey.

Management suggestions: the location of the subpopulations noted above should be revisited, and an assessment made of the number of individuals and regeneration levels as a baseline for future monitoring. Gazettement of the proposed Mefou National Park would provide a secure station for this taxon in the wild.

RUBIACEAE (M. Cheek)

Chassalia manningii O. Lachenaud ined.

EN B2ab(iii)

Range: Nigeria (Benin Province, Usonigbe FR, 3 coll.), Cameroon (SW: S Bakundu FR, 2 coll. C: Mefou proposed National Park, 1 coll.).

Here assessed as EN since three locations (above, AOO 12 km² with 4 km² cells, threats as below).

Habitat: lowland evergreen forest; sea-level–700 m alt.

Threats: nearly 90% of the original forest of Nigeria has been lost to timber extraction and agriculture, including in Forest Reserves; the species may no longer survive in Nigeria. These same threat also occur in S Bakundu.

Management suggestions: Mefou proposed NP should be upgraded to full NP status to protect this and the other threatened species that it contains. A census of basic populational data of this species at Mefou would provide a baseline for future monitoring and gauge needs for management interventions.

Chazaliella obovoidea Verdc. subsp. ***villosistipula*** Verdc.

VU B2ab(iii)

Range: Cameroon (S: Bipindi; Yaoundé-Deng Deng; Batouri-Djampiel; Lolodorf; Kribi-Lolodorf 40 km; NE Sangmélima. C: Mefou proposed National Park; E: Kadei R (left bank)) and Congo (Brazzaville) (Odzala National Park).

Here assessed as Vulnerable since nine locations (above, AOO 36 km² with 4 km² cells, threats as below).

Unusual in the genus is being so small, shrubby and deciduous. The stipules are not always villose!

Habitat: semi-deciduous and evergreen forest; 350–700 m alt.

Threats: clearance of forest especially semi-deciduous, for logging, followed by agriculture: cocoa, pineapples, food-crops.

Management suggestions: formal ratification of the Mefou proposed National Park would provide the first protected area for this shrub. Within the Park it is rather common and easily viewed, even characteristic of semi-deciduous forest remnants. However, globally it is threatened on current data. Survey of basic population parameters would serve as a baseline for future monitoring.

Hymenocoleus nervopilosus Robbr. var. *orientalis* Robbr.

EN B2ab(iii)

Range: Cameroon (S: Ébolowa-Kribi at Kianke R.; 4–5 km E Kribi. C: Mefou proposed National Park) and Gabon (Cristal Mts at Nkam; Makokou; Libreville; Mouila).

Here assessed as Vulnerable since six locations are known (above, AOO 24 km^2 with 4 km^2 cells, threats as below).

Habitat: lowland evergreen forest; 600–700 m alt.

Threats: clearance of forest for timber and/or agriculture.

Management suggestions: several records exist from Mefou proposed National Park so this is a logical centre to conserve the plant if only governmental approval for the National Park is achieved. Baseline population data is needed against which to monitor in future and to gauge need for management intervention.

Morinda mefou Cheek

VU B2ab(iii)

Range: Cameroon (S: Station du Cacaoyer de N'Koemvone; Bipindi; C: Mefou proposed National Park) and Gabon (Route Owendo–Ikoi Komo; Moanda–Franceville; Région des Abeilles) and Congo (Brazzaville) (Environs de Brazzaville).

Morinda mefou is formally named in this book, where it is assessed as Vulnerable VU B2ab(iii) since it is only known from six locations (see specimens cited above) equating to an area of occupancy of 24 km^2 using the recommended 4 km^2 cells, and since the species is threatened by forest clearance for agriculture, for example at Mefou itself.

Habitat: Lowland evergreen semi-deciduous to submontane forest; either forest edge or secondary; 350–600 m alt.

Threats: clearance of forest for timber and/or agriculture.

Management suggestions: It is to be hoped that this and other Red Data species at Mefou proposed National Park identified in this volume will be incorporated into a future management plan for the Park and protected for the future. The first step, in this instance, is to refind the original plant, survey for additional individuals in the area, establish actual and possible threats to those individuals so that they may be addressed, evaluate actual regeneration levels in the wild, begin seed collection for propagation in a nursery and future planting out in suitable safe sites to bolster the wild population.

Poecilocalyx setiflorus (R. Good) Bremek.

VU B2ab(iii)

Range: Cameroon (C: Mefou proposed National Park; 35 km SE Mbalmayo and 28 km S of Atozok village on Soa River), Gabon (Etéké, distr. Mimongo; Belinga; Mouila; Moukalaba-Doudou National Park; Rabi-Kounga; Massif du Chaillu) and Angola (Cabinda).

Here assessed as Vulnerable since nine locations are known (AOO 36 km^2 with 4 km^2 cells) and threats as below.

Distinguished from the similar *P. schumannii* in the stipule being entire and acute, not multifid, secondary nerves 7–10 pairs not 11–12 pairs.

Habitat: evergreen riverine forest; 600–700 m alt.

Threats: clearance of forest for agriculture and timber.

Management suggestions: this is a relatively common understorey shrub in *Gilbertiodendrom dewevrei* forest in Mefou proposed National Park, but globally it is rare. Formal ratification of the National Park would help to protect this species from extinction. So long as its habitat is protected at Mefou it will be secure. Public education at Mefou for visitors might help understanding of rare plant species in Cameroon.

Pseudosabicea sthenula N.Hallé

VU B2ab(iii)

Range: Cameroon (C: Mefou Proposed National Park) and Gabon (Akoga; 37 km SW Makokou; Mimongo-Mbigou; Lopé National Park; Boyan-Oyan; Koumameyong; Minkébé; St. Germain).

Here *Pseudosabicea sthenula* is assessed as Vulnerable since 10 locations (above, AOO 40 km^2 with 4 km^2 cells) are known with threats as below.

Habitat: lowland evergreen forest; 710 m alt.

Threats: road clearance and widening (Mefou) since the site is along an overgrown sideroad which if cleared might damage the population.

Management suggestions: official recognition of Mefou as a National Park would safeguard Cameroon's only known population of this species. At Mefou the species is locally abundant and so long as it is protected and monitored, if likely to persist.

Tricalysia amplexicaulis Robbr.

VU A2c,d

Range: Cameroon endemic (SW: S Bakundu. C: Mt Kala; Nkolbisson; Mbam-Minkom; Makak; Mefou proposed National Park. Littoral & S: just E of Kribi, Kienke R.; Campo-Ma'an).

The main distribution of *T. amplexicaulis* is Yaoundé to Kribi and Campo. More than 10 sites are known. At the Mefou proposed National Park it occurs in undisturbed evergreen forest areas, but only rarely and this forest is under severe threat from agricultural clearance (Cheek pers. obs.). At S Bakundu large areas have suffered from agricultural operations., while at Mt Kala slash and burn agriculture is also a threat. It is here estimated that over 30% of habitat at its known sites has been lost in the last 3 generations/100 years (understorey shrubs in forest are believed to be long-lived) and losses are ongoing so *T. amplexicaulis* is here assessed as VU A2c,d.

Habitat: lowland evergreen forest; sea-level to c. 700 m alt.

Threats: see above. At Mefou proposed National Park several plants were found by us in a forest patch destined for a new chimpanzee enclosure. Chimps are very destructive of forest habitat when in enclosures, so this posed a threat to these rare plants. Fortunately it was possible to move the plants to a safe location nearby.

Management suggestions: marking, publicising, protecting and monitoring clusters of plants in protected areas such as at Mefou proposed National Park is advisable to avoid accidental destruction of this often very local species.

Note: Robbrecht indicated that four specimens are close to *T. amplexicaulis* but differ in having smaller, shortly petiolate leaves. These need more research to elucidate their taxonomic status.

STERCULIACEAE (M. Cheek)

Cola letouzeyana Nkong.

VU B2ab(iii)

Range: Cameroon endemic (SW: Mt Cameroon, Bonja-Onge. C: Mefou proposed National Park; Mbam-Minkom. S: Kribi-Lolodorf; Boga, 30 km N Eséka. Littoral: Hikoa Mahouda, Ébolowa-Mbam; Ebo).

Here *Cola letouzeyana* is assessed as Vulnerable since 8 locations are known (above, AOO 36 km^2 with 4 km^2 cells) and threats are as below.

Habitat: lowland evergreen forest including *Gilbertiodendron dewevrei* dominated; 200–700 m alt.

Threats: forest clearance for agriculture, both plantation (Mt Cameroon) and slash and burn (S and C Regions).

Management suggestions: formal recognition of the Mefou National Park would protect a major centre for this species since it is relatively frequent in *Gilbertiodendron* forest here. Considerable data on the population at Mefou already exists and could be a basis for future monitoring. Studies on rates of regeneration would be valuable. A record of this species in Bakossi requires confirmation.

Leptonychia subtomentosa K. Schum.

EN B1a, b(iii)

Range: Cameroon endemic (C: Yaoundé area at Mefou proposed National Park; Ottotomo Reservoir; Colline Kombeng, 8 km SSE, Matomb).

The ten specimens known for *Leptonychia subtomentosa* derive from four locations: the proposed Mefou National Park (five collections), Colline Kombeng, Ottotomo Reservoir and an unknown location in Yaoundé (the type collection). These all fall in Yaoundé's SW quadrant within an area of about 50 × 50 km, giving an extent of occurrence of less than 250 km^2. Yaoundé is the capital city of Cameroon with a population of over a million people and steadily expanding. It is very likely that the original Zenker and Staudt collections from over a century ago were made on one of the city's now deforested hills. Even at Mefou proposed National Park destruction of natural habitat continues, albeit at a smaller scale than in the wider area. Accordingly, *L. subtomentosa* is here assessed as EN B1a, b (iii) according to the criteria of IUCN (2001), that is, Endangered.

Leptonychia subtomentosa is easily identified by the long length and persistence of the stem hairs. No other *Leptonychia* species of central Africa known to the authors has these features.

Habitat: Within semi-deciduous forested areas: submontane forest in hilly areas and also along rivers in *Gilbertiodendron dewevrei* gallery forest; 550–880 m alt.

Threats: see above.

Management suggestions: since the most recent records derive from the Mefou proposed National Park, this is the logical place to focus conservation efforts. Basic populational data should be gathered to inform future monitoring and to assess needs for intervention.

VITACEAE (M. Cheek)

Cyphostemma cameroonense Desc.

EN B2ab(iii)

Range: Cameroon endemic (C: Nkolbison; Mefou proposed National Park; Nanga Eboko. E: Bertoua).

Here *Cyphostemma cameroonense* is assessed as Endangered since there are only four locations (above, AOO 16 km^2 with 4 km^2 cells) and threats as below. EOO is calculated as 11168 km^2.

Habitat: semi-deciduous forest; c. 700 m alt.

Threats: the habitat of this species has been logged in most parts of its range, often followed by cultivation of cacao or manioc (pers. obs. Cheek & Onana 2002, 2004).

Managment suggestions: since the most recent record derives from the Mefou proposed National Park, this is the logical place to focus conservation efforts. Basic populational data should be gathered to inform future monitoring and to assess needs for intervention. Propagation and reintroduction may be needed.

MONOCOTYLEDONAE

ZINGIBERACEAE (M. Cheek)

Aulotandra kamerunensis Loes.

EN B2ab(iii)

Range: Cameroon endemic (S: Bipindi; Bodi, 20 km SW Eséka; Ebianemeyong et Nyabesan. C: Mefou proposed National Park).

Here *Aulotandra kamerunensis* is assessed as Endangered since only four locations are known (above; AOO 16 km^2 with 4 km^2 cells) with threats as below.

First published in (1909), based on *Zenker* 36908 from Bipindi collected in 1908, the species was not seen again until 1973 near Eséka (*Letouzey* 12504). It was then recorded from the Nyabesan Campo area by Tchoutou (2004) but we have not seen a specimen. The Mefou record, (*Tadjouteu* 630 4/2004), is only the fourth global location for this very rare species. After his specimen was identified at Kew by David Harris, Tadjouteu returned to Mefou to attempt rediscovery of the plant but sadly it was not refound.

This is the only continental African species of the genus, the other five all occurring in Madagascar.

Habitat: evergreen *Gilbertiodendron dewevrei* forest; c. 650 m alt.

Threats: clearance of forest for cultivation.

Management suggestions: the best hope for the survival of this species is Mefou proposed National Park, should it be gazetted. Tadjouteu should be engaged to attempt to refind this species there during April, when, in flower it is conspicuous. This species has potential as an ornamental plant and there is certainly potential to multiply the species in cultivation with a view to introducing it to the wild.

REFERENCES

Harvey, Y., Pollard, B.J., Darbyshire, I., Onana, J.-M. & Cheek, M. (eds) (2004). The Plants of Bali Ngemba Forest Reserve, Cameroon: A Conservation Checklist. Royal Botanic Gardens, Kew, UK. iv + 154 pp.

Tchoutou, M.P.G. (2004). Plant Diversity in a Central African Rainforest: implications for Biodiversity Conservation in Cameroon. PhD thesis, University of Wageningen.

FIG. 9 *Funtumia elastica* (Apocynaceae) by W.E. Trevithick

FIG. 10 *Impatiens irvingii* (Balsaminaceae) by W.E. Trevithick

100

FIG. 11 *Kigelia africana* (Bignoniaceae) by W.E. Trevithick

FIG. 12 *Newbouldia laevis* (Bignoniaceae) by W.E. Trevithick

101

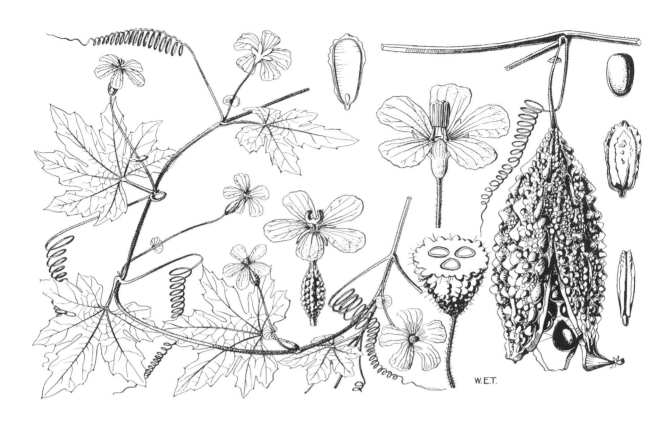

FIG. 13 *Momordica charantia* (Cucurbitaceae) by W.E. Trevithick

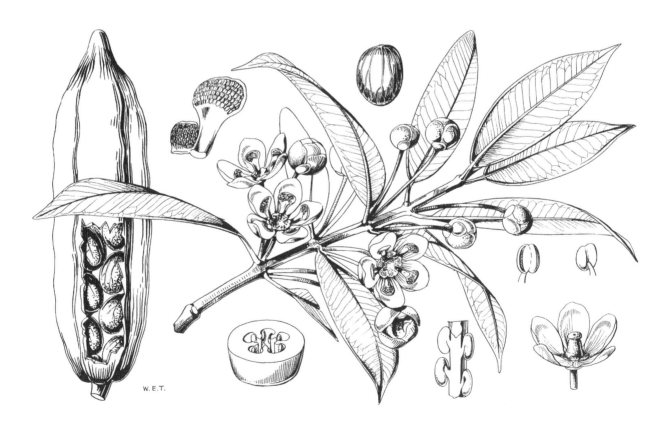

FIG. 14 *Allanblackia floribunda* (Guttiferae) by W.E. Trevithick

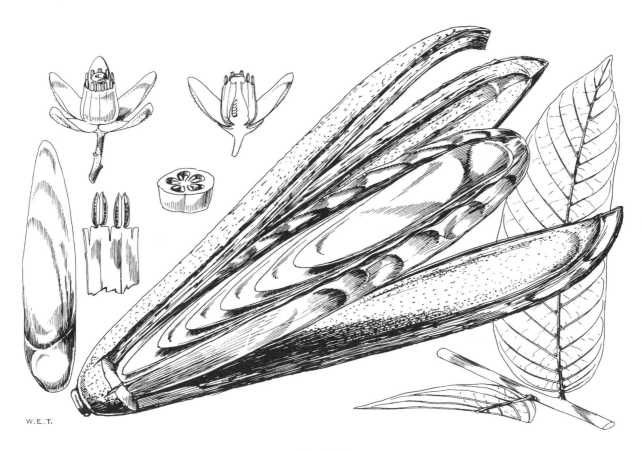

FIG. 15 *Entandrophragma utile* (Meliaceae) by W.E. Trevithick

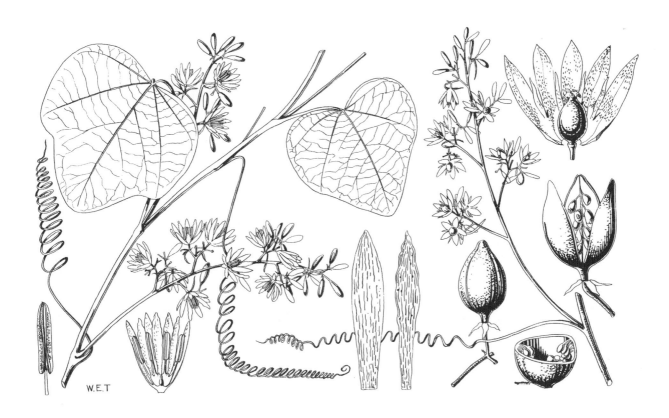

FIG. 16 *Adenia cissampeloides* (Passifloraceae) by W.E. Trevithick

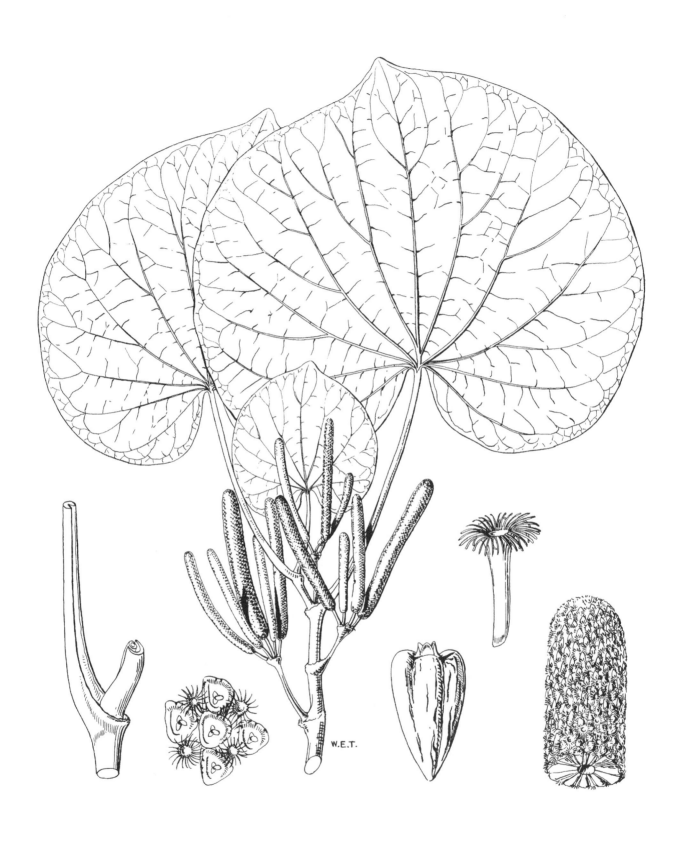

FIG. 17 *Piper umbellatum* (Piperaceae) by W.E. Trevithick

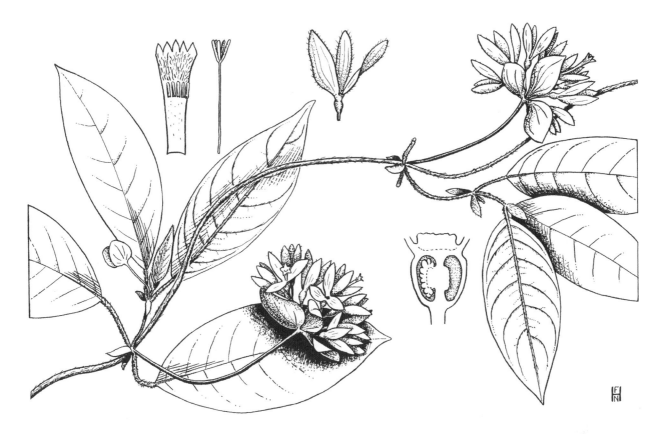

FIG. 18 *Sabicea calycina* (Rubiaceae) by F. Nigel Hepper

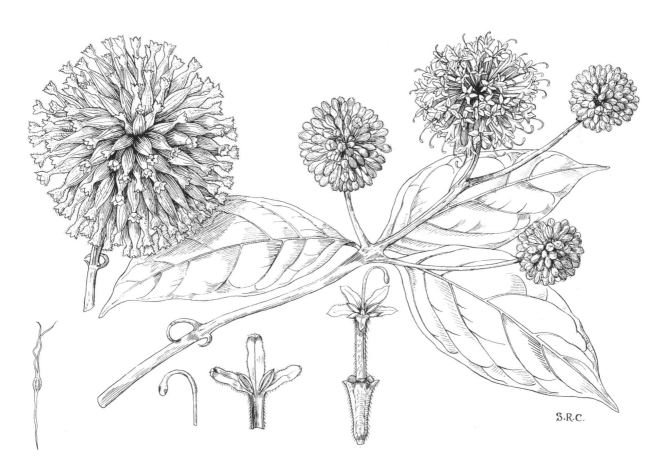

FIG. 19 *Uncaria africana* (Rubiaceae) by Stella Ross-Craig

Fɪɢ. 20 *Paullinia pinnata* (Sapindaceae) by W.E. Trevithick

Fɪɢ. 21 *Bacopa crenata* (Scrophulariaceae) by W.E. Trevithick

FIG. 22 *Grewia pubescens* (Tiliaceae) by W.E. Trevithick

BIBLIOGRAPHY

Angiosperm Phylogeny Group II (2003). An update of the Angiosperm Phylogeny Group classification for the orders and families of flowering plants: APG II. *Botanical Journal of the Linnean Society* 141: 399–436.

Angiosperm Phylogeny Group III (2009). An update of the Angiosperm Phylogeny Group classification for the orders and families of flowering plants: APG III. *Botanical Journal of the Linnean Society* 161: 105–121.

Breton, R. (1979). Ethnies et langues: 31–33. In Laclavère, G. & Loung J F.(eds.) Atlas de République Unie du Cameroun. Editions j.a. Paris.

Brummitt, R.K. & Powell, C.E., eds. (1992). Authors of Plant Names. Royal Botanic Gardens, Kew, U.K.

Brummitt, R.K. (1992). Vascular Plant Families and Genera. Royal Botanic Gardens, Kew, U.K.

Brunel, J. F. (1987). Sur le genre *Phyllanthus* L. et quelques genres voisins de la tribu des Phyllantheae Dumort. (Euphorbiaceae, Phyllantheae) en Afrique intertropicale et à Madagascar. Thèse doctorale, Université Louis Pasteur, Strasbourg, France.

Cable, S. & Cheek, M. (eds) (1998). The Plants of Mount Cameroon: A Conservation Checklist. Royal Botanic Gardens, Kew, UK. lxxix + 198 pp.

Cheek, M. (2008). The Importance Of The Cross-Sanaga River Interval For Plants. Nigerian Field 72 (2): 93–100 (2008)

Cheek, M., Onana, J.-M. & Pollard, B.J. (eds) (2000). The Plants of Mount Oku and the Ijim Ridge, Cameroon: A Conservation Checklist. Royal Botanic Gardens, Kew, UK. iv + 211 pp.

Cheek, M. (April 2001). Kew Scientist 19: 3. "Cameroonian Minister Visit". RBG, Kew, UK.

Cheek, M., Pollard, B.J., Darbyshire, I., Onana, J.-M. & Wild, C. (eds) (2004). The Plants of Kupe, Mwanenguba and the Bakossi Mountains, Cameroon: A Conservation Checklist. Royal Botanic Gardens, Kew, UK. iv + 508 pp.

Cheek. M., Harvey, Y. & Onana, J.-M. (eds) (2010). The Plants of Dom, Bamenda Highlands, Cameroon: A Conservation Checklist. Royal Botanic Gardens, Kew, UK. iv + 162 pp.

Convention on Biological Diversity (2002). Convention on Biological Diversity: Text and Annexes. United Nations Environment Programme, Montreal, Canada.

Cook, F.M. (1995). Economic Botany Data Collection Standard. RBG, Kew. 146 pp.

Govaerts, R., Frodin, D. G. and Radcliffe-Smith, A. (2000). *World checklist and bibliography of Euphorbiaceae (with Pandaceae)*, vols. 1–4. Royal Botanic Gardens, Kew, UK.

Harvey, Y., Pollard, B.J., Darbyshire, I., Onana, J.-M. & Cheek, M. (eds) (2004). The Plants of Bali Ngemba Forest Reserve, Cameroon: A Conservation Checklist. Royal Botanic Gardens, Kew, UK. iv + 154 pp.

Harvey, Y., Tchiengué, B. & Cheek, M. (eds) (2010). The Plants of Lebialem Highlands (Bechati-Fosimondi-Besali), Cameroon: A Conservation Checklist. Royal Botanic Gardens, Kew, UK. 170 pp.

Hepper, F.N. (1963) *Morinda*, pp. 188–190 in Hepper (Ed.) Flora of West Tropical Africa Ed. 2,2. Crown agents, London

Hoffmann, P. and Cheek, M. (2003). Two new species of *Phyllanthus* (Euphorbiaceae) from south west Cameroon. *Kew Bulletin* 58: 437–446.

IUCN (2001). IUCN Red List Categories and Criteria. Version 3.1. IUCN, Gland, Switzerland. 30 pp.

IUCN (2003). Guidelines for Application of IUCN Red List Criteria at Regional Levels: Version 3.0. IUCN Species Survival Commission. IUCN, Gland, Switzerland and Cambridge, UK.

Kathriarachchi, H., Samuel, R., Hoffmann, P., Mlinarec, J., Wurdack, K.J., Ralimanana, H., Stuessy, T.F. and Chase, M.W. (2006). Phylogeny of the tribe Phyllantheae (Phyllanthaceae) based on nrITS and plastid *matK* sequence data. *American Journal of Botany* 93: 637–655.

Keay, R.W.J. & Hepper, F.N. (eds) (1954–1972). Flora of West Tropical Africa, 2nd ed., 3 vols. Crown Agents, London.

Letouzey, R. (1968). Les Botanistes au Cameroun. Flore du Cameroun 7. Museum National d'Histoire Naturelle, Paris.

Letouzey, R. (1968). Étude Phytogéographique du Cameroun, Edition Lechevalier, Paris.

Letouzey, R, (1976). Contribution de la botanique au problème d'une éventuelle langue pigmée. Selaf – Paris.

Letouzey, R. (1980). Les Botanistes Hollandais au Cameroun. Miscellaneous Papers 19: 245–252. Landbouwhogeschool, Wageningen, The Netherlands.

Letouzey, R. (1985). Notice de la carte phytogéographique du Cameroun au 1: 500,000. IRA, Yaoundé, Cameroon.

Letouzey, R. & Mouranche, R. (1952). Ekop du Cameroun. Publication n°14 du Centre technique forestier tropical. Centre technique forestier tropical. Nogent Sur-Marne. France.

Mabberley, D.J. (1997). The Plant Book. Second edition. Cambridge University Press, U.K.

Manning, S.D. (1996). Revision of *Pavetta* Subgenus *Baconia* (Rubiaceae: *Ixoroideae*) in Cameroon. Annals of the Missouri Botanical Garden, 83(1): 87–150.

Mdida, M.C. (2006). Patrimoine culturel: 70. In Ben Yamed, D. (ed.) Atlas de l'Afrique. Cameroun. Les Editions J.A. Paris.

Petit, E, (1962) Rubiaceae Africanae IX. pp. 173–198. Bull. Jard. Bot. Brux. 32.

Ralimanana, H., and Hoffmann, P. Taxonomic revision of *Phyllanthus* (Phyllanthaceae) in Madagascar and the Comoro Islands I: Synopsis and Subgenera *Isocladus*, *Betsileani*, *Kirganelia* and *Tenellanthus*. *Kew Bulletin*, in press.

Savolainen, V., Fay, M.F., Albach, D.C., Backlund, A., van der Bank, M., Cameron, K.M., Johnson, S.A., Lledó, M.D., Pintaud, J.-C., Powell, M., Sheahan, M.C., Soltis, D.E., Soltis, P.S., Weston, P., Whitten, W.M., Wurdack, K.J. and Chase, M.W. (2000). Phylogeny of the eudicots: A nearly complete familial analysis based on *rbcL* gene sequences. *Kew Bulletin* 55: 257–309.

Tchoutou, M.P.G. (2004). Plant Diversity in a Central African Rainforest: implications for Biodiversity Conservation in Cameroon. PhD thesis, University of Wageningen.

Verdcourt, B.V. (1976). Rubiaceae (Part 1). Flora of Tropical East Africa, Crowns Agents, London.

Vorontsova, M.S., Hoffmann, P., Kathriarachchi, H., Kolterman, D.A., and Chase, M.W. (2007a). *Andrachne cuneifolia* (Phyllanthaceae; Euphorbiaceae s.l.) is a *Phyllanthus. Botanical Journal of the Linnean Society* 155: 519–525.

Vorontsova, M.S., Hoffmann, P., Maurin, O. and Chase, M.W. (2007b). Molecular phylogenetics of tribe Poranthereae (Phyllanthaceae; Euphorbiaceae sensu lato). *American Journal of Botany* 94: 2026–2040.

Webster, G. L. (1956). A monographic study of the West Indian species of *Phyllanthus*. *Journal of the Arnold Arboretum* 37: 91–122, 217–268, 340–359.

Webster, G. L. (1957). A monographic study of the West Indian species of *Phyllanthus*. *Journal of the Arnold Arboretum* 38: 51–80, 170–198, 295–373.

Webster, G. L. (1958). A monographic study of the West Indian species of *Phyllanthus*. *Journal of the Arnold Arboretum* 39: 49–100, 111–212.

Webster, G. L. (2002). Three new sections and a new subgenus of *Phyllanthus* (Euphorbiaceae). *Novon* 12: 290–298.

White, F. (1983). The Vegetation of Africa: a descriptive memoir to accompany the Unesco/AETFAT/UNSO vegetation map of Africa. Unesco, Switzerland.

Wurdack, K. J., Hoffmann, P., Samuel, R., de Bruijn, A., van der Bank, M. and Chase, M. W (2004). Molecular phylogenetic analysis of Phyllanthaceae (Phyllanthoideae pro parte, Euphorbiaceae sensu lato) using plastid *rbcL* DNA sequences. *American Journal of Botany* 91: 1882–1900.

A. Evergreen *Gilbertiodendron dewevrei* forest near Ndanan 1, Oct. 2002, by Yvette Harvey
B. Canopy view, evergreen forest, near Ndanan 1, Feb. 2006, by Martin Cheek
C. Semi-deciduous forest, tree with leafless crown, Dec. 2001, by Benedict Pollard
D. Open stream with *Caldesia reniformis* (Alismataceae) flanked by fragments of semi-deciduous forest, Dec. 2001, by Benedict Pollard
E. Semi-deciduous forest replaced by pineapple plantation, Dec. 2001, by Benedict Pollard
F. The first botanical team at Mefou, Oct. 2002, by Martin Cheek

Plate 1

A. *Phyllanthus kidna* (Euphorbiaceae), *Cheek* 13102, Feb. 2006, by Martin Cheek. Base of trunk
B. *Phyllanthus kidna* (Euphorbiaceae), *Cheek* 13102, Feb. 2006, by Martin Cheek. View from ground of canopy
C. *Phyllanthus kidna* (Euphorbiaceae), *Cheek* 13102, Feb. 2006, by Martin Cheek. Bark and slash
D. *Phyllanthus kidna* (Euphorbiaceae), *Cheek* 13102, Feb. 2006, by Martin Cheek. Victor Ngueben with fruiting branch
E. *Phyllanthus kidna* (Euphorbiaceae), *Cheek* 13102, Feb. 2006, by Martin Cheek. Section of immature fruit
F. *Phyllanthus kidna* (Euphorbiaceae), *Cheek* 13102, Feb. 2006, by Martin Cheek. Fruiting branch

Plate 2

A. *Brillantaisia vogeliana* (Acanthaceae), Oct. 2002, by Yvette Harvey
B. *Greenwayodendron suaveolens* (Annonaceae), *Cheek* 11224, Oct. 2002, by Martin Cheek
C. *Pararistolochia macrocarpa* var. *macrocarpa* (Aristolochiaceae), *Cheek* 11227, Oct. 2002, by Martin Cheek
D. *Ritchiea albersii* (Capparaceae), *Cheek* 11083, Oct. 2002, by Martin Cheek
E. *Myrianthus preussii* subsp. *preussii* (Cecropiaceae), *Cheek* 11118, Oct. 2002, by Martin Cheek
F. Sebsebe Demissew and *Maytenus gracilipes* subsp. *gracilipes* (Celastraceae), *Gosline* 410, Oct. 2002, by Martin Cheek

Plate 3

A. *Diospyros sp.* (Ebenaceae), *Cheek* 13100, Oct. 2002, by Martin Cheek
B. *Macaranga saccifera* (Euphorbiaceae), *Cheek* 11237, Oct. 2002, by Martin Cheek
C. *Allanblackia floribunda* (Guttiferae), *Cheek* 11159, Oct. 2002, by Martin Cheek
D. *Garcinia mannii* (Guttiferae), *Cheek* 11239, Oct. 2002, by Martin Cheek
E. *Clerodendrum globuliflorum* (Labiatae), *Cheek* 11122, Oct. 2002, by Martin Cheek
F. *Clerodendrum splendens* (Labiatae), *Cheek* 11485, March 2004, by Martin Cheek

Plate 4

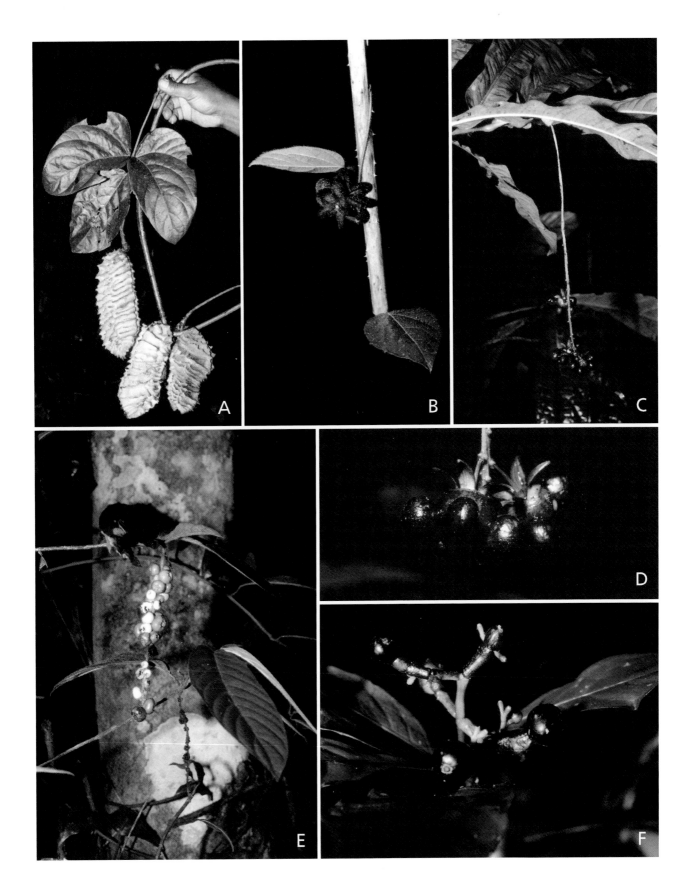

A. *Mucuna sloanei* (Leguminosae-Papilionoideae), *Cheek* 11137, Oct. 2002, by Martin Cheek
B. *Synclisia scabrida* (Menispermaceae), *Cheek* 11079, Oct. 2002, by Martin Cheek
C. *Campylospermum elongatum* (Ochnaceae), *Cheek* 11014, Oct. 2002, by Martin Cheek
D. *Campylospermum elongatum* (Ochnaceae), *Cheek* 11014, Oct. 2002, by Martin Cheek. Close up of fruit.
E. *Bertiera aethiopica* (Rubiaceae), *Cheek* 11062, Oct. 2002, by Martin Cheek
F. *Chassalia pteropetala* (Rubiaceae), *Cheek* 11249, Oct. 2002, by Martin Cheek

Plate 5

A. *Chassalia subherbacea* (Rubiaceae), *Cheek* 11084, Oct. 2002, by Martin Cheek
B. *Chazaliella macrocarpa* (Rubiaceae), *Cheek* 11247, Oct. 2002, by Martin Cheek
C. *Sabicea venosa* (Rubiaceae), *Cheek* 11216, Oct. 2002, by Martin Cheek
D. *Schumanniophyton magnificum,* (Rubiaceae), *Cheek* 11124, Oct. 2002, by Martin Cheek
E. *Sherbournia hapalophylla,* (Rubiaceae), *Cheek* 11163, Oct. 2002, by Martin Cheek
F. *Tricalysia amplexicaulis* (Rubiaceae), *Cheek* 11208, Oct. 2002, by Martin Cheek

Plate 6

A. *Uncaria africana* (Rubiaceae), *Cheek* 11217, Oct. 2002, by Martin Cheek
B. *Zanthoxylum buesgenii* (Rutaceae), *Cheek* 11814, March 2004, by Martin Cheek
C. *Quassia africana* (Simaroubaceae), *Cheek* 11136, Oct. 2002, by Martin Cheek
D. *Cola letouzeyana* (Sterculiaceae), *Cheek* 11192, Oct. 2002, by Martin Cheek
E. *Cola pachycarpa* (Sterculiaceae), *Cheek* 11085, Oct. 2002, by Martin Cheek
F. *Ancistrocarpus densispinosus* (Tiliaceae), *Cheek* 11002, Oct. 2002, by Martin Cheek

Plate 7

A. *Desplatsia dewevrei* (Tiliaceae), *Cheek* 11113, Oct. 2002, by Martin Cheek
B. *Marantochloa filipes* (Marantaceae), *Cheek* 11139, Oct. 2002, by Martin Cheek
C. *Trachyphrynium braunianum* (Marantaceae), *Cheek* 11135, Oct. 2002, by Martin Cheek
D. *Aerengis stelligera* (Orchidaceae), *Cheek* 11259, Oct. 2002, by Martin Cheek
E. *Calamus deëratus* (Palmae), photo. record, Feb. 2006, by Martin Cheek
F. *Laccosperma acutiflorum* (Palmae), *Cheek* 11158, Oct. 2002, by Martin Cheek

Plate 8

READ THIS FIRST:
EXPLANATORY NOTES TO THE CHECKLIST

Yvette Harvey

Herbarium, Royal Botanic Gardens, Kew, Richmond, Surrey, TW9 3AE, UK

Before using this checklist, the following explanatory notes to the conventions and format used should be read.

The checklist is compiled in an alphabetical arrangement: species within genera, genera within families and families within the groups Dicotyledonae, Monocotyledonae, Gnetopsida, Lycopsida (fern allies) and Filicopsida (true ferns), following Kubitzki, in Mabberley (1997: 771–781).

Identifications and descriptions of the species were carried out on a family-by-family basis by both family specialists and general taxonomists; these authors are credited at the head of each account. As a general rule, if two authors are listed, the primary author is responsible for the determinations and the second author for the compiling of the account, including writing of descriptions, distributional data and conservation assessments.

As the incomplete Flore du Cameroun (1963–) is a particularly relevant source of information on the plants of the checklist area, a reference to the volume and year of publication is listed at the head of each family account where available.

The families and genera accepted here follow Brummitt (1992), with recent updates on the Kew Vascular Plant Families and Genera database.Within the checklist each taxon in the family account is treated in the following manner:

TAXON NAME

The species name adopted follows the most recent taxonomic work available. Author abbreviations follow the standards of Brummitt & Powell (1992) (www.ipni.org).

Species names not validly published at the time of publication of the checklist are noted as "in press" or "ined." depending upon the extent to which the publication process has advanced, the former indicating that the protologue has been accepted for, but is awaiting, publication and the latter that the species concept is firmly established but that the protologue is not yet submitted for publication.

Not all names listed are straightforward binomials with authorities. A generic name followed by "sp." generally indicates that the material was inadequate to name to species, for example *Sorindeia* sp. (Anacardiaceae). Use of "sp. 1", "sp. 2" and "sp. A", "sp. B" etc. generally indicate unmatched specimens which may be new to science or may prove to be variants of a currently accepted species; these taxa usually require additional material in order to confirm identity, for example *Trichoscypha sp. 1 of Mefou* (Anacardiaceae), or new taxa for which sufficient material is available, but are awaiting formal description. Unless otherwise stated, or inferred from the distribution, these provisional names are applicable only to the current checklist, thus "sp. 1 of Mefou" indicates "sp. 1 of The Plants of Mefou Proposed National Park, Yaoundé, Cameroon: A Conservation Checklist". The use of "sp. nov." is a firm statement that the taxon is new to science but awaiting formal description; sufficient material may or may not be available for this process, for example *Psychotria longicalyx* O.Lachenaud sp. nov. A generic name followed by "cf.": indicates that the specimens cited should be compared with the associated specific epithet, for example Darbyshire 247 (*Polystachya cf. modesta* Rchb.f.) should be compared with *Polystachya modesta* Rchb.f. This is an indication of doubt (sometimes due to poor material), suggesting that the taxon is close to (but may differ from) the described taxon. The terms "aff.", indicating that the taxon has affinity to the subsequent specific epithet, and "vel. sp. aff.", indicating that the specimen refers to the taxon listed or a closely allied entity, are applied in a similar fashion. These uncertainties are generally explained in the taxon's "Notes" section (see below).

TAXON REFERENCE

The majority of species referred to within the checklist are found in the 2nd edition of Flora of West Tropical Africa (F.W.T.A.: Keay 1954–58; Hepper 1963–72), the standard regional flora. Only species names which do not occur in F.W.T.A. are given a reference here; if no reference is cited the taxon name is currently accepted and occurs in F.W.T.A. The references listed are not necessarily the place of first publication of the name; rather, we have tried where possible to use widely accessible publications which provide useful information on that taxon, such as a description and/or distribution and habitat data. The reference is recorded immediately below the taxon name. In the case of scientific journals, we list the journal name, volume and page numbers and date of publication, with recording of volume part number where it aids in access to the publication. In the case of books, we list the surname of the author(s), the book title, the page number for the taxon in question and the year of publication. Journal and book titles are often abbreviated in the interest of economy of space. Several notable publications are:

Fl. Cameroun

Flore du Cameroun (1963–) Muséum National d'Histoire Naturelle, Paris, France & MINREST, Yaoundé, Cameroon

Fl. Gabon

Flore du Gabon (1961–) Muséum National d'Histoire Naturelle, Paris, France.

F.T.E.A.

The Flora of Tropical East Africa (1952–) Crown Agents, London & A.A.Balkema, Lisse, Netherlands.

SYNONYMS

In some instances, names used in F.W.T.A. have been superseded and are thus reduced to synonymy; these are listed below the accepted name, with the prefix "Syn.". Names listed in synonymy in F.W.T.A. are not recorded here. Other important synonyms are, however, recorded.

TAXON DESCRIPTION

The short descriptions provided for each taxon are based primarily upon the material cited in order to provide the most accurate representation for field botanists working within the checklist area. However, where necessary, they are supplemented by extracted details from the descriptions in F.W.T.A., Fl. Cameroun and the cited taxonomic works. The descriptions are not exhaustive or necessarily diagnostic; rather, they aim to list the key characters to enable field identification of live or dried material, thus microscopic or complex characters are referred to only when they are essential for identification. Where two or more taxa closely resemble one another, a comparative description may be used, by for example stating "…herb… resembling *Phaulopsis ciliata* but …"; such comparisons are only made to other taxa occurring within the checklist area.

Several abbreviations are used in the descriptions, most notably dbh (referring to "diameter at breast height", being a standard measure of the diameter of a tree trunk), the use of "c." as an abbreviation for "approximately", "±" meaning "more or less". In addition, in a few accounts, where male and female flowering parts are repeatedly referred to, the symbols "♀" and "♂" are applied for "male" and "female" respectively. Secondary also appears as 2° in some accounts.

HABITAT

The habitat, recorded at the end of the description, is derived mainly from the field notes of the cited specimens and therefore does not necessarily reflect the entire range of habitats for that taxon; rather, those in which it has been recorded within the checklist area. Habitat information is taken from published sources only where field data is not available, for example, where the only specimens recorded were not available to us, but are cited in F.W.T.A.. Altitudinal measurements are derived from barometric altimeters carried by the collectors. Altitudinal ranges, listed together with habitat, are generated directly from the database of specimens from the checklist area, and thus do not necessarily reflect the entire altitudinal range known for the taxon. Where no altitudinal data was recorded with the specimens, it is omitted.

DISTRIBUTION

For the sake of brevity, country ranges are generally recorded for each taxon rather than listing each separate country, for example "Sierra Leone to Uganda" is taken to include all or most of the intervening countries. Only where taxa are recorded from only two or three, rarely more, countries within a wide area of occurrence are the individual countries listed, for example "Sierra Leone, Cameroon & Uganda". For more widespread taxa, a more general distribution such as "tropical Africa" or "pantropical" is recorded. Where a species is alien to the checklist area, its place of origin is noted, together with its current distribution. Several country abbreviations are used:

Guinea (Bissau): former Portuguese Guinea.

Guinea (Conakry): the Republic of Guinea, or former French Guinea.

CAR: Central African Republic.

Congo (Brazzaville): the Republic of Congo or former French Congo.

Congo (Kinshasa): the Democratic Republic of Congo, or former Zaïre.

Abbreviations for parts of the country are also used (N: north, S: south (or southern), E: east, W: west, C: central), with the exception of South Africa. Where appropriate, Equatorial Guinea is divided into Bioko, Annobón (both islands) and Rio Muni (mainland), and the Angolan enclave of Cabinda north of the Congo River is recorded separately from Angola itself (south of the river).

In addition to country range, a chorology, largely based upon the phytochoria of White (1983) but with modifications to reflect localised centres of endemism in W Africa, is recorded in square brackets. The main phytochoria used are:

upper Guinea: broadly the humid zone following the Guinean coast from Senegal to Ghana.

lower Guinea: separated from upper Guinea by the "Dahomey Gap", an area of drier savanna-type vegetation, that reaches the atlantic coast. Lower Guinea represents the humid zone from Nigeria to Gabon, including Rio Muni, Cabinda, and the wetter parts of western Congo (Brazzaville).

Congolian: the basin of the River Congo and its tributaries, from eastern Congo (Brazzaville) and southern CAR, through Congo (Kinshasa) and to Uganda, Zambia and Angola.

Cameroon endemic: for those taxa restricted to Cameroon, a subdivision of lower Guinea. Taxa endemic to montane western Cameroon are however recorded under W Cameroon Uplands unless they are endemic to the checklist area, when they are listed as a narrow endemic (see below).

narrow endemic: for those taxa restricted to the checklist area, a subdivision of W Cameroon Uplands.

These phytochoria are variously combined where appropriate. Taxa with ranges largely confined to the Guineo-Congolian phytochorion, but with small outlying populations in wet forest in, for example, west Tanzania or northern Zimbabwe, are here recorded as Guineo-Congolian rather than tropical African, as the latter would provide a more misleading representation of the taxon's true phytogeography. A range of othier chorologies are used for more widespread species, including "tropical Africa", "tropical & S Africa", "tropical Africa & Madagascar", "palaeotropics" (taxa from tropical Africa and Asia or some other Old World region), "amphi-Atlantic" (taxa from tropical Africa and S America), "pantropical" and "cosmopolitan". If these terms are used in the distribution, no separate phytochorion is listed.

For some taxa, such as those native to one area of the tropics but widely cultivated elsewhere, the chorology is difficult to define and is thus omitted. Both distribution and chorology are omitted for taxa where there is uncertainty over its identification.

CONSERVATION ASSESSMENT

The level of threat of future extinction on a global basis is assessed for each taxon that has been fully identified and has a published name, or for which publication is imminent, under the guidelines of the IUCN (2001). Under the heading "IUCN:", each taxon is accredited one of the following Red List categories:

LC: Least Concern   NT: Near-threatened

VU: Vulnerable   EN: Endangered

CR: Critically Endangered

Those taxa listed as VU, EN or CR are treated in full within the chapter on Red Data species, where the criteria for assessment are recorded. Those listed as LC or NT are not treated further in this publication, but it is recommended that further investigation of the threats to those taxa recorded as NT are made. Undescribed taxa, or those with an uncertain determination, are not assessed. In addition, we do not consider it appropriate to assess taxa from the poorly-known genus *Anchomanes* (*Araceae*) for which species delimitation is currently poorly understood and thus for which conservation assessments would be somewhat meaningless; this genus should be revisited once a full taxonomic revision has been completed.

SPECIMEN CITATIONS

Specimens from the checklist area are recorded below the distribution and conservation assessment, with the following information:

Location: Ndanan 1 and Ndanan 2 are a rough indicators as to the actual location, which can be provided by reference to the cited specimen.

Collector and number: within each location, collectors are ordered alphabetically and, together with the unique collection number, are underlined. Only the principal collector is listed here, thus for example, some of the collections listed under "Harvey" may originally have been recorded as "Harvey, Nana, F., Nana, V., Kruger, Wandayi".

Phenology: this information is derived directly from the Cameroon specimens database at Kew and is thus dependent upon recording of such information at the time of collection or at the point of data entry onto the database; if this was not done no phenological information is listed. Collections are recorded as flowering (fl.), fruiting (fr.) or sterile (st.) where applicable.

Date: the month and year of collection is recorded for each specimen; within each collector from each location, collections are ordered chronologically.

The specimens cited are of herbarium material derived from a variety of sources; the chapter on collectors in the checklist area provides further information. In addition, a very few confident sight records of uncollected taxa are also included, also photographic records where no specimen was obtained.

NOTES

Items recorded in the notes field at the end of the taxon account include:

Taxonomic notes, for example in the case of new or uncertain taxa, how they differ from closely related species.

Notes on the source of specimen data, for example for those specimens recorded in Fl. Cameroun, or for specimens not seen by the author(s) of the family account.

ETHNOBOTANICAL INFORMATION

Local names and uses are listed for each taxon where appropriate; these are derived largely from local residents and field assistants and are reproduced here with their consent. Local names are in the Ewondo-Beti language. For each local name or use listed, a source is attributed, usually by reference to the collector and number of the specimen from which the information was derived. The sources of the name or use, where recorded, were the local guides who accompanied the teams. In circumstances where the ethnobotanical information was provided verbally with no attached specimen, or where it is known by the author(s) of the family account, the terms *"fide"*, *"pers. comm."* (personal communication) or *"pers. obs."* (personal observation) are used to attribute the information to a source. In the cases where no source is attributed, they are a well known use throughout Cameroon.

The layout of the information on local uses follows the convention of Cook (1995), listing level 1 state categories of use in capital letters, followed by the level 2 state categories, then the specific use. In order to comply with guidelines set by the Convention on Biological Diversity (2002), the detailed uses of medicinal plants are not listed here; only the level 1 state "MEDICINES" is recorded.

REFERENCES

Brummitt, R.K. (1992). *Vascular Plant Families and Genera*. Royal Botanic Gardens, Kew, U.K.

Brummitt, R.K. & Powell, C.E., eds. (1992). *Authors of Plant Names*. Royal Botanic Gardens, Kew, U.K.

Convention on Biological Diversity (2002). *Convention on Biological Diversity: Text and Annexes*. United Nations Environment Programme, Montreal, Canada.

Cook, F.M. (1995). *Economic Botany Data Collection Standard*. RBG, Kew. 146 pp.

IUCN (2001). *IUCN Red List Categories and Criteria. Version 3.1*. IUCN, Gland, Switzerland. 30 pp.

Keay, R.W.J. & Hepper, F.N. (eds) (1954–1972). *Flora of West Tropical Africa*, 2nd ed., 3 vols. Crown Agents, London.

Mabberley, D.J. (1997). *The Plant Book*. Second edition. Cambridge University Press, U.K.

White, F. (1983). The Vegetation of Africa: a descriptive memoir to accompany the Unesco/AETFAT/ UNSO vegetation map of Africa. Unesco, Switzerland.

ANGIOSPERMAE

DICOTYLEDONAE

ACANTHACEAE

I. Darbyshire (K)

Acanthus montanus (Nees) T.Anderson

Erect herb, to 1.5m; stems woody at base, hirsute at nodes and upwards; leaves subcoriaceous, oblanceolate in outline, 25–30 × 6.5–13.5cm, pinnately lobed with spinose teeth, upper surface variegated, scabrid beneath, cystoliths absent; petiole to 8mm; inflorescence a dense terminal raceme to 14cm, bracts broadly ovate, 2.5cm, spinosely toothed, glabrescent; calyx glumaceous, dorsal sepal to 3cm long; corolla white or pink, with purple streaking, splitting near base into a single 5-lobed lip held below the stamens, 3.5 × 3cm; stamens 4, anthers monothecous, densely white-hairy on one side; capsule flattened-ellipsoid, 4-seeded. Forest and secondary growth; 600m.
Distr.: Benin to CAR, Angola and NW Zambia [Guineo-Congolian].
IUCN: LC
Mefou Proposed National Park: Etuge 5261 3/2004.

Ascotheca paucinervia (T.Anderson ex C.B.Clarke) Heine

Fl. Gabon 13: 210 (1966).
Syn. *Rungia paucinervia* (T.Anderson ex C.B.Clarke) Heine
Decumbent herb, to 70cm; stems dark red, tomentose; leaves ovate, 10–16 × 4.5–7cm, base obtuse to rounded, apex acuminate, margin subentire, cystoliths present; petiole to 8.5cm, tomentose; inflorescence a strobilate terminal spike c. 4.5cm, 1-sided, one bract of a pair sterile; peduncle 7.5cm; fertile bracts broadly obovate, 7 × 5mm, shallowly emarginate but with central apiculum, ciliate, pale green with deep red margins; calyx enclosed within bracts, of 5 subulate lobes; corolla shortly exserted, white with purple markings, 2-lipped; stamens 2, thecae offset, lower one tailed, both dehiscing by basal pores; capsule stipitate, 4-seeded. Forest and disturbed forest; 700m.
Distr.: SE Nigeria to Gabon [lower Guinea].
IUCN: NT
Ndanan 1: Harvey 159 10/2002.

Asystasia gangetica (L.) T.Anderson subsp. ***micrantha*** (Nees) Ensermu

Proc. XIII Plenary Meet. AETFAT, Zomba, Malawi 1: 343 (1994).
Trailing or straggling herb, to 1m; stems sparsely pubescent; leaves ovate, c. 7 × 3cm, base subcordate or attenuate and abruptly rounded, apex acuminate, margin subentire, surfaces sparsely pubescent or glabrous, cystoliths present; petiole to 1.5cm; inflorescence a lax 1-sided axillary spike to 14cm, 6–11-flowered; bracts lanceolate, c. 1mm, ciliate; calyx lobes subulate, c. 4mm, pubescent; corolla 2-lipped, white with purple markings on throat and palate; tube 1.3 × 0.8cm, with expanded throat, puberulent; stamens 4, bithecous; ovary and base of style pubescent; capsule stipitate, 2cm, puberulent, 4-seeded, seeds angular, grit-like. Farmbush and forest edges; 700m.
Distr.: Senegal to Djibouti and Ethiopia, S to South Africa, Madagascar and tropical Arabia [tropical Africa, Madagascar & Arabia].
IUCN: LC
Ndanan 1: Harvey 88 10/2002.

Asystasia macrophylla (T.Anderson) Lindau

Shrub, to 3m; leaves oblanceolate, 27–37 × 10–12cm, base long-cuneate, apex caudate, margin subcrenulate, glabrous, cystoliths present; petiole 2.5cm; inflorescence a slender terminal spiciform thyrse to 24cm, 6–10 flowers per cymule; bracts triangular, 2mm, pubescent; calyx lobes subulate, 3.5mm, pubescent; corolla 2-lipped, white, drying red-black, tube 3.5 × 1cm with an elongate expanded throat; stamens 4, bithecous, pistil glabrous; capsule stipitate, to 3.3cm, glabrous, 4-seeded. Forest; 700–720m.
Distr.: SE Nigeria & Bioko to Gabon [lower Guinea].
IUCN: NT
Mefou Proposed National Park: Cheek in MF 2 3/2004; Nana 45 3/2004; **Ndanan 1:** Cheek 11190 10/2002; 11491 3/2004; 11843 3/2004; Onana 2868 3/2004.

Brillantaisia lamium (Nees) Benth.

Erect perennial herb, to 1.5m; stems subglabrous or tomentose upwards, hairs to 4mm; leaves ovate, 8.5–11 × 5–6cm, base cordate or truncate, apex acuminate, margin subentire, surfaces pilose, cystoliths present; petiole to 6.5cm, tomentose; inflorescence a lax panicle, glandular-hairy throughout; main axis bracts ovate or elliptic, 9 × 8mm at midpoint; pedicels to 1.3cm; calyx lobes linear, 8mm; corolla purple or blue, 2.5cm long, 2-lipped, tube 7–10mm, upper lip hooded, lower lip shortly 3-lobed; stamens 2, bithecous; capsule linear, 30–35mm long, glabrous, over 10-seeded. Farmbush & forest margins; 700m.
Distr.: Guinea (Conakry) to Angola & W Tanzania [Guineo-Congolian].
IUCN: LC
Ndanan 1: Harvey 127 10/2002.

Brillantaisia vogeliana (Nees) Benth.

Annual or perennial herb, to 1.5m; stems largely glabrous; leaves broadly ovate, 7–20 × 5–12cm, base truncate to subcordate, margin dentate, apex (sub)acuminate, hairy mainly along principal veins, cystoliths present; petiole to 10cm, upper portion often with irregularly toothed wing; panicle lax, 5–20cm long, many-flowered, glandular hairy; main axis bracts ovate, leafy, reducing upwards, sometimes caducous; calyx lobes linear-lanceolate ± 1cm long; corolla blue

or purple, 2-lipped, tube white 5–7mm, lips 10–18mm, upper lip hooded, lower lip shortly 3-lobed; stamens 2, held in upper lip, bithecous; capsule linear, 17–23mm long, over 10-seeded. Forest & forest margins; 700m.
Distr.: Ghana to Sudan & W Kenya [Guineo-Congolian].
IUCN: LC
Ndanan 1: Cheek 11023 10/2002.

Dicliptera elliotii C.B.Clarke
Weakly decumbent or erect herb, to 1m; stems glabrescent; leaves ovate, 2–8.5 × 1–4cm, base attenuate, apex (sub)acuminate, margin entire, largely glabrous, cystoliths present; petiole to 2cm; inflorescences axillary, umbellate, often several pedunculate umbels per axil; flowers enclosed within paired broadly ovate bracts 7–9mm long, pilose; calyx hidden within bracts, lobes subulate; corolla minute, 5.5–7.5mm, white, 2-lipped, resupinate; stamens 2, thecae superposed; capsule shortly stipitate, 4.5–5mm, pubescent, 4-seeded. Understorey of *Raphia* swamp; 670m.
Distr.: Sierra Leone to W Tanzania and Angola (Cabinda) [Guineo-Congolian].
IUCN: LC
Ndanan 2: Darbyshire 222 fl., fr., 3/2004.

Elytraria marginata Vahl
Herb, to 30cm, decumbent; stems pubescent; leaves in terminal rosette, obovate or oblanceolate, 8–13 × 3–4cm, base cuneate, apex acute or acuminate, margin subentire, upper surface sparsely pilose, cystoliths absent; petiole to 1cm; inflorescence a dense slender terminal spike to 13cm; bracts rigid, ovate, 0.5cm, apiculate, margins ciliate; corolla white, c. 5mm, 2-lipped, often cleistogamous; stamens 2. bithecous; capsule beaked, c. 5mm long, glabrous, over 10-seeded, not held on hooks. Forest & farmbush; 658–700m.
Distr.: Guinea (Conakry) to Angola & Uganda [Guineo-Congolian].
IUCN: LC
Mefou Proposed National Park: Etuge 5306 3/2004; **Ndanan 1:** Harvey 198 10/2002; Onana 2905 3/2004; **Ndanan 2:** Cheek 11149 10/2002.

Eremomastax speciosa (Hochst.) Cufod.
Fl. Gabon 13: 30 (1966); Kew Bull. 44: 69 (1989).
Syn. *Eremomastax polysperma* (Benth.) Dandy
Syn. *Clerodendrum eupatorioides* Baker
Herb, to 2m; stems robust, tomentose towards apex, hairs to 1.5mm; leaves broadly ovate, 5–7 × 3–5cm, base truncate to subcordate, apex acuminate, margin crenate-dentate, surfaces pubescent along veins, cystoliths present; petiole to 5.7cm; inflorescence a many-flowered terminal panicle; bracts leafy but reducing up the axis, finely pilose; calyx lobes linear, c. 1.5cm, tomentose; corolla 3–5cm long with tube 3cm, blue to mauve, 5 lobes held in a single lip below the stamens; stamens 4, bithecous; capsule linear, 17mm, glandular towards apex, usually over 4-seeded. Forest & farmbush; 700m.

Distr.: Guinea (Conakry) to SW Ethiopia & NW Tanzania, Madagascar [Guineo-Congolian].
IUCN: LC
Ndanan 2: Cheek 11144 10/2002.
Local Name: Dibielog (Ewondo) (Cheek 11144).
Uses: RELIGIOUS USES — ritual (Cheek 11144).

Hypoestes aristata (Vahl) Soland. ex Roem. & Schult.
Erect or scrambling herb, to 2m; stems pubescent, hairs white; leaves ovate, 4–18.5 × 1–8cm, base acute, apex gradually acuminate, margin subentire, veins, margins and upper surface ± pilose, cystoliths present; petiole 1.5–7cm; inflorescences of dense axillary fascicles; flowers enclosed in paired basally-fused bracts lanceolate, to 1.3cm long with arista to 5–13mm; calyx enclosed within bracts, of 4 subulate lobes; corolla c. 2cm long, pink with white base, 2-lipped, resupinate; stamens 2, monothecous; capsule stipitate, 11–17mm long, glabrous, 4-seeded. Secondary forest & farmbush; 600–710m.
Distr.: Nigeria to Ethiopia, S to South Africa [tropical Africa].
IUCN: LC
Mefou Proposed National Park: Cheek in RP 9 3/2004; Etuge 5205 3/2004; **Ndanan 1:** Cheek 11583 3/2004.

Justicia tenella (Nees) T.Anderson
Small creeping or decumbent herb, to 20cm; stems white-pubescent; leaves ovate, c. 2 × 1.3cm, base obtuse, apex acute or obtuse, margin subentire, largely glabrous, cystoliths present; petiole c. 1.3cm; inflorescence a dense spike 1–2cm long, lateral in clusters of 2–3; peduncle to 1.7cm; bracts obovate-orbicular, 2.5mm diam., pale green with paler margin, imbricate; corolla minute, 2-lipped, lower lip purple with white margin; stamens 2, thecae offset, the lower tailed; capsule shortly stipitate c. 3mm long, glabrous, 4-seeded. Farmbush, grassland and forest margins; 600–700m.
Distr.: tropical Africa & Madagascar [tropical Africa & Madagascar].
IUCN: LC
Mefou Proposed National Park: Etuge 5238 3/2004; **Ndanan 1:** Harvey 214 10/2002.

Lepidagathis alopecuroides (Vahl) Griseb.
Decumbent herb, to 20cm; stems pubescent or pilose when young; leaves ovate, elliptic or lanceolate, 3–6.5 × 0.5–2.5cm; base attenuate or cuneate, margin entire, apex obtuse to subattenuate, sparsely hairy or glabrous, cystoliths present; petiole to 15mm; inflorescence a dense subcapitate to cylindrical terminal thyrse; bracts elliptic or lanceolate, 3.5–7.5mm, with mixed glandular and eglandular hairs; calyx divided into 5 lobes of unequal width; corolla minute, 5–7.5mm, white to pink with darker markings, 2-lipped; stamens 4, bithecous; capsule flattened oblong-elliptic, 4mm, 4-seeded; seeds with appressed hairs. *Raphia* swamp; 600–710m.

Distr.: Senegal to Sudan & Angola, Tanzania, also neotropics [tropical Africa & neotropics].
IUCN: LC
Mefou Proposed National Park: <u>Cheek in RP 40</u> 3/2004; <u>Etuge 5204</u> 3/2004.

Mendoncia gilgiana (Lindau) Benoist

Slender woody twiner, to 10m; leafy stems pubescent; leaves ovate-elliptic, 6.5–9.5 × 2.5–4cm, base rounded or shallowly cordate, apex acute to acuminate, apiculate, margin subcrenulate, young leaves pubescent, later glabrescent, cystoliths absent; petiole 1.8–4.5cm, pubescent; inflorescence an axillary fascicle, 1–4 flowers per node; pedicel to 2cm, pubescent; flowers enclosed within paired ovate bracteoles 1.6 × 1cm, apex acuminate, densely pubescent; corolla white, tube 1.7cm long, limb of 5 spreading lobes; stamens 4, included, bithecous; fruit a green-black ellipsoid drupe c. 1cm long, enclosed within the persistent bracteoles. Forest & clearings; 700m.
Distr.: Cameroon to Congo (Kinshasa), S Sudan to NW Tanzania [Guineo-Congolian].
IUCN: LC
Ndanan 1: <u>Cheek in MF 398</u> 10/2002; <u>Harvey 85</u> 10/2002.

Nelsonia canescens (Lam.) Spreng.

Creeping or decumbent herb; stems densely villose; leaves ovate, 3–3.5 × 2–3cm, base truncate to attenuate, apex acute, margin subcrenulate, surfaces densely pubescent particularly when young, cystoliths absent; petiole to 2cm, villose; inflorescence a cylindrical terminal spike to 12cm long, densely white-hairy throughout; bracts ovate, c. 5 × 2mm, acuminate, imbricate; calyx enclosed within bracts, of 4 lobes of unequal width; corolla pink to purple, c. 5–10mm, limb subregular; stamens 2, bithecous; capsule shortly beaked, 4–7mm, over 10-seeded, these not held on hooks. Farmbush & disturbed environments; 710m.
Distr.: palaeotropical, introduced in tropical America, the Pacific & Australia [pantropical].
IUCN: LC
Mefou Proposed National Park: <u>Cheek in RP 18</u> 3/2004.

Phaulopsis ciliata (Willd.) Hepper

Symb. Bot. Ups. 31: 103 (1995).
Syn. *Phaulopsis falsisepala* C.B.Clarke F.W.T.A. 2: 399 (1963).
Perennial herb, to 1.3m; stems sparsely hairy mainly on angles; leaves ovate(-elliptic), 5.5–15 × 2–5cm, base attenuate, apex acuminate, margin entire or crenulate, sparsley hairy mainly on veins beneath, cystoliths present; petiole 1–6cm; inflorescences terminal and axillary strobilate spikes, 1-sided; fertile bracts broadly elliptic or orbicular, 7.5–11mm long, often brown or purple-tinged, long-ciliate with stiff yellowish eglandular hairs; sterile bracts much narrower; calyx of five lobes of unequal width, the anterior pair obliquely

oblanceolate with blunt apex, posterior lobe elliptic or obovate; corolla white with purple markings, 9.5–11mm, 2-lipped; stamens 4, anthers of longer pair shortly exserted, all bithecous; capsule elliptic, 4-seeded. Forest, forest margins & secondary thicket; 710m.
Distr.: Senegal to Bioko, Cameroon, Gabon, Congo (Brazzaville), Congo (Kinshasa), CAR, Chad and Sudan [Guineo-Congolian].
IUCN: LC
Ndanan 1: <u>Cheek 11573</u> 3/2004.

Phaulopsis imbricata (Forssk.) Sweet subsp. *poggei* (Lindau) M.Manktelow

Symb. Bot. Ups. 31: 138 (1995).
Syn. *Phaulopsis poggei* Lindau Fl. Gabon 13: 48 (1966).
Perennial herb, erect or often straggling, to 1m; resembling *P. ciliata* but stem and leaves more hairy, bracts pale green, with mixed glandular and eglandular hairs, the latter not so stiff; anterior calyx lobes narrower, subulate or narrowly spathulate; corolla 7–11mm long, anthers included within corolla tube. Forest clearings, secondary scrub, rough grassland & farmbush; 700–710m.
Distr.: Guinea Highlands, Ghana, Nigeria to Ethiopia, S to Angola & Zambia [Guineo-Congolian].
IUCN: LC
Ndanan 1: <u>Cheek 11754</u> 3/2004; **Ndanan 2:** <u>Darbyshire 220</u> fl., 3/2004.

Phaulopsis talbotii S. Moore

Herb, resembling *P. imbricata* subsp. *poggei* but annual, leaves more oblique at base, glandular hairs on the inflorescence having a red gland-tip, anterior pair of calyx lobes very narrow, linear with sharply acute apex, and corolla only 5.5–6.5mm long. Forest, secondary forest and farmbush; 710m.
Distr.: Senegal to Ghana, then scattered E to S Sudan, also in South America and West Indies [Guineo-Congolian & neotropical].
Ndanan 1: <u>Cheek 11857</u> 3/2004.
Note: whilst common in upper Guinea, this species is apparently rare and highly scattered east of the Dahomey interval, Darbyshire ix.2010.

Pseuderanthemum tunicatum (Afzel.) Milne-Redh.

Herb, to 60cm; stems glabrous, often woody towards base; leaves elliptic or oblanceolate, 10–17 × 3.5–6cm, base cuneate, apex acuminate, margin irregularly crenulate, cystoliths present; petiole to 2cm; inflorescence a slender (sub)terminal spike to 10–24cm long; peduncle to 9cm, sessile cymules widely spaced; bracts minute; calyx lobes subulate, to 3.5mm; corolla (bluish-)white with pink or purple markings on limb, tube long-cylindrical, to 14mm, limb of 5 subequal lobes; stamens 2, bithecous; capsule stipitate, to 2cm, glabrous, 4-seeded. Forest; 670–710m.
Distr.: W Africa to Congo (Kinshasa) [Guineo-Congolian].
IUCN: LC

Mefou Proposed National Park: <u>Cheek in RP 70</u> 3/2004; <u>Nana 53</u> 3/2004; **Ndanan 1:** <u>Cheek 11544</u> 3/2004; <u>Darbyshire 184</u> fl., fr., 3/2004; <u>Tadjouteu 598</u> 3/2004; **Ndanan 2:** <u>Darbyshire 293</u> fr., 3/2004.

Rhinacanthus virens (Nees) Milne-Redh.

Herb, or subshrub, to 1.3m; stems puberulent; leaves ovate-elliptic, 5.5–17 × 2.5–6cm, base acute, apex acuminate, margin subentire, glabrous or sparsely pilose, cystoliths present; petiole to 3.3cm; inflorescence a ± lax, few-flowered panicle to 27cm long; peduncles wiry, puberulent towards apices; bracts linear-lanceolate to 2mm long; calyx lobes lanceolate, 4mm, glandular-puberulent; corolla white to mauve, to 1.7cm, 2-lipped, tube narrowly cylindrical, often long, limb much shorter, upper lip triangular, lower lip 3-lobed; stamens 2, thecae offset, lower theca minutely tailed; capsule stipitate, c. 1.1cm, puberulous, 4-seeded. Forest; 710m.
Distr.: Guinea (Conakry) to Uganda & NW Tanzania [Guineo-Congolian].
IUCN: LC
Ndanan 1: <u>Cheek 11914</u> 3/2004.

Whitfieldia elongata (P.Beauv.) De Wild. & T.Durand

Woody scrambler or scandent shrub, to 3m; stems subglabrous; leaves elliptic-ovate, 10–28 × 4–11cm, base attenuate, apex acuminate, margin subcrenulate, glabrous, cystoliths present; petiole to 2cm; inflorescence a lax terminal raceme, occasionally with lateral racemes at base together forming a panicle, to 17.5 × 13cm; bracts green-white, narrowly ovate, to 1–4cm long; bracteoles as bracts but elliptic, c. 15 × 7mm; calyx white, lobes linear, 23–40 × 3.5mm, glandular puberulent; corolla white, tube cylindrical, 33–50 × 2mm, lobes elliptic, 18–28mm long, often reflexed; stamens 4, bithecous; capsule stipitate, 2-seeded. Forest & forest margins; 710m.
Distr.: Nigeria to Kenya, S to Tanzania, Zambia & Angola [tropical Africa].
IUCN: LC
Ndanan 1: <u>Cheek 11898</u> 3/2004.

ALANGIACEAE

M. Cheek (K)

Fl. Cameroun 10 (1970).

Alangium chinense (Lour.) Harms

Tree, 5–18m, glabrescent; leaves alternate, ovate, c. 10 × 6cm, acuminate, obliquely rounded, entire, digitately 5-nerved, scalariform; inflorescences cymose, axillary, c. 20-flowered, 4cm; peduncle 1.5cm; flowers orange, 1.1cm; fruit ellipsoid, 1cm, fleshy. Forest margins and farmbush; 710m.
Distr.: palaeotropical, excluding upper Guinea [palaeotropics].
IUCN: LC
Mefou Proposed National Park: <u>Cheek in RP 87</u> 3/2004.

AMARANTHACEAE

C.C. Townsend (K), E. Fenton & Y.B. Harvey (K)

Fl. Cameroun 17 (1974).

Aerva lanata (L.) Juss. ex Schult.

Scrambling herb, 0.1–2m; branched from the base; stem and branches with whitish soft hairs; leaves elliptic-lanceolate to ± round, 10–50 × 5–35mm; sessile, or with petioles to 2cm; spikes 0.5–2 × 0.3–0.4cm, solitary or clustered; flowers female, bisexual and ± functionally male; tepals ± densely lanate dorsally, 0.75–1.75mm, inner 3 shorter to narrower; capsule c. 1mm, round, compressed. Roadside, disturbed ground; 710m.
Distr.: widespread from Sierra Leone to South Africa and Madagascar, in Asia from Arabia to New Guinea [tropical Africa and tropical Asia].
IUCN: LC
Ndanan 1: <u>Cheek 11825</u> 3/2004.

Alternanthera sessilis (L.) R.Br.

Scrambling herb, to 0.3m; rooting at the nodes; stems red-green, with a narrow line of whitish hairs, and tufts of white hairs in branch and leaf axils; leaves lanceolate to obovate-spathulate, 1–4 × 0.5–2cm, apex shortly acuminate, base attenuate, glabrous; sessile or petioles to 4mm; inflorescences sessile clusters, c. 0.5cm diam.; bracts white, to 1mm, glabrous; bracteoles to 1.5mm; tepals 1.5–2.5mm, white, glabrous; fruit 2–2.5mm long, compressed, glabrous. River margins; 650m.
Distr.: pantropical.
IUCN: LC
Ndanan 2: <u>Darbyshire 206</u> fl., 3/2004.

Amaranthus dubius C.Mart. ex Thell.

Erect herb, to 0.5m; stem branched, angular, glabrous in lower parts; leaves ovate or rhomboid-ovate, 4–7 × 1.5–2.5cm, apex mucronate, base cuneate; flowers green; lower inflorescences axillary clusters, 1–10cm diam., towards apex, inflorescences branched spikes to 15 × 1cm; lower clusters of flowers female, spikes with few male flowers at apex; bracts and bracteoles deltoid-ovate with reddish awn; perianth segments 5; female flowers c. 1.5–2.75mm long; fruit a capsule subequalling the perianth, ovoid urceolate. Disturbed ground; 660m.
Distr.: tropical Africa & tropical America [tropical Africa & tropical America].
IUCN: LC
Ndanan 2: <u>Darbyshire 331</u> fl., 4/2004; <u>335</u> fl., 4/2004.

Celosia isertii C.C.Towns.

F.T.E.A. Amaranthaceae: 15 (1985); Fl. Zamb. 9(1): 34 (1988).
Climbing or scrambling herb, to 2m; stems glabrous; leaves drying green-black, alternate, ovate, 6.8–9.3 × 3–5cm, apex acuminate-apiculate, base attenuate then abruptly truncate, margin irregularly undulate, glabrous

or scabridulous, veins occasionally purplish below; petiole c. 2.5cm; inflorescence spikes axillary or terminal, occasionally branched, 2.5–10cm; peduncle to 3.5cm, glabrescent; bracteoles and tepals ovate-elliptic, white, the latter 3–4mm long; style bifid. Forest & forest edge; 700m.
Distr.: tropical African highlands [afromontane].
IUCN: LC
Ndanan 1: Cheek 11253 10/2002; Harvey 172 10/2002.

Cyathula prostrata (L.) Blume var. *prostrata*
Scrambling herb, to 1.5m; stem and branches bluntly 4-angled, subglabrous to densely pilose; leaves 3–6(–21) × 2.5–4.5(–7)cm, rhombic to rhombic-ovate, apex acute, base cuneate to cuneate-attenuate, subglabrous to pilose; subsessile to petiolate (petiole to 1cm); spikes terminal, initially to c. 2cm, then elongating to 18cm or more; peduncles to 1mm, shorter than the bract; outer tepals ± white pilose; spines of modified flowers sharply uncinate. Farmbush, swamp and forest; 660–710m.
Distr.: pantropical.
IUCN: LC
Mefou Proposed National Park: Cheek in RP 62 3/2004; **Ndanan 1:** Cheek 11577 3/2004; **Ndanan 2:** Darbyshire 324 fl., 4/2004.
Note: Cheek 11577, has much larger leaves (to 21 × 7cm) than the other collections made here (to 8 × 5cm), Harvey iii.2010.

ANACARDIACEAE

H. Fortune-Hopkins (K) & F. Breteler (WAG)

Lannea welwitschii (Hiern) Engl.
Tree 20–25m; leaves 30–40cm, 2–3-jugate; leaflets leathery, drying dark brown below, ovate, c. 10 × 5cm, acuminate, obtuse, nerves 8 pairs, raised; petiolule 1cm; petiole 15cm; inflorescence terminal or from upper 2–3 axils, 7–20cm, branches 2–7cm long; flowers white 2mm; fruit discoid, 8mm, foveolate, fleshy. Forest; 710m.
Distr.: Ivory Coast to Congo (Kinshasa) [Guineo-Congolian].
IUCN: LC
Mefou Proposed National Park: Cheek in RP 32 3/2004.

Pseudospondias microcarpa (A.Rich.) Engl. var. *microcarpa*
Tree, 5–30m; broken stems scented of mango; leaves often only 3–4-jugate, pinnate, to 30cm long; leaflets 3–12 pairs, to 20 × 8cm, oblique, acuminate, obtuse to rounded; petiolule 0.5cm; petiole to 8cm; inflorescence c. 20 × 12cm, highly diffuse; flowers 4mm, greenish-yellow; stamens 8; fruit ellipsoid, 2–5cm, red. Farmbush and secondary forest; 550m.
Distr.: Senegal to Malawi [tropical Africa].
IUCN: LC
Mefou Proposed National Park: Etuge 5191 3/2004.

Sorindeia africana (Engl.) van der Veken
Bull. Jard. Bot. Natl. Belg. 24: 242 (1959); Enum. Pl. Afr. Trop. 2: 229 (1992).
Syn. *Sorindeia nitidula* Engl.
Syn. *Sorindeia gilletii* De Wild.
Tree, 5–7m; exudate white; leaves 1–3-jugate, the c. 35cm, apical leaflets drying brown, leathery, larger, 21 × 8cm, acuminate, unequally obtuse, lateral nerves c. 9 pairs, matt, margin revolute, undulate; petiolule 0.5cm; petiole 8cm; inflorescence terminal, c. 35 × 35cm; flowers white, 2mm; fruits ovoid 1.5cm. Forest; 710m.
Distr.: Nigeria to Angola [lower Guinea & Congolian].
IUCN: LC
Ndanan 1: Cheek 11585 3/2004.

Sorindeia grandifolia Engl.
Stout tree, often leaning 1.5(–4)m; leaves matt dark green above and below, 30–60cm, 1–3-jugate, oblong-elliptic, to 25(–44) × 10(–14)cm, shortly acuminate, obliquely rounded, lateral nerves 8–10 pairs, yellow below; petiolule 0.5–1.5cm; petiole 7–15cm; inflorescence cauliflorous, pendulous, 4–30cm, 1–8-branched; pink in bud; flowers 2–3mm; fruit ellipsoid, 2cm. Forest; 600–720m.
Distr.: Guinea-Conakry to S Sudan & São Tomé, N of equator (Guineo-Congolian) [lower Guinea].
IUCN: LC
Mefou Proposed National Park: Cheek in MF 36 3/2004; Etuge 5222 3/2004; Nana 55 3/2004; **Ndanan 1:** Onana 2854 3/2004.
Note: Cheek in plot voucher series MF36 is probably this taxon but fertile material needed to be sure, Hopkins iii.2010.

Sorindeia sp. of Mefou
Tree seedling, to 10cm; leaves opposite, 3-jugate, leaflets ovate, 7 × 2.5cm, glabrous, attenuate, tip to 1.5cm; petiole 1.7cm. Lowland evergreen forest near slow-flowing swamp-edged water course; 700m.
Mefou Proposed National Park: Cheek in MF 61 3/2004.

Trichoscypha acuminata Engl.
Tree, 15m; leaves 1.2m, c. 15-jugate; middle leaflets leathery oblong-lanceolate c. 20–25 × 5–10cm, acuminate, rounded, c. 15 nerve pairs; petiole 15cm; inflorescence cauliflorous, 15–30 × 6cm; flowers 2–3mm, pink; fruit red, 3cm. Forest; 700m.
Distr.: S Nigeria to Congo (Kinshasa) [lower Guinea & Congolian].
IUCN: LC
Ndanan 1: Cheek in MF 391 10/2002.
Note: Cheek in plot voucher series MF391 is probably this taxon, Hopkins iv.2010.

Trichoscypha sp. 1 of Mefou
Tree, to 5m; exudate white; leaves alternate; lateral leaflets to 30 × 7cm, oblong, apex acute, glabrous, petiolules to 1.5cm, apical leaflet 30 × 9.5cm,

oblanceolate, apex acute; petiole to 5cm. Forest; 700m.
Ndanan 1: Cheek in MF 307 10/2002.
Note: plot voucher series MF307 is close to *T. acuminata* Engl. and *T. arborea* (A.Chev.) A.Chev. but not an exact match with either, Hopkins iii.2010.

Trichoscypha sp. 2 of Mefou

Tree seedling, to 10cm; stem glabrous; leaves alternate, ovate, 8–9 × 2.5–3cm, glabrous, attenuate, tip to 1.5cm; petiole 1.5–2.3cm, scattered hairs. Lowland evergreen forest near slow-flowing swamp-edged water course; 700m.
Mefou Proposed National Park: Cheek in MF 60 3/2004.
Note: Cheek in plot voucher series MF60 has an upper surface of blade that suggests *Trichoscypha mannii* Hook.f. or if not, then *T. lucens* Oliv. Adult material needed, Hopkins iii.2010.

Trichoscypha sp. probably *bijuga* Engl.

Tree, 7m; leaflets to 18 × 5.8cm, oblong, apex attenuate, tip to 2cm, glabrous except for hairs on lower midrib; petiolule to 1mm, scattered hairs. Forest; 700m.
Ndanan 1: Cheek in MF 372 10/2002.
Note: plot voucher MF372 is sterile. More material needed to confirm. Hopkins x.2010

ANCISTROCLADACEAE

C. Couch (K) & M. Cheek (K)

Ancistrocladus letestui Pellegr.

Bull. Soc. Bot. France 98: 18 (1951).
Woody climber, to 7m; stem with hooked branchlets; leaves simple, alternate, 9.5–21.5 × 3–4.5cm, lanceolate-oblanceolate, entire, glabrous, stipulate, reticulate venation; flowers in terminal panicles, hermaphrodite, 5-merous; fruit indehiscent, unevenly winged. Forest; 600m.
Distr.: Cameroon and Gabon [lower Guinea].
IUCN: LC
Mefou Proposed National Park: Etuge 5274 3/2004.
Note: Etuge 5274 is sterile, specific identification uncertain, Cheek xi.2010.

ANNONACEAE

G. Gosline (K), H. Fortune-Hopkins, X. van der Burgt (K) & M. Cheek (K)

Annickia affinis (Exell) Versteegh & Sosef

Syst. & Geogr. Pl. 77(1): 95 (2007).
Tree, to 30m; wood and sap bright yellow; leaves elliptic-oblong, 5–22 × 2–7cm, 7–14 pairs secondary nerves, inferior face appressed pubescent; flowers solitary, extra-axillary; petals 3 opposite sepals, lanceolate, densely appressed pubescent, triangular cross-section, 2–3 × 0.6–1.2cm; fruit with numerous ovoid monocarps, 2.5–3 × 1cm; stipes 2–3cm; pedicel 1.5–2cm. Secondary forest; 600–700m.

Distr.: Nigeria to Congo (Kinshasa) [lower Guinea & Congolian].
IUCN: LC
Mefou Proposed National Park: Etuge 5139 3/2004;
Ndanan 1: Cheek in MF 253 10/2002.
Note: distinguished from *Annickia chlorantha* by obverse of leaf with simple hairs all oriented towards apex and midrib on upper surface glabrous, Gosline vi.2010.
Local Name: Mfol (Etuge 5139).
Uses: MEDICINES (Etuge 5139).

Anonidium mannii (Oliv.) Engl. & Diels var. *mannii*

Tree, to 30m; leaves elliptic-obovate to obovate-oblong, 20–45 × 7–18cm, sharply acuminate, acumen 2–3cm long, 10–20 pairs secondary nerves; inflorescence pendant from trunk or bare branches, cymes to 2–3m long; flowers hermaphordite or male; petals 6 greenish-yellow, elliptic to obovate 2.5–5 × 1.5–4cm; fruits syncarpous, large cylindric-ovoid, 25–50 × 10–30cm, to 10kg, surface reticulate-areolate. Forest; 700m.
Distr.: Ghana to Congo (Kinshasa) [Guineo-Congolian].
IUCN: LC
Ndanan 1: Cheek 11248 10/2002; **Ndanan 2:** Gosline 420 10/2002.

Artabotrys insignis Engl. & Diels

Woody climber; stems drying nearly black, glabrous; leaves oblanceolate 4–10 × 2–4cm, glabrous, base cuneate, acumen linear, 1–2cm long, 6–8 lateral nerves looping and joining 3–5mm from margin; petioles 3–5mm long, thin; hooks thin, 12–15mm long. Forest; 710m.
Distr.: Sierra Leone to Cameroon, Gabon and Congo (Kinshasa) [Guineo-Congolian].
IUCN: LC
Ndanan 1: Cheek 11641 3/2004.
Note: more material needed to confirm identification of Cheek 11641, Gosline vi.2101.

Friesodielsia enghiana (Diels) Verdc.

Fl. Gabon 16: 240 (1969).
Syn. *Oxymitra obanensis* (Baker f.) Sprague & Hutch.
F.W.T.A. 1: 45 (1954).
Woody climber, to 30m or understorey shrub, 1–2m when young; shoots densely rusty hirsute; leaves oblong-lanceolate 8–34 × 3.5–8cm, acuminate, leaf undersurface strikingly blue-grey, 11–20 pairs secondary nerves, nerves and lamina hirsute; inflorescence extra-axillary in 2–3-flowered fascicles; external petals papery, elliptic-ovate, 15–25 × 12–20mm, rufus tomentose, inner petals coriacious, smaller, glabrescent; fruit pedicel 2–3cm long, monocarps ellipsoid or moniliform, rufus hirsute. Forest; 710m.
Distr.: Sierra Leone to Congo (Kinshasa) [Guineo-Congolian].

IUCN: LC
Ndanan 1: Cheek 11499 3/2004.

Greenwayodendron suaveolens (Engl. & Diels)
Verdc.
Adansonia (Sér. 2) 9: 90 (1969); F.T.E.A. Annonaceae:
67 (1971); Keay R.W.J., Trees of Nigeria: 23 (1989).
Syn. *Polyalthia suaveolens* Engl. & Diels
Tree, to 10–25m; bark smooth pale to dark grey; leaves
oblong-elliptic 4–12 × 3–5cm acute to long-acuminate
apex, 5–13 lateral nerves; flowers polygamous, solitary
or in 2–8-flowered fascicles; petals subequal linear-
oblong or lanceolate 10–28 × 1.5–2.5mm finely
pubescent; fruiting pedicels 0.3–2.2cm long; stipes
5–8mm long; monocarps c. 8–10, globose, 1.2–1.8 ×
0.9–1.5cm, 1–2 seeded; seeds compressed-hemiglobose,
c 1cm diam., strongly wrinkled with a circumferential
groove. Forest; 700m.
Distr.: Nigeria to Uganda [Guineo-Congolian].
IUCN: LC
Mefou Proposed National Park: Cheek in MF 27
3/2004; 66 3/2004; 78 3/2004; **Ndanan 1:** Cheek 11224
10/2002; Cheek in MF 343 10/2002.
Note: LeThomas does not recognise Verdcourt's
separation of *Greenwayodendron* from *Polyalthia*.

Hexalobus crispiflorus A.Rich.
Tree, to 30m; trunk deeply channelled, bark brown,
fissured; leaves elliptic to ovate-lanceolate, 7.5–20cm ×
3–8cm, 12–17 pairs secondary nerves; flowers solitary
or 2–3 in axils; sepals 3, 1–2 × 0.5–1cm; petals pale
yellow, united at base, crispate borders, linear 2.5–8 ×
0.5–1.5cm; anthers numerous; carpels 6–10; fruits on a
very thick pedicel 1–2cm long; mericarps oblong, 8–9 ×
4–5cm. Forest; 700m.
Distr.: Guinea to Congo (Kinshasa) & Sudan [Guineo-
Congolian].
IUCN: LC
Ndanan 1: Cheek 11064 10/2002.

Monanthotaxis angustifolia (Exell) Verdc.
J. Bot. 75: 163 (1937); Kew Bull. 25: 21 (1971).
Syn. *Enneastemon angustifolius* Exell
Liane; twigs grey; leaves obovate to oblanceolate 8–15
× 4–6cm, acumen brief, obverse distinctly glaucous
with appressed white hairs, 8–12 pairs lateral nerves;
petioles 5–6mm. Forest, near river; 710m.
Distr.: SE Nigeria & Cameroon [lower Guinea].
IUCN: DD
Ndanan 1: Cheek 11657 3/2004.
Note: this species is known from Nigeria and
neighbouring Cameroon. More material is needed to
confirm identification of Cheek 11657, Gosline vi.2010.

Neostenanthera myristicifolia (Oliv.) Exell
Tree, to 1m tall; dbh 10cm; leaves obovate 7–30 ×
3–12cm, base rounded, 12–17 pairs lateral nerves;
petioles 6–12mm; flowers 10–30mm long, exterior

petals much longer than interior; mericarps distributed
spherically; pedicel 1.5–3cm long, thick; stipes
1.5–3cm long, thin; mericarps numerous, ellipsoid 9–15
× 5–9mm, mucronate; 1 seed per mericarp. River bank;
700–720m.
Distr.: Nigeria to CAR and Congo (Kinshasa) [lower
Guinea & Congolian].
IUCN: LC
Ndanan 1: Cheek 11965 3/2004; Onana 2870 3/2004.

Uvariastrum pierreanum Engl.
Tree, to 10(–30)m; leaves lanceolate to oblanceolate
7–15 × 2–4cm, base cuneate, acumen linear to 2cm,
lateral nerves 8–12 pairs, oblique, joining 0.5–1cm
from margin; petioles 5–6mm long; pedicel 2–6cm;
flower buds pyramidal 2–3 × 1–2cm with winged sides;
fruiting pedicel 8mm thick; mericarps cylindrical, 3–10
× 2–5cm, smooth and tomentose. Lowland forest near
swamp; 700m.
Distr.: Ghana to CAR and Congo (Kinshasa) [Guineo-
Congolian].
IUCN: LC
Mefou Proposed National Park: Cheek in MF 86
3/2004.

Uvariodendron connivens (Benth.) R.E.Fr.
Tree, to 13m; leaves elongate-obovate to 40 × 12cm,
abruptly acuminate, glabrous, c. 20 pairs lateral veins;
flowers axillary on young shoots, solitary; pedicel c.
1.5cm; sepals green c. 1cm; petals purple ovate c. 3 ×
2cm tomentellous; mericarps glabrous, ellipsoid, c. 4 ×
3cm. Forest understorey; 700–710m.
Distr.: SE Nigeria, Bioko & Cameroon [lower Guinea].
IUCN: NT
Ndanan 1: Cheek 11839 3/2004; Cheek in MF 380
10/2002; Tadjouteu 566 3/2004.

Uvariopsis solheidii (De Wild.) Robyns & Ghesq.
Ann. Soc. Sci. Bruxelles, Ser. B liii: 321 (1933).
Tree, to 10m; bark smooth, mottled-grey; leaves obovate
17–28 × 5–9cm, rounded at base, briefly acuminate,
8–11 pairs ascendant lateral nerves; petiole 3–4mm;
monoecious, flowers pendant from burrs on lower part of
trunk; male flower pedicels c 1.5cm long; buds conical;
petals free, 7–10 × 2.5–4mm; female flower pedicels
8–16cm long; buds conical; petals free, 15–23 × 5–7mm;
fruiting pedicel thin, 13–16cm long; mericarps ellipsoid-
cylindrical, 1.5–5.5 × 1.5–3cm, attenuate at base, apex
apiculate. Evergreen forest, near river; 700–710m.
Distr.: Cameroon, CAR, Gabon, Congo (Brazzaville),
Congo (Kinshasa) [lower Guinea & Congolian].
IUCN: NT
Mefou Proposed National Park: Cheek in MF 42
3/2004; 43 3/2004; **Ndanan 1:** Cheek 11606 3/2004.

Xylopia acutiflora (Dunal) A.Rich.
Shrub, or small tree; leaves ovate-elliptic to lanceolate
4.5–10 × 2–3.7cm, obtusely acuminate, obtuse or

rounded at base, thinly pilose beneath, glabrescent, midrib above hirsute; flowers solitary, sessile; outer petals c. 4.5cm long; fruits scarlet. Forest; 710m.
Distr.: Sierra Leone to Congo (Kinshasa), Zambia and CAR [Guineo-Congolian].
IUCN: LC
Ndanan 1: Cheek 11690 3/2004.

Xylopia aethiopica (Dunal) A.Rich.

Tree, to 30m; lightly buttressed to 1m; leaves elliptic 7–20 × 3–9cm, more or less long-acuminate, 7–8 pairs lateral nerves, not very apparent; flowers axillary, solitary or 2–6-fasciculate, needle-like 4–5cm long; pedicel thin, 0.5–1cm long; mericarps to 30, subsessile, long-cylindrical, 5–6 × 0.5cm, dehiscing to reveal bright red interior and black seeds. Forest and farms; 700m.
Distr.: Senegal to Sudan & Mozambique [tropical Africa].
IUCN: LC
Ndanan 1: Cheek 11057 10/2002; Gosline 423 10/2002.
Uses: FOOD ADDITIVES — Unspecified Parts — used as a pepper spice.

Xylopia hypolampra Mildbr.

Fl. Gabon 16: 181 (1969).
Tree, to 40m; dbh 90cm; bole cylindric whitish grey; leaves evenly ranked 1cm apart, lanceolate 4–9 × 1.5–2cm, underside covered with long silky hairs giving an iridescent sheen; flowers axillary, 2–3cm long; fruits subsessile; mericarps sessile, obovoid, 2–5 × 1.5–3cm. Forest; 700–710m.
Distr.: Cameroon, Gabon, Congo (Brazzaville), Congo (Kinshasa), Angola (Cabinda) and CAR [Guineo-Congolian].
IUCN: LC
Ndanan 1: Cheek 11487 3/2004; **Ndanan 2:** Gosline 411 10/2002.

Xylopia rubescens Oliv.

Tree, to 20(–30)m; numerous stilt roots to 5m; branches reddish-brown; leaf oblong to elliptic, 7–23 × 3–8cm, cuneate at base and decurrent on petiole and twig, more or less abruptly acuminate, coriacious and drying reddish below; petiole 5–18mm; flowers axillary; exterior petals 2.5–4cm long, much longer than interior ones; fruits with pedicel 1–1.5cm long; up to 11 moniliform mericarps, 7–11cm long; Swamp forest; 663–710m.
Distr.: Liberia to Angola, Zambia and Sudan [Guineo-Congolian].
IUCN: LC
Mefou Proposed National Park: Cheek in RP 88 3/2004; Etuge 5305 3/2004; **Ndanan 1:** Cheek 11579 3/2004; 11999 3/2004.

APOCYNACEAE

Y.B. Harvey (K), J.-M. Onana (YA), H. Fortune-Hopkins (K), X. van der Burgt (K) & E. Fenton (K)

Alstonia boonei De Wild.

Tree, 10–40m, glabrous; branches and leaves whorled; leaves obovate to oblanceolate, 7–25 × 3–11cm, subsessile, lateral nerves 30–55 pairs; panicles terminal; corolla white, 9–15mm including lobes 3–6mm, twisting to right; follicles paired, 50 × 0.5cm, villous. Forest; 710m.
Distr.: Senegal to Tanzania [tropical Africa].
IUCN: LC
Mefou Proposed National Park: Cheek in RP 60 3/2004.
Note: although initially thought to be *A. congensis* Engl., owing to the long petioles (>5mm) plot voucher RP60 is more likely to be *A. boonei*. Flowers/fruits required to confirm this, Harvey ix.2010.

Baissea axillaris (Benth.) Hua

Liana, densely brown pubescent; internodes c. 1.5cm; leaves ovate-oblong, c. 4.5 × 2.5cm, apex acute, base cordate or truncate, 4–5 pairs lateral nerves; petiole 2mm; panicles axillary, 1–2cm, 2–6-flowered; corolla twisting to right, 5mm, white; fruit not seen. Forest; 710m.
Distr.: Senegal to Congo (Kinshasa) [Guineo-Congolian].
IUCN: LC
Ndanan 1: Cheek 11732 3/2004.

Callichilia bequaertii De Wild.

Meded. Land. Wag. 78(7): 12 (1978).
Syn. *Callichilia macrocalyx* Schellenb.
Shrub, 2m; leaves elliptic or obovate to 26 × 11cm, subacuminate, lateral nerves 13; petiole 2cm; axillary panicles c. 10cm, pedulous, 2–4-flowered; calyx 3cm, divided to base; corolla white, 12cm, including tube 6cm. Forest; 700–710m.
Distr.: Nigeria to Congo (Kinshasa) [lower Guinea & Congolian].
IUCN: LC
Ndanan 1: Cheek 11522 3/2004; 11704 3/2004; **Ndanan 2:** Cheek 11094 fr., 10/2002.

Dictyophleba leonensis (Stapf) Pichon

Liana seedling; branches hirsute; stipules laciniate, pilose; leaf-blades to 18 × 9.5cm, obovate, apex shortly acuminate, base rounded to subtruncate, glabrous upper surface (hirsute midrib and secondary veins), scattered hairs on lower surface, more densely so on veins; 5–6 pairs of secondary veins; petioles 1.5–2cm, densely hirsute. Evergreen forest at river bank; 710m.
Distr.: Guinea (Conakry) to SW Cameroon [Guineo-Congolian].
IUCN: LC
Ndanan 1: Cheek 11656 3/2004.

Dictyophleba setosa B.de Hoogh

Bull. Jard. Bot. Nat. Belg. 59(1–2): 220 (1989).
Tendrillous liana; stems with long chestnut coloured hairs; stipules numerous, to 40 × 1mm, hirsute; leaf-blades oblanceolate, 12–13.5 × 4–4.5cm, apex acuminate (to 2.5cm), base truncate, upper surface glabrous below except midrib and veins, lateral veins to 10 pairs; petiole 1–1.2cm, hirsute; inflorescence tendrilled and winding, cymes grouped into terminal panicles; bracts filiform, to 1cm; calyx to 2mm; corolla white, hyperocrateriform, to 1.5cm, tube cylindrical, lobes narrowly linear, contorted; fruit a many-seeded berry. Evergreen forest near river; 710m.
Distr.: Cameroon & Gabon [lower Guinea].
IUCN: VU
Ndanan 1: Cheek 11621 3/2004.

Funtumia africana (Benth.) Stapf

Tree, resembling *F. elastica*, but leaf domatia hairy, without pits; sepals 1.5–4mm; corolla lobes usually longer than tube; fruit fusiform, to 32 × 1cm, acute. Forest; 720m.
Distr.: Senegal to Mozambique [tropical Africa].
IUCN: LC
Ndanan 1: Onana 2886 3/2004.

Funtumia elastica (Preuss) Stapf

Tree, to 15m, glabrous; stem drying black, smooth; leaves elliptic to 19 × 7cm, acumen cuneate, 1.5cm, base acute-decurrent, lateral nerves c. 10 pairs, domatial pits to 1.5mm, margin undulate; fascicles axillary, 2cm; sepals 3–5mm; corolla white, c. 9mm, twisted to right, lobes shorter than tube; fruit with paired follicles, patent, c. 17 × 2cm, obtuse, angled, seeds hairy. Forest; 700m.
Distr.: Guinea (Conakry) to Uganda [Guineo-Congolian].
IUCN: LC
Ndanan 1: Cheek in MF 364 10/2002.
Note: provides excellent quality rubber according to the scientific literature.

Landolphia congolensis (Stapf) Pichon

Syn. *Cyclocotyla congolensis* Stapf
Liana, glabrous; leaves thinly coriaceous, slightly grey-green above, green and often black dotted below, elliptic, c. 11 × 3.5cm, acumen ligulate, 1cm, cuneate, lateral nerves 4–7 pairs below; petiole 0.5–1cm; flowers single, axillary; corolla white, 3cm; fruit narrowly turbinate, 6 × 2.5cm. Forest; 700–710m.
Distr.: Cameroon to Congo (Kinshasa) [lower Guinea & Congolian].
IUCN: LC
Ndanan 1: Cheek 11728 3/2004; Cheek in MF 401 10/2002.

Landolphia dulcis (Sabine) Pichon

Syn. *Landolphia dulcis* (Sabine) Pichon var. *barteri* (Stapf) Pichon
Liana, pilose when young; leaves papery to leathery,

ovate to obovate, 4–22 × 1.8–11cm, acumen up to 2.5cm, base cuneate to auriculate, glandular dotted below, lateral nerves 4–7 pairs; petiole 2–17mm; inflorescence axillary, 1–3 per axil, 1–7-flowered, 1.5 × 1.5cm or morel; peduncle 0–5.5mm; calyx lobes to 4 × 2mm; corolla tube 7–20 × 0.9–2mm, tomentose to glabrous; corolla lobes 5–22 × 1.5–4mm; fruit 2–10 × 2–6cm. Forest; 710m.
Distr.: Senegal to Gabon [upper & lower Guinea].
IUCN: LC
Ndanan 1: Cheek 11580 3/2004.

Landolphia hirsuta (Hua) Pichon

Liane, hirsute when young; leaves oblong, 8–11 × 4–5cm, apex acute, base subtruncate, glabrescent above, hirsute along the nerves below, lateral nerves 6–8 pairs; petiole c. 1.5cm; flowers in compact, sessile, axillary clusters; calyx pubescent, c. 2.5mm long, sepals ovate-oblong; corolla small, tube pubescent, to 6mm, lobes linear, 6 × 1.5mm; fruit globose, 5–8cm diam.; seeds c. 2mm long, in a yellow pulp. Forest; 710m.
Distr.: Senegal to Cameroon [Guineo-Congolian].
IUCN: LC
Ndanan 1: Cheek 11816 3/2004.

Landolphia leptantha (K.Schum.) Persoon

Agric. Univ. Wag. Papers 92(2): 120 (1992).
Syn. *Aphanostylis leptantha* (K.Schum.) Pierre
Climbing shrub, to 3m; leaves oblong, 10–15 × 8.5–7cm, coriaceous, apex acuminate, base acute to subacute, nerves 7–8 pairs; petiole 3.5–7cm; cymes few to 12-flowered, minutely puberulous; pedicels to 2.5mm; calyx c. 1mm long, sepals ovate; corolla 8–10mm, tube urceolate, to 2mm, lobes obtuse, 5–7mm; fruit cylindric-oblong, 5 × 2cm, orange mottled brown; seeds 12–15mm long. Forest; 600–710m.
Distr.: SE Nigeria, Cameroon and Gabon [lower Guinea].
IUCN: NT
Mefou Proposed National Park: Etuge 5266 3/2004; **Ndanan 1:** Cheek 11991 3/2004; Tadjouteu 585 3/2004; **Ndanan 2:** Cheek 11923 3/2004; Darbyshire 315 fl., 3/2004.

Landolphia robustior (K.Schum.) Persoon

Agric. Univ. Wag. Papers 92(2): 172 (1992).
Syn. *Anthoclitandra robustior* (Schumann) Pichon
Liana, glabrous; leaves drying brown-black, narrowly elliptic, c. 16 × 5cm, acumen 0.5cm, base obtuse, lateral nerves c. 8 pairs; petiole 1cm; fascicles axillary, dense, 1cm; corolla 1cm, white; fruit ellipsoid, 4 × 1.5cm, smooth. Forest; 700m.
Distr.: Nigeria to Angola [Guineo-Congolian].
IUCN: LC
Mefou Proposed National Park: Cheek in MF 37 3/2004; **Ndanan 1:** Cheek 11220 10/2002.

Landolphia stenogyna Pichon

Mem. Inst. Franc. Afr. Noire No. 35 (Monogr. Landolph.): 151 (1953).

Climbing liana; stems glabrous; leaves elliptic, 8–10 ×
3.5–4.5cm, glabrous, apex acuminate, base cuneate,
9–10 pairs of lateral nerves; petiole to 1cm, glabrous;
fruit globose, 7 × 6cm, green; pedicel to 5mm; many-
seeded. Secondary forest; 700m.
Distr.: Cameroon & Gabon [Guineo-Congolian].
IUCN: NT
Ndanan 1: Onana 2200 10/2002.

Landolphia sp. 1 of Mefou
Liana seedling, to 2m; stem minutely haired; leaves
oblong, 8–12.5(–15) × 3–4(–5)cm, glabrous, apex
acuminate, base rounded, 14–16 pairs of lateral nerves;
petiole to 5mm, glabrous. Lowland evergreen forest
near swamp; 700m.
Mefou Proposed National Park: Cheek in MF 10
3/2004; 56 3/2004.

Landolphia sp. 2 of Mefou
Liana seedling; stem glabrous; branches with few
scattered hairs; leaves ovate, 9–12 × 2.5–4cm, glabrous,
apex acuminate, acumen to 2cm, base rounded, to 14
pairs of lateral veins; petioles to 5mm, glabrous. Lowland
evergreen forest near swamp edged river; 700m.
Mefou Proposed National Park: Cheek in MF 77
3/2004.

Motandra guineensis (Thonn.) A.DC.
Liana, puberulent; leaves papery, elliptic-oblong or
oblanceolate, to 12 × 3.5cm, acumen 1–2cm, base
rounded, nerves c. 4 pairs; petiole 0.5cm; panicles
terminal, rachis to 12cm; corolla twisting to right,
white, 5mm; fruit bifid or entire, fleshy, narrowly ovoid,
c. 8 × 1.5cm, apex pointed, densely hairy, winged.
Forest; 710m.
Distr.: Guinea to Uganda [Guineo-Congolian].
IUCN: LC
Ndanan 1: Cheek 11804 3/2004.
Note (Congo Brazzaville): Cheek 15513 is close to this
taxon, except it has petioles that are too thick, Hopkins,
xi.2009.

Orthopichonia sp. of Mefou
Liana seedling; stem glabrous; leaves oblong, to 13 ×
14cm, glabrous, apex acuminate, acumen to 2.5cm,
base cuneate, 40+ pairs of parallel lateral nerves;
petiole to 7mm. Lowland evergreen forest near swamp-
edged river; 700m.
Mefou Proposed National Park: Cheek in MF 89
3/2004.

Pleiocarpa bicarpellata Stapf
Shrub, to 8m; leaves elliptic to oblong, c. 10 × 4cm,
acumen 1.5cm, ligulate, base acute or cuneate, lateral
nerves 19–25 pairs; petiole 4–14mm; inflorescence
10–15-flowered; calyx lobes ovate, to 3mm; corolla
white, tube 9–15mm long, lobes narrowly oblong to
broadly ovate, 5–11mm long; anthers up to 1.3mm from

throat; fruit with 2 carpels, each obovoid or ellipsoid, to
20 × 12mm. Forest; 550–710m.
Distr.: Nigeria to Kenya [tropical Africa].
IUCN: LC
Mefou Proposed National Park: Etuge 5192 3/2004;
Nana 74 3/2004; **Ndanan 1:** Cheek 11572 fr., 3/2004;
11614 3/2004; 11789 3/2004; 11832 3/2004; Onana
2912 3/2004; **Ndanan 2:** Darbyshire 250 fr., 3/2004.

Rauvolfia caffra Sond.
Syn. *Rauvolfia macrophylla* Stapf
Tree, 15–30m; trunk smooth, grey; leaves in whorls, 4,
oblanceolate, to c. 30 × 9cm, subacuminate, decurrent-
sessile; petiole to 2cm, lateral nerves c. 30 pairs;
inflorescence as *R. vomitoria*; fruit with carpels
completely united. Farmbush; 700–710m.
Distr.: tropical and subtropical Africa [tropical &
subtropical Africa].
IUCN: LC
Ndanan 1: Cheek 11576 3/2004; Cheek in MF 285
10/2002; Harvey 90 10/2002.

Rauvolfia mannii Stapf
Shrub, 1–3m, glabrous; leaves opposite or in whorls of
3, elliptic, c. 9(–15) × 3(–7)cm, acumen 1cm, cuneate-
decurrent, lateral nerves 10 pairs; petiole 1cm; panicle
1.5cm, 3–10-flowered; corolla 8mm, white with 5 red
spots, rarely pink; berry bilobed, rarely entire. Forest;
600–710m.
Distr.: tropical Africa [tropical Africa].
IUCN: LC
Mefou Proposed National Park: Etuge 5164 fl.,
3/2004; Nana 29 3/2004; **Ndanan 1:** Cheek 11055
10/2002; 11870 3/2004; Harvey 92 10/2002; **Ndanan
2:** Darbyshire 246 fr., 3/2004.

Rauvolfia vomitoria Afzel.
Shrub or tree, 2–8m; stems greyish-white; leaves in
whorls of 3, membranous-papery, elliptic c. 20 × 7cm,
subacuminate, cuneate, lateral nerves 12 pairs; petiole
2cm; panicles puberulent, 10cm; corolla white, 8mm;
fruit with 2 ovoid berries 8mm. Forest & savanna;
600–720m.
Distr.: Senegal to Uganda [Guineo-Congolian].
IUCN: LC
Mefou Proposed National Park: Etuge 5122 3/2003;
5282 3/2004; **Ndanan 1:** Onana 2883 3/2004.
Uses: MEDICINES (Onana 2883).

Saba comorensis (Bojer) Pichon
Bull. Jard. Bot. Nat. Belg. 59: 190 (1989).
**Syn. *Saba florida* (Benth.) Bullock F.W.T.A. 2: 61
(1963).**
Liane, to 40m; trunk to 15cm diam.; leaf-blade ovate-
elliptic, 4–25 × 3.5–12cm, apex obtuse, base rounded or
subcordate; petiole 7–20mm; inflorescence a many-
flowered congested cyme; peduncle 2–6(–30)cm;
pedicels 1–7mm; corollas white with a yellow or orange

throat, tube greenish, 16–34mm, lobes 20–40 × 4–12mm; fruit yellow or orange (sub-)globose, to 11cm diam.; seeds to 1cm diam. Forest; 550m.
Distr.: tropical Africa [tropical Africa].
IUCN: LC
Mefou Proposed National Park: <u>Etuge 5186</u> 3/2004.

Strophanthus gracilis K.Schum. & Pax
Meded. Land. Wag. 82(4): 78 (1982).
Climbing shrub, to 12m; branches minutely puberulous when young, then glabrescent; leaves elliptic-oblong, 5–10 × 3–4cm, ± acuminate, rounded at base, coriaceous, scattered hairs; petiole 4–7mm; cymes terminal, peduncled, few to 10-flowered; peduncle 12–27mm; pedicels to 17mm; calyx c. 1.5cm; corolla yellow or orange with brown or purple streaks and spots, puberulous, lobes ovate, with filiform yellow tails, 10–13mm long. Secondary forest, roadsides; 710–720m.
Distr.: Nigeria, Cameroon & Gabon [lower Guinea].
IUCN: NT
Ndanan 1: <u>Cheek 11709</u> 3/2004; <u>Onana 2887</u> 3/2004.

Strophanthus mortehanii De Wild.
Meded. Land. Wag. 82(4): 109–113 (1982).
Stem pubescent when immature, glabrous with age; leaves suborbicular, 6.5–9.5 × 5–7cm, few scattered hairs above, densely pubescent below, apex acuminate, base rounded, c. 6 pairs of lateral nerves; petiole to 5mm, pubescent; flowers in 6+-flowered terminal cyme; pedicels to 5mm; calyx tube to 2mm, lobes narrowly lanceolate, 10 × 1.5mm; corolla tube 12–118mm, expanding in upper two thirds, lobes narrowly triangular, to 2.5cm long; stamens partly exserted. Riverside; 710m.
Distr.: Cameroon, Equatorial Guinea (Rio Muni), Gabon, N Congo (Kinshasa) [Congolian].
IUCN: LC
Ndanan 1: <u>Cheek 11878</u> fl., 3/2004.

Tabernaemontana brachyantha Stapf
Tree, 8–15m, glabrous; leaves elliptic c. 30 × 17cm, acumen 0.7cm, obtuse, lateral nerves c. 12 pairs; petiole 1cm; panicle c. 20 × 16cm; peduncle 7–23cm; corolla white, tube 1.2–1.8cm; fruit bicarpellate, carpels divided to base, each depressed, subglobose or transversely oblong, 4–5 × 5.5–8 × 4–6cm, sutured, wall 5–15mm thick; containing a mass of ellipsoid seeds 12mm. Forest; 710m.
Distr.: SE Nigeria to Congo (Kinshasa) [lower Guinea].
IUCN: LC
Ndanan 1: <u>Cheek 11790</u> 3/2004.
Note: remarkable for the short corolla tube.

Tabernaemontana crassa Benth.
Tree, 6–20m, resembling *T. brachyantha* and perhaps not reliably distinguishable in fruit; peduncle 3.5–11cm; corolla tube 3.5–6cm; carpels obliquely subglobose, 5–12 × 4–11 × 4–10cm, wall 9–40mm. Forest; 700–720m.
Distr.: Sierra Leone to Congo (Kinshasa) [Guineo-Congolian].
IUCN: LC
Mefou Proposed National Park: <u>Cheek in MF 57</u> 3/2004; **Ndanan 1:** <u>Cheek 11995</u> 3/2004; <u>Harvey 117</u> 10/2002; <u>Onana 2891</u> 3/2004.
Note: Harvey 117 is tentatively placed in this taxon, Onana x.2009.

Tabernaemontana eglandulosa Stapf
Small, night-flowering liana; stem 1–3cm diam.; anisophyllous or not; leaves subelliptic, 4–22 × 1.5–10cm, base and apex variable, lateral nerves 4–10 pairs; petioles 4–50mm; intrapetiolar stipules; inflorescence congested, to 14 × 10cm, 3–12-flowered; pedunculate; sepals 1.5–7mm; corolla tube 2.5–6cm; carpels subglobose, to 8 × 7 × 5cm, yellow, with two blunt angles. Forest; 720m.
Distr.: Benin to Congo (Kinshasa). [Lower Guinea & Congolian].
IUCN: LC
Ndanan 1: <u>Onana 2850</u> 3/2004.

Tabernaemontana penduliflora K. Schum.
Shrub, 3–5m, resembling *T. brachyantha*, but leaves c. 18 × 8cm, acumen 1cm, lateral nerves 10–12 pairs; petiole 1cm; fascicles 3–10-flowered; corolla tube 1.2cm; fruzit not seen. Forest; 700–720m.
Distr.: Nigeria to Congo (Kinshasa) [lower Guinea & Congolian].
IUCN: LC
Mefou Proposed National Park: <u>Cheek in MF 11</u> 3/2004; **Ndanan 1:** <u>Cheek 11493</u> 3/2004; <u>11566</u> 3/2004; <u>11686</u> 3/2004; <u>11687</u> 3/2004; <u>Harvey 163</u> 10/2002; <u>Onana 2867</u> 3/2004.
Local Name: Obaetoan (Beti) (Cheek 11566).

Voacanga africana Stapf
Tree, 7–12m; stems greyish-white; leaves obovate, c. 22 × 10cm, rounded, decurrent-sessile, lateral nerves c. 15 pairs; inflorescence erect, 6–15cm, c. 20-flowered; corolla white, tube 8mm, lobes patent, 20mm; fruit pendulous, follicles 2, each 4cm, green mottled white, splitting to show black seeds in orange mass, wall 1cm thick. Forest; 700–720m.
Distr.: Senegal to Tanzania [Guineo-Congolian].
IUCN: LC
Ndanan 1: <u>Cheek 11000</u> 10/2002; <u>11483</u> 3/2004; <u>11571</u> 3/2004; <u>11724</u> 3/2004; <u>Cheek in MF 405</u> 10/2002; <u>Onana 2865</u> 3/2004; <u>Tadjouteu 583</u> 3/2004.
Local Name: Obaetoan (Beti) (Cheek 11571).

Voacanga thouarsii Roem. & Schult.
Syn. *Voacanga obtusa* K.Schum.
Tree, to 15m; branchlets ± glabrous; leaf-blade narrowly obovate, 6–25 × 2–9cm, apex obtuse or rounded, base attenuate, glabrous; petiole 8–25mm, ± glabrous; inflorescence up to 20cm corymbs, few-

flowered; peduncle to 20cm; bracts ovate; pedicels to 15mm; flowers yellow; corolla tube to 23mm, lobes obcordate, 20–30 × 30–45mm, apex emarginate; fruits spotted, 2 free mericarps, 4–10cm diam.; seeds dark brown, to 1cm. *Raphia* swamp; 720m.
Distr.: Senegal to Sudan and Angola and Malawi to S Africa, Madagascar [tropical Africa].
IUCN: LC
Ndanan 1: <u>Onana 2825</u> 3/2004.

ARALIACEAE

M. Cheek (K)

Fl. Cameroun 10 (1970).

Polyscias fulva (Hiern) Harms

Tree, to 30m; leaves to 80cm long, imparipinnate; leaflets 3–12 paired, coriaceous, to 17 × 7.5cm, lanceolate-ovate; inflorescence of compound racemosely arranged racemes; pedicels to 5mm; petals greenish to creamy white; fruit broadly ovoid to subglobose, 3–6 × 2–5mm, ribbed. Submontane forest on slopes of hills; 700–710m.
Distr.: tropical Africa.
IUCN: LC
Ndanan 1: <u>Cheek 11241</u> 10/2002; <u>11560</u> 3/2004.

Schefflera barteri (Seem.) Harms

Tree, 4–20m, hemiepiphyte; leaves c. 7 digitately compound, 35cm, leaflets elliptic, 7–13 × 7cm, acumen 0.5cm, obtuse-decurrent, lateral nerves c. 6 pairs; petiolules to 3cm; stipule intrapetiolar; umbels of 15 spikes 30cm; partial peduncles numerous, 1cm; pedicels 5mm; flowers white, 2mm; stamens 6, erect; fruit 5mm, red. Forest; 600m.
Distr.: Guinea (Conakry) to Congo (Kinshasa) [Guineo-Congolian].
IUCN: LC
Mefou Proposed National Park: <u>Etuge 5208</u> 3/2004.

ARISTOLOCHIACEAE

J.-M. Onana (YA), L. Pearce (K), M. Cheek (K) & Y.B. Harvey (K)

Pararistolochia ceropegioides (S.Moore) Hutch. & Dalziel

Adansonia (Sér. 2) 17: 482 (1978).
Climber, 5–15m; stem 8-shaped in section, 1–2cm diam., flexible; leaves leathery, ovate, c. 13 × 8cm, 3-nerved, acuminate, rounded-truncate; inflorescence cauliflorous, 1–2m from ground, 3–8-flowered; flowers 5–7cm; perianth tube pouched at base, pale brown outside; tepals 3, equal, 1–2cm, slightly splayed, throat orange; fruit fleshy, ellipsoid, c. 5–6 × 3–4cm, 5-ridged, base and apex truncate; pedicel 3–4cm. Forest; 720m.
Distr.: Cameroon & Gabon [lower Guinea].
IUCN: NT
Ndanan 1: <u>Onana 2888</u> 3/2004.

Pararistolochia macrocarpa (Duch.) Poncy var. *macrocarpa*

Adansonia (Sér. 2) 17: 488 (1978).
Syn. *Pararistolochia tribrachiata* (S.Moore) Hutch. & Dalziel
Syn. *Pararistolochia flos-avis* (A.Chev.) Hutch. & Dalziel
Climber, to 25m; stem fissured, 8-shaped in section, to 1.5cm diam.; leaves leathery, ovate, c. 10–15(–18) × 5–7(–11)cm, 3-nerved, acuminate, base rounded; petiole 1.5–3.5(–5.5)cm; inflorescence cauliflorous, 1–3-flowered; flowers scented of urine; peduncle to 2cm; pedicel to 6.5cm; perianth tube pouched at base, 7–9cm long, inner surface pale brown, throat white (or with purple blotches) with black hairs; tepals 3, 5.5–7cm, narrowly triangular to ovate, ventral one 1/2 size of laterals and inrolled from apex by 1/3 length, colour varies from pale brown with dull purple markings to deep purple with white or black markings; fruit cylindrical, 30 × 3cm, 6-sided. Along river banks, *Raphia* swamp and secondary forest; 600–720m.
Distr.: Sierra Leone, Ivory Coast, Ghana, S Nigeria, Cameroon, Equatorial Guinea, CAR, Gabon, Congo (Brazzaville), Congo (Kinshasa) [upper & lower Guinea].
IUCN: LC
Mefou Proposed National Park: <u>Etuge 5255</u> 3/2004;
Ndanan 1: <u>Cheek 11227</u> 10/2002; <u>Onana 2826</u> 3/2004.

Pararistolochia cf. *promissa* (Mast.) Keay

Climber, with twisting petioles, glabrous; leaves alternate, estipulate; blade ovate or elliptic, c. 11.5 × 5cm, acumen 1cm, base rounded, 3-nerved from base, basal laterals extending half the length of the leaf, other laterals c. 5 pairs; petiole 3cm, terete. Forest; 700m.
Ndanan 1: <u>Cheek in MF 273</u> 10/2002.
Note: plot voucher MF273 is immature. Fertile material needed to identify fully, Pearce ix.2010.

ASCLEPIADACEAE

D. Goyder (K)

Cynanchum adalinae (K.Schum.) K.Schum. subsp. *adalinae*

Ann. Missouri Bot. Gard. 83: 291 (1996).
Slender twiner; latex white; leaves opposite, to c. 7.5 × 5cm, ovate with cordate base, glabrous; inflorescences extra-axillary, subsessile with sweetly scented flowers; corolla rotate, creamish green, lobes 3.5–4mm long; corona white, 2.5–3mm long, obscuring the gynostegium. Margins of secondary forest; 710m.
Distr.: Ivory Coast to Congo (Kinshasa) [Guineo-Congolian].
IUCN: LC
Ndanan 1: <u>Cheek 11747</u> 3/2004.

Marsdenia latifolia (Benth.) K. Schum.

Omlor, Gen. Rev. Marsdenieae (Asclepiadaceae): 75 (1998).

Syn. *Gongronema latifolium* Benth.
Slender to robust woody scrambler; latex white; leaves opposite, to c. 12 × 7cm, ovate to ovate-oblong, pubescent; inflorescences extra-axillary with cream or yellow-green flowers scattered along the axes; corolla lobes c. 2–2.5mm long; corona lobes fleshy and with an apical tongue; follicles single, subcylindrical. Margins of evergreen forest or scrub; 600–720m.
Distr.: Widespread in C and W Africa, to Zambia, Angola [tropical Africa].
IUCN: LC
Mefou Proposed National Park: Etuge 5248 3/2004; **Ndanan 1:** Cheek 11054 10/2002; Cheek in MF 325 10/2002; Onana 2881 3/2004; 2902 3/2004; **Ndanan 2:** Cheek 11157 10/2002.
Local Name: Zole (Ewondo) (Cheek 11157).

Mondia whitei (Hook.f.) Skeels

Liana, to 4m; white exudate; stems pubescent; leaves opposite, ovate, 14–17.5 × 10–12cm, acumen 1.5cm, base cordate, 5 nerves on each side of the midrib, entire, abaxial surface densely pubescent, adaxial surface sparsely pubescent; petiole (3–)6–8cm; inflorescence a panicle to 15cm; corolla 0.7cm, outside pale green, inside yellow–dark red; 700m.
Distr.: Guinea (Bissau), Guinea, Sierra Leone, Liberia, Ghana, Togo, Benin, Nigeria, Cameroon, Equatorial Guinea, Congo (Kinshasa), Burundi, Mozambique, Malawi, Zambia, Zimbabwe, Angola, South Africa [Guineo-Congolian].
IUCN: LC
Ndanan 1: Cheek 11024 10/2002.

Secamone afzelii (Schultes) K.Schum.

Slender woody climber; latex white; leaves opposite, to c. 7 × 3cm, oblong-elliptic, glabrous; inflorescences terminal or extra-axillary, lax, well-branched with minute yellow or orange flowers; corolla lobes c. 1mm long; follicles slender, paired, 5–7cm long. Forest; 700m.
Distr.: widespread across C and W Africa [tropical Africa].
IUCN: LC
Ndanan 1: Cheek in MF 229 10/2002.
Note: plot voucher series MF229 is tentatively placed in this taxon, Goyder vi.2010.

Telosma africana (N.E.Br.) N.E.Br.

Fl. Cap. (Harvey) 4(1,5): 776 (1908).
Woody climber; latex clear, often sparse; leaves opposite, to c. 10 × 5cm, oblong to ovate; inflorescences extra-axillary, subsessile; flowers green, sometimes tinged with red, with long hairs in throat; corolla with narrow tube 5–8mm long; lobes 5–8mm long, twisted. Scrub & forest margins; 600–720m.
Distr.: widespread in tropical and subtropical Africa [tropical & subtropical Africa].
IUCN: LC

Mefou Proposed National Park: Onana 2803 3/2004; **Ndanan 1:** Cheek 11815 3/2004; Onana 2889 3/2004; Tadjouteu 622 fl., 3/2004; **Ndanan 2:** Etuge 5116 3/2003.
Note: erroneously published in F.W.T.A. as *Telosma africanum* (N.E.Br) Colville (F.W.T.A. 2: 97 (1963)), although Coville described the genus, he never made this combination. Goyder x.2010.

Tylophora cameroonica N.E.Br.

Twining climber, to 7m, glabrous; leaves opposite, suborbicular, 14.5–19.5 × 10.5–15.5cm, apex very shortly cuspidate, base rounded-subcordate; petiole 3–4.5cm; inflorescence a panicle to 10cm; corolla 1–2mm, pink-purple. Forest; 700m.
Distr.: Cameroon, Congo (Kinshasa), Uganda [Guineo-Congolian].
IUCN: NT
Ndanan 1: Gosline 421 10/2002.
Note: Gosline 421 is tentatively placed in this taxon.

Tylophora conspicua N.E.Br.

Slender woody twiner; latex cloudy; leaves opposite, to c. 20 × 13cm, ovate-oblong, glabrous above; inflorescences extra-axillary, zig-zag, with white to dark red flowers at the angles; corolla rotate, fleshy, 13–18mm across; corona lobes fleshy; follicles mostly single, to 10cm long. Evergreen or deciduous forest, often near streams; 600m.
Distr.: widely distributed across tropical Africa [tropical Africa].
IUCN: LC
Mefou Proposed National Park: Etuge 5271 3/2004.

Tylophora sylvatica Decne.

Slender woody twiner; latex clear; leaves 3–9 × 1.5–5cm, ovate to oblong, apex acute and somewhat attenuate, base cordate, rarely truncate, glabrous or subglabrous; inflorescences extra-axillary, up to c. 20cm long with clusters of minute flowers scattered at intervals of 1–3cm along the simple or branched axes; corolla reddish brown or dull purple, glabrous, 3–5mm across; corona lobes maroon, c. 0.5mm high, fleshy; follicles mostly occuring singly, if paired then held at 180º, 6–9.5cm long. *Alchornea* swamp near river; 710m.
Distr.: [tropical Africa].
IUCN: LC
Ndanan 1: Cheek 11696 3/2004; 11785 3/2004.

BALANITACEAE

M. Cheek (K) & Y.B. Harvey (K)

Balanites wilsoniana Dawe & Sprague

Tree, to 50m; trunk to 1.2m diam., sometimes buttressed, fluted; bark yellowish to light brown; spines to 15cm; branchlets from swollen nodes, branchlet-spines to 1cm, borne 5mm above the axil, simple or forked, 5–6mm from base; leaflets ovate to ovate-elliptic, to 3.5 × 7.5cm, slightly bilaterally asymmetric, acuminate, tip to 2cm; petiolule to 6mm; petioles to 1cm; inflorescence 2–8-flowered umbellate cyme; pedicels to 1.3cm; sepals elliptic to ovate, 3–6 × 1.5–3.7mm, acute; petals pale yellowish-green, oblong-elliptic, 6–10 × 1.5–2.5mm, villous within, acute; fruit brown to yellow, ovoid or ellipsoid, 8–12 × 3.5–7cm, one fruit per inflorescence. Disturbed areas within forest; 720m.

Distr.: Ivory Coast west to Uganda and southwards to Congo (Kinshasa) [tropical Africa].

IUCN: LC

Ndanan 1: Onana 2843 3/2004.

Note: Onana 2843 is a tree seedling, Cheek vii.2006.

BALANOPHORACEAE

J.-M. Onana (YA), E. Fenton (K) & Y.B. Harvey (K)

Fl. Cameroun 33 (1991).

Thonningia sanguinea Vahl

Underground root parasite, leafless, achlorophyllose; flower-like inflorescences on forest floor, pink or red, c. 5cm diam.; outer bracts petal-like, stiff, 2cm; flowers minute, numerous, yellow, resembling stamens. Forest floor; 693–710m.

Distr.: Sierre Leone to Zambia [Guineo-Congolian].

IUCN: LC

Ndanan 1: Cheek 11904 3/2004; Onana 2927 3/2004.

BALSAMINACEAE

M. Cheek (K), X. van der Burgt (K) & Y.B. Harvey (K)

Fl. Cameroun 22 (1981).

Impatiens irvingii Hook.f.

Herb, to 1m; stem succulent, tinged pink; leaves spirally arranged, 2–15 × 1–4cm, lanceolate to ovate or oblanceolate, subglabrous, dark green above, sometimes reddening at apex, lighter below, margins serrate; petioles 1–10mm; flowers red-purple to pink with white throat; bracts 2–4mm, linear-lanceolate; pedicels 2–7mm; lateral sepals 4–6mm; dorsal petal 10–15 × 12–18mm; lateral united petals 17–26mm, variable in size; ovary glabrous; fruit 14–18 × 4–6mm, fusiform, glabrous. Forest beside river, *Raphia* swamp; 600–710m.

Distr.: Guinea to Cameroon, Gabon, Congo (Brazzaville), Congo (Kinshasa), Angola (Cabinda), CAR, Uganda, Sudan, Tanzania, Malawi and Zambia [tropical Africa].

IUCN: LC

Mefou Proposed National Park: Etuge 5130 fl., 3/2004; **Ndanan 1:** Cheek 11779 3/2004; Harvey 100 10/2002.

Note: Cheek 11779 and Harvey 100 are tentatively placed in this taxon. They are a little too small to be *I.*

irvingii. The leaves of *I. irvingii* can be narrow but the leaves of this collection are also very smooth and do not look like any other specimen of *I. irvingii* at K. *Impatiens irvingii* was collected at Mefou (Etuge 5130) and that collection looks different from the others, van der Burgt viii.2009.

Impatiens niamniamensis Gilg

Terrestrial, epiphytic or epilithic herb, 0.3–1m, glabrous; leaves alternate, lanceolate, c. 20 × 8cm; flowers in axillary fascicles of 2–4; peduncle sessile; pedicels c. 4cm; flowers red, with white hood, face concave, c. 3 × 1.5cm; lateral sepals 2mm; spur recurved, erect. Forest; 710m.

Distr.: Cameroon to NW Tanzania [lower Guinea & Congolian].

IUCN: LC

Ndanan 1: Cheek 11746 3/2004.

Impatiens sp. aff. *mannii* Hook.f. of Mefou

Succulent herb, 10–150cm tall; stems and leaves glabrous; leaves alternate, internodes c. 4cm; leaf-blade elliptic, c. 7.5 × 3.5cm, gradually acuminate, base cuneate-decurrent, margin crenate, teeth alternating with mucra; flowers two per axil; peduncle absent, all parts with scattered patent red hairs; pedicel 5–8mm; flower pink, 9mm long, spur 7mm, filiform, curved downwards. River edge; 710m.

Ndanan 1: Cheek 11589 3/2004.

Note: this species keys out in Fl. Cameroun as *I. mannii* but is a very different species being restricted to river edges where it grows with the lower part submerged in water. *Impatiens mannii* sensu stricto is a submontane species of the Cameroon Highlands with flowers which are white, barred purple and glabrous. Since Grey-Wilson lists other collections from the Mefou region (e.g. De Wilde, 1148, N'Kolbison; Letouzey 1964, raphiale du Nia, 40km SW Nanga Eboko) it seems likely that two species have been united in one. More work is needed to find the correct name with which to resurrect the pink, hairy riverine species. *Impatiens adenopus* Gilg (type of Lolodorf, *Staudt* 205) seems the most likely. Cheek x.2010.

BEGONIACEAE

M.S.M. Sosef (WAG)

Netherlands Centre for Biodiversity Naturalis (section NHN), Biosystematics group, Wageningen University Gen. Foulkesweg 37, 6703 BL Wageningen, The Netherlands

Begonia macrocarpa Warb.

Erect, unbranched terrestrial herb; stem c. 30cm, juicy; leaf blade obliquely obovate or oblong, c. 10 × 4.5cm, acuminate, base unequally obtuse, slightly serrate; petiole 0.5–3cm; stipules obovate, 1.2cm, serrate; inflorescence axillary, peduncle 1–2cm; tepals white, 5–10mm; fruit

strongly 3-winged, 1.5 × 1.5cm. Forest; 658m.
Distr.: Guinea (Conakry) to Angola [upper & lower Guinea].
IUCN: LC
Mefou Proposed National Park: <u>Etuge 5298</u> 3/2004.

Begonia potamophila Gilg
Agric. Univ. Wag. Papers 94(1): 171 (1994).
Terrestrial herb, 40cm; lacking aerial stems; blade broadly elliptic, c. 20 × 12cm, sparsely hairy above, acumen 1cm; peduncle 15cm, few-flowered; perianth 1cm, red, turning orange outside, yellow inside; fruit ellipsoid, narrowly winged, hairy. Forest, rocks; 660m.
Distr.: Cameroon to Congo (Brazzaville) [lower Guinea].
IUCN: NT
Ndanan 2: <u>Darbyshire 268</u> fr., 3/2004.

BIGNONIACEAE

J.-M. Onana (YA), C. Couch (K), E. Fenton (K), X. van der Burgt (K) & Y.B. Harvey (K)

Fl. Cameroun 27 (1984).

Crescentia cujete L.
Fl. Zamb. 8(3): 61 (1988).
Tree, 7–10m; dbh 30cm diam; leaves oblanceolate, apex mucronate, base attenuate to acute, 7–11 nerves on each side of the midrib, entire, adaxial surface glabrous, abaxial surface pubescent on the veins; flower buds green, 1.5cm; fruit spherical to ovoid-elliptic, 13–20cm diam, hard-shelled, indehiscent. Cultivated; 600m.
Distr.: native to Mexico and N Central America, extensively cultivated in the tropics.
IUCN: LC
Mefou Proposed National Park: <u>Etuge 5234</u> 3/2004.
Uses: MATERIALS — Unspecified Materials — widely cultivated for the fruits once used as containers (Cheek pers. comm.). ENVIRONMENTAL USES — Unspecified Environmental Uses — cultivated tree (Etuge 5234).

Fernandoa ferdinandi (Welw.) Milne-Redh.
Fl. Gabon 27: 26, pl. 4 (1985).
Tree, to 30m; dbh 50cm; leaves 5–11-foliolate; leaflets 2–7 × 1–3.5cm, obovate to oblong, apex acute, base asymmetric (cuneate), puberulent; petiolules to 2–3mm; panicles short, 1–5-flowered; pedicels to 2–3cm; calyx campanulate, 2–3-lobed, 1.5–2.2 × 2–3cm, puberulent; corolla orange, campanulate, 6–9 × 2–4.5cm (at throat); tube 4–6cm; lobes 1.5–2.5cm, puberulent exterior, glabrous within; stamens reaching end of tube. Forest; 720m.
Distr.: Cameroon, Gabon, Angola [lower Guinea].
IUCN: VU
Ndanan 1: <u>Onana 2875</u> 3/2004.
Note: Onana 2875 is represented by fallen 'yellow' flowers, Harvey vi.2010.

Kigelia africana (Lam.) Benth.
Tree, to 15m, glabrous; leaves opposite, imparipinnate, to 50cm, 5–6-jugate, leaflets oblong-elliptic, 11–20 × 3.5–8cm, apex acuminate, base acute to rounded; petiole 8–25cm; petiolules 3–6mm; panicles pendent, lax, to 40cm, lateral branches to 7cm; peduncle to 20cm; pedicels 1.2cm; calyx cupular, 2.8 × 2.5cm, lobes triangular, 0.6–1cm; corolla campanulate, 6(–8)cm, dark-red; fruit sausage-shaped, 20–50(–100)cm. Forest & farmbush; 700–710m.
Distr.: Senegal, Gambia, Mali, Guinea (Conakry), Sierra Leone, Liberia, Ivory Coast, Ghana, Togo Republic, N Nigeria, S Nigeria, Cameroon [tropical Africa].
IUCN: LC
Ndanan 1: <u>Cheek 11669</u> 3/2004; <u>Harvey 151</u> 10/2002.

Markhamia lutea (Benth.) K.Schum.
Tree, to 20m, glabrous; stems lenticellate; pseudostipules orbicular, 2cm; leaves 4-jugate, to 30cm; leaflets oblong-elliptic, 11–14 × 4–5.5cm, acuminate, base asymmetric-acute; petiole to 9cm; petiolules c. 5mm; panicles to 8cm, few-branched, 6–10-flowered; calyx lepidote, c. 2cm, beaked in bud, split on one side; corolla 6cm, yellow with red lines in throat; fruit linear, to 50 × 1.5cm, valves single-nerved. Forest & farmbush; 700–710m.
Distr.: Ghana to Congo (Kinshasa), Uganda, Kenya, Tanzania [tropical Africa].
IUCN: LC
Mefou Proposed National Park: <u>Cheek in RP 86</u> 3/2004; **Ndanan 1:** <u>Cheek in MF 334</u> 10/2002.

Markhamia tomentosa (Benth.) K. Schum. ex Engl.
Tree, 3–10m; dbh 25–40cm; stems pubescent; leaves compound, imparipinnate, 5–13 leaflets; leaflets elliptic, 9–12.5 × 3–4.3cm, acumen 1cm, base attenuate, 6–8 nerves on each side of the midrib, entire, abaxial surface densely pubescent, adaxial surface sparsely pubescent; inflorescence a panicle, hairy; pedicels 0.5–1cm; bracts tomentose; flowers yellow; corolla 3–4cm; fruits linear, flattened, 30–110cm, with dense bronze-coloured hairs. Semi-deciduous forest; 700m.
Distr.: Guinea (Bissau), Guinea, Sierra Leone, Ivory Coast, Burkina Faso, Cameroon, Equatorial Guinea, Gabon, Congo (Brazzaville), Congo (Kinshasa), Zambia, Angola [upper & lower Guinea].
IUCN: LC
Ndanan 1: <u>Cheek 11009</u> 10/2002.
Note: 21 specimens are known from Cameroon alone (Fl. Cameroun).

Newbouldia laevis (P.Beauv.) Seeman ex Bureau
Tree, to 15m, glabrous; leaves to 30cm, 4-jugate; leaflets subsessile, coriaceous, oblong-elliptic, 12–15.5 × 5–6.5cm, acumen 1.5cm, base asymmetric-acute, margin serrate to subentire; petiole to 11cm; panicles racemose, to 8cm, dense; pedicels 7mm; calyx to 2cm, splitting on one side; corolla 5cm, pink to purple and

white; fruit linear, to 34 × 2.3cm, valves 3-nerved; seeds biwinged, 1 × 4.5cm. Forest & cultivated; 600–710m.
Distr.: Senegal to Congo (Kinshasa) [Guineo-Congolian].
IUCN: LC
Mefou Proposed National Park: Etuge 5217 3/2004; 5232 3/2004; **Ndanan 1:** Cheek 11822 3/2004.

BOMBACACEAE

C. Couch (K) & J.-M. Onana (YA)

Fl. Cameroun 19 (1975).

Ceiba pentandra (L.) Gaertn.

Deciduous spiny tree, to 60m; triangular buttress 2 × 2m; leaves alternate palmately compound; leaflets c. 7, normally elliptic-oblong, c. 17 × 4cm, sessile; inflorescence fasciculate on leafless wood; flowers 3cm, white; pedicels 4cm; fruit pendulous, c. 20cm long, valves falling to reveal white fibres. Forest and farmbush; 600–710m.
Distr.: tropical America & Africa [amphi-Atlantic].
IUCN: LC
Mefou Proposed National Park: Cheek 13122 2/2006; Etuge 5215 3/2004; **Ndanan 1:** Cheek 11067 10/2002; 11761 3/2004.

BORAGINACEAE

X. van der Burgt (K)

Cordia dewevrei De Wild. & T.Durand
Bull. Soc. Roy. Bot. Belg. 33(2): 37 (1899); F.T.A. 4(2): 9 (1906); Fl. Afr. Cent. Boraginaceae: 8 (1971). Shrub or tree, to 4m; branchlets pilose; leaves elliptic, 5.5–15 × 5–6cm, apex acute, base cuneate to rounded, upper surface subglabrous, lower surface pilose; petiole 5–13mm; cymes terminal, many flowered; calyx tubular-campanulate, to 0.8cm, longitudinally ridged, 5-lobed; corolla tubular-campanulate, tube yellow, lobes orange, to 0.5cm; fruit ovoid, to 1.4 × 0.8cm, orange. Evergreen forest near river; 710m.
Distr.: Cameroon and Congo (Kinshasa) [lower Guinea & Congolian].
IUCN: NT
Ndanan 1: Cheek 11684 3/2004.

BURSERACEAE

J.-M. Onana (YA), H. Fortune-Hopkins (K) & X. van der Burgt (K)

Canarium schweinfurthii Engl.
Tree, to 40m; copious exudate, drying whitish; young branches and young leaves rusty-pubescent; leaves clustered towards branch apices, c. 30–60cm long, imparipinnate, 8–12-jugate; leaflets oblong-ovate, 5–20 × 3–5.5cm, base cordate, lateral nerves pubescent below; petiole flat on upper side towards base; panicles to 30cm long; pedicels c. 3mm; flowers cream-white, c. 1cm; fruits ellipsoid, c. 3.5 × 2cm, epicarp glabrous, purple-black, endocarp trigonous. Forest; 700m.
Distr.: Senegal to Sudan & Tanzania [Guineo-Congolian].
IUCN: LC
Ndanan 1: Cheek in MF 236 10/2002; 332 10/2002; 385 10/2002; 390 10/2002; 410 10/2002; **Ndanan 2:** Cheek 11145 10/2002.
Note: plot voucher series MF390 is probably this taxon, Hopkins & van der Burgt iv.2010.
Local Name: Abel (Ewondo) (Cheek 11145).

Santiria trimera (Oliv.) Aubrév.
Tree, to 20m; laterally-flattened stilt roots, glabrous; stem strongly smelling of turpentine when cut; leaves imparipinnate, 2–3-jugate, leaflets oblong-elliptic, 10.5–16.5 × 4–6.5cm, acumen narrow, to 2.2cm long, base acute to obtuse, lateral nerves 7–10 pairs; petiole 9.5cm; petiolules 1cm; panicles axillary, to 9cm; fruit green-black, asymmetrically oblate, 1.2 × 2 × 1.5cm, style-remains lateral; fruiting pedicel 0.8cm. Forest; 700m.
Distr.: Sierra Leone to Congo (Kinshasa) [Guineo-Congolian].
IUCN: LC
Mefou Proposed National Park: Cheek in MF 118 3/2004.

CACTACEAE

M. Cheek (K) & X. van der Burgt (K)

Rhipsalis baccifera (J.M.Mill.) Stearn
Syn. *Rhipsalis cassutha* Gaertn.
Epiphyte, pendulous, 45cm; stems terete, grey-green, succulent, glabrous, branching at intervals of c. 10cm; flower buds red, 1cm; petals white at anthesis; fruit globose 8mm, white tinged-red. Forest; 550–700m.
Distr.: tropical America & Africa [amphi-Atlantic].
IUCN: LC
Mefou Proposed National Park: Etuge 5199 3/2004; Onana 2810 3/2004.

CAMPANULACEAE

M. Thulin (UPS)

Lobelia gilletii De Wild.
Fl. Afr. Cent. Lobeliaceae: 41 (1985); Fl. Gabon 40: 40 (2010).
Prostrate, mat-forming herb; stems glabrous, rooting; leaf-blade ovate, 5–10 × 5–10mm, apex rounded, base cordate, margin dentate or crenate, sparsely pubescent on the upper face; petiole 2–7mm, glabrous; flowers simple, axillary; pedicel 5–12mm; calyx 1–2mm;

corolla white, 7–10mm; fruit obovoid, 3–5mm, indehiscent. *Raphia* swamp; 600–700m.
Distr.: Cameroon, Gabon, Congo (Brazzaville) & Congo (Kinshasa) [Congolian].
IUCN: EN
Mefou Proposed National Park: <u>Etuge 5185</u> fl., 3/2004; **Ndanan 2:** <u>Darbyshire 189</u> 3/2004.
Note: in view of the indehiscent fruit, once placed in a separate genus, as *Pratia gilletii* (De Wild.) E.Wimm., Cheek v.2010.

CAPPARACEAE

M. Cheek (K), E. Fenton (K), X. van der Burgt (K), C. Couch (K) & Y.B. Harvey (K)

Fl. Cameroun 29 (1986).

Cleome afrospina Iltis
Fl. Cameroun 29: 40 (1986).
Erect herb; leaves palmately 7-foliolate; leaflets to 10cm, secondary nerves 17–28, not less than 17; stipular spines; capsules 9–12mm wide, not 4mm; fruiting gynophore as long as pedicel, not twice as long; petals 0.9–1.6cm, not to 2.3cm. Secondary forest; 600–710m.
Distr.: Cameroon to Congo (Kinshasa) [lower Guinea & Congolian].
IUCN: LC
Mefou Proposed National Park: <u>Etuge 5163</u> 3/2004; **Ndanan 2:** <u>Cheek 11926</u> 3/2004.

Cleome sp. of Mefou
Erect herb, to 60cm, glabrous; stem glabrous, occasional spines opposite petioles; leaves palmately 5-foliolate; leaflets to 11cm; petiole to 12cm. *Raphia* swamp; 710m.
Mefou Proposed National Park: <u>Cheek in RP 67</u> 3/2004.
Note: similar to *Cleome speciosa* but differing in leaf venation, van der Burgt viii.2010.

Ritchiea albersii Gilg & Benedict
Shrub or tree, to 11(–20)m; leaves simple to 5-foliolate; leaflets elliptic or oblong-elliptic, to 20 × 8cm, base cuneate; petiolule 2–5mm long; inflorescence terminal; pedicels 1.5–3.5cm; sepals lanceolate or ovate, 12–20 × 5–10mm; petals 4, to 3.5cm, clawed; gynophore 2–3cm long; fruits cylindrical, to 5 × 2.5cm with 6 longitudinal grooves. Forest; 600–700m.
Distr.: SE Nigeria to E Africa [afromontane].
IUCN: LC
Mefou Proposed National Park: <u>Etuge 5183</u> fr., 3/2004; **Ndanan 2:** <u>Cheek 11083</u> 10/2002; <u>Darbyshire 219</u> fr., 3/2004.
Note: Darbyshire 219 is tentatively placed in this taxon (*Ritchiea* cf. *albersii*) (field determination, i.2009). The Mefou material is discordant being a subshrub 0.5m with unifoliolate leaves at low altitude, Cheek xi.2010.

CARYOPHYLLACEAE

J.-M. Onana (YA)

Drymaria cordata (L.) Willd.
Straggling, slightly viscid herb; stems to 90cm; leaves broadly ovate, 1–2.5cm long; inflorescence a terminal cyme; flowers few, white. Roadside weed; 660m.
Distr.: pantropical [pantropical].
IUCN: LC
Ndanan 2: <u>Darbyshire 330</u> 4/2004.

CECROPIACEAE

M. Cheek (K), X. van der Burgt (K) & E. Fenton (K)

Fl. Cameroun 28 (1985).

Musanga cecropioides R.Br. ex Tedlie
Tree, to 30m with an umbrella-like crown; stipule sheaths 10–20cm, deep red outside, silky within, caducous; leaves 12–15-digitately divided almost to base, less so when immature, lobes oblanceolate, c. 25 × 6cm, felty grey-puberulent below, dark green above, lateral nerves numerous, prominent below; male flowers in numerous globose heads 4mm on a many-branched panicle; female inflorescence dense, ellipsoid, 2cm on a peduncle to 12cm; fruit ellipsoid, yellow-green. Farmbush & forest; 700–710m.
Distr.: Guinea (Conakry) to Uganda [Guineo-Congolian].
IUCN: LC
Ndanan 1: <u>Cheek 11575</u> 3/2004; <u>Cheek in MF 411</u> 10/2002.

Myrianthus arboreus P.Beauv.
Tree, 7(–25)m; leaves alternate, palmately compound; leaflets 7, elliptic oblong c. 30 × 10cm, acute or acuminate, serrate, concolorous; petiole robust, 30cm; male inflorescence c. 20 × 15cm; fruit compound, ellipsoid, c. 15 × 10cm, yellow, juicy. Forest, farmbush; 700m.
Distr.: Guinea (Conakry) to W Tanzania [Guineo-Congolian].
IUCN: LC
Ndanan 1: <u>Cheek in MF 336</u> 10/2002.

Myrianthus preussii Engl. subsp. *preussii*
Treelet, 2–4(–9)m; leaves alternate, palmately compound; leaflets 5(–7), bright-white below, obovate, long-acuminate; petioles slender, 15cm; petiolules c. 3cm; male inflorescence c. 6 × 4cm; fruit globose, c. 5cm diam. Forest; 600–700m.
Distr.: SE Nigeria & Cameroon [lower Guinea].
IUCN: NT
Mefou Proposed National Park: <u>Etuge 5161</u> fl., 3/2004; **Ndanan 2:** <u>Cheek 11118</u> 10/2002.
Local Name: Engakom (Cheek 11118).
Uses: FOOD — Infructescences — fruits edible. MEDICINES (Cheek 11118).

CELASTRACEAE

G. Gosline (K), X. van der Burgt (K) &
M. Cheek (K)

Fl. Cameroun 19 (1975) & 32 (1990).

Campylostemon angolense Welw. ex Oliv.
Woody climber, to 10m; with resinous threads,
glabrous; wood with rays; bark reddish and sloughing;
young shoots four-winged; leaves 3–13 × 2–5cm,
chartaceous, elliptic, acumen caudate to 3cm long, 5–8
pairs secondary nerves, margin slightly dentate; petiole
5–10mm long; inflorescence an axillary dichotomous
cyme 1.5–5cm long; flowers without a disc, stellate,
5–6mm diam.; petals thin; 5 stamens. Forest; 700m.
Distr.: Sierra Leone to Bioko, Cameroon, Congo
(Kinshasa), Angola, CAR and Uganda [Guineo-
Congolian].
IUCN: LC
Mefou Proposed National Park: Cheek in MF 91
3/2004; **Ndanan 1:** Cheek in MF 215 10/2002.

Cuervea macrophylla (Vahl) R.Wilczek ex N.Hallé
Fl. Gabon 29: 224 (1986); Fl. Cameroun 32: 184 (1990).
Syn. *Hippocratea macrophylla* Vahl
Woody climber, to 10–20m; opposite branched; bark
thick, stems round, twigs smooth, green; leaves ovate
10–27 × 7–17cm, acumen minimal or absent, densely
reticulate beneath, 4–6 pairs lateral nerves; petiole
1–2cm long; infloresecence lax, strictly bifurcate;
flowers 1cm or more in diam.; petals 5, yellowish-
white, entire, membranaceous; fruits 8–15cm long, 3
triangular to ovate mericarps 8–11 × 10–11cm; 4–6
wingless seeds per carpel. Forest; 640–700m.
Distr.: Guinea to Cameroon, Gabon, Congo
(Brazzaville), Congo (Kinshasa), Angola and CAR
[Guineo-Congolian].
IUCN: LC
Mefou Proposed National Park: Onana 2808 fl.,
3/2004; **Ndanan 2:** Darbyshire 292 fl., 3/2004.

Loeseneriella apocynoides (Welw. ex Oliv.)
N.Hallé ex J.Raynal var. *apocynoides*
Fl. Gabon 29: 247 (1986).
Woody climber, to 10–30m; young branches reddish-
brown puberulent, plant climbing partly via paired
twigs 2–3cm long; leaves c. 8 × 5cm, 5–6 secondary
veins; inflorescence a dichotomous cyme; peduncles
and bracts tomentose; flowers 3–8mm diam.,
puberulent; fruiting mericarps elliptic, to 7 × 4cm.
Forest, riverbanks; 640–720m.
Distr.: Guinea (Conakry) to Tanzania [Guineo-
Congolian].
IUCN: LC
Ndanan 1: Cheek 11775 3/2004; Onana 2873 3/2004;
Ndanan 2: Darbyshire 287 fl., 3/2004.

Maytenus gracilipes (Welw. ex Oliv.) Exell subsp.
gracilipes
Syn. *Maytenus serrata* (Hochst. ex A.Rich.) R.Wilczek
var. *gracilipes* (Welw. ex Oliv.) R.Wilczek Fl. Cameroun
19: 22 (1975); Symb. Bot. Ups. 25: 79 (1985).
Tree, or shrub, to 5m; spiny; leaves 8–16 × 3–17cm,
elliptic, base strongly attenuate, apex angled, not
acuminate, secondary nerves 10–12 pairs; inflorescence
a loose axillary cyme 5–14cm long; flowers tiny, c. 2 ×
2mm; fruit a tri-locular capsule, red-violet when fresh,
1.5–2 × 1cm; 2–4 seeds per locule. Forest; 600–700m.
Distr.: Cameroon to Angola [lower Guinea].
IUCN: LC
Mefou Proposed National Park: Etuge 5136 3/2004;
Ndanan 2: Gosline 410 10/2002.

Salacia dusenii Loes.
Small tree, or woody climber, to 15m; no resinous
threads; bark dark to light grey; leaves oblanceolate to
ovate, 7–16 × 4–8cm, acumen to 2cm, lateral nerves
6–8 pairs; petiole 6–10mm; flowers sessile; pedicels
7–20mm, thickening slightly towards the top; fruit
globular, to 3cm diam. Forest; 640–710m.
Distr.: Nigeria to Congo (Kinshasa) [lower Guinea &
Congolian].
IUCN: LC
Ndanan 1: Cheek 11591 3/2004; 11997 3/2004;
Ndanan 2: Darbyshire 304 fl., 3/2004.

Salacia lehmbachii Loes. var. *lehmbachii*
Shrub, to 2m; without resinous threads; stems
quadrangular, slightly pustulose; leaves ovate to 19 ×
9cm, c. 5 secondary veins, tertiary veins perpendicular
to midrib across leaf; petiole 5 × 10mm; inflorescence
axillary short pedunculate; flowers 5–8mm diam., red to
orange; fruit long and acuminate or turbinate to 7 ×
4cm. Forest; 679m.
Distr.: Cameroon to Congo (Kinshasa) [Guineo-
Congolian].
IUCN: LC
Ndanan 1: Onana 2913 fl., 3/2004.

Salacia loloensis Loes. var. *sibangana* N.Hallé
Fl. Gabon 29: 82 (1986); Fl. Cameroun 32: 60 (1990).
Syn. *Salacia* sp. A sensu Hepper & Blakelock
Tree, or shrub, 1.5–3m; no resinous threads, slight
ridges decurrent on branches; leaves large, oblanceolate
22–30 × 8–12cm, acumen linear to 3cm, 15–20 pairs
lateral nerves, margin toothed; inflorescence sessile in
axis with 1–5 flowers; pedicels filiform, 5–12mm long;
buds subglobose 2.5mm diam., green-black; flowers
purplish; fruits fusiform to 5cm long, orange. Forest;
680–700m.
Distr.: SE Nigeria and SW Cameroon [western
Cameroon].
IUCN: NT
Mefou Proposed National Park: Nana 48 3/2004;
Ndanan 2: Darbyshire 301 fl., 3/2004.

Salacia longipes (Oliv.) N.Hallé var. *camerunensis* (Loes.) N.Hallé

Fl. Gabon 29: 66 (1986); Fl. Cameroun 32: 49 (1990).
Syn. *Salacia camerunensis* Loes.
Woody climber, to 15m; no resinous threads, branchlets green; leaves 5–18 × 2–8cm drying green, acumen linear to 1.5cm, 6–12 pairs secondary nerves; petiole 2–12mm; inflorescence with a peduncle 4–5cm long, carrying a fascicle to 10mm broad; pedicels 9–14mm; buds 2–3mm diam.; flowers to 10mm diam., orange to yellow; fruits variable, oblong, ellipsoid or turbinate 3–6.5 × 2–4cm; 2–6 seeds. Forest; 600–710m.
Distr.: Liberia to Congo (Kinshasa) [Guineo-Congolian].
IUCN: LC
Mefou Proposed National Park: Etuge 5166 fl., 3/2004; 5172 fl., 3/2004; Onana 2806 fl., 3/2004; **Ndanan 1:** Cheek 11852 fl., 3/2004; Gosline 408 fr., 10/2002; 440 10/2002.

Salacia pallescens Oliv.

Fl. Cameroun 32: 60 (1990).
Shrub, to 3m; branches brown, lenticular; no resinous threads; leaves elliptic 7–16 × 2.5–6cm, toothed, margins undulate, 8–12 pairs secondary nerves; inflorescence sessile fascicles of 3–8; pedicel 3–13mm; flowers 3–4 × 4–6mm, violet-black; fruit globose, 1.5–3cm diam., orange-red with violet markings; 1–3 seeds. Forest; 640m.
Distr.: Guinea (Conakry) to Zimbabwe [Guineo-Congolian].
IUCN: LC
Ndanan 2: Darbyshire 290 fl., 3/2004.

Salacia pyriformioides Loes.

Fl. Gabon 29: 57 (1986).
Woody climber, to 7m; no resinous threads, twigs olive when dried; leaves elliptic 8–25 × 5–12cm, acumen 5–15mm, olive green when dried, secondary nerves 6–9 pairs arcuate; petiole canaliculate; inflorescence sessile on stems, few to many flowers; pedicels long filiform, 9–40mm; flower yellow, 4.5mm diam.; fruits ellipsoid, 3–7 × 3–5cm; numerous seeds. Forest; 600m.
Distr.: Nigeria to Congo (Kinshasa) [lower Guinea & Congolian].
IUCN: NT
Mefou Proposed National Park: Etuge 5264 3/2004.
Note: a widespread but rare species, known over much of its range from only early to mid-twentieth century collections.

Thyrsosalacia nematobrachion Loes.

Fl. Gabon 29:21 (1986); Fl. Cameroun 32:12 (1990).
Shrub, to 2m or climber, to 6m; leaves opposite, ovate to obovate 5–20 × 2–9cm, base cuneate, apex short acuminate, 6–8 lateral nerves ascendent, obverse light green, fine reticulation on both surfaces; inflorescence axillary, a thin raceme c. 10cm long carrying delicate

panicles sometimes with both flowers and fruits; buds globose, 1–2mm diam. at tips of very thin pedicels 1–2cm long; sepals translucent; flowers 4–9mm diam., pale orange-yellow; mature fruits 4–10 × 2–3cm, attenuate at base and rostrate; 1–3 seeds. Forest understorey near river banks and swamps; 650–700m.
Distr.: Cameron, Gabon [lower Guinea].
IUCN: LC
Mefou Proposed National Park: Etuge 5295 fr., 3/2004; **Ndanan 1:** Cheek 11051 fr., 10/2002; Tadjouteu 582 3/2004; 624 fl., fr., 4/2004; **Ndanan 2:** Darbyshire 214 fl., fr., 3/2004.

CHRYSOBALANACEAE

G. Prance (K), Y.B. Harvey (K) & M. Cheek (K)

Fl. Cameroun 20 (1978).

Chrysobalanaceae sp. 1 of Mefou

Tree, only known from seedlings 0.5 × 1m; stem with long, white hairs, both appressed and spreading; stipules brown, parallel to stem, 15 × 1mm, margin with curved glandular teeth; leaf-blade elliptic, 12–18 × 4–7cm, acumen slender, 1.5–2cm, base abruptly rounded, or obtuse, margin with glandular teeth, lateral nerves 9–11 pairs, tertiary nerves scalariform, lower surface with long, spider-web appressed hairs, upper surface with gland pair at base on each side of the midrib, glands raised, circular, 1mm diam., black; petiole 4mm. Evergreen forest/*Raphia* swamp boundary; 710m.
Ndanan 1: Cheek 11628 3/2004; 11912 3/2004.
Note: determined by Prance ix.2008 as "not matched"; confirmed by Breteler ix.2010 as Chrysobalanaceae and not Dichapetalaceae. Possibly a new, undescribed species, but equally likely to be the unknown juvenile stage of a species already known to science. Flowering material needed, Cheek ix.2010.

Chrysobalanaceae sp. 2 of Mefou

Tree, known only from seedlings up to 6m tall; stems dark brown, with long white hairs 1–2mm, spreading and appressed, moderately sparse; stipules strap-like, 5–7 × 1mm, brown, margin entire, held at 45° from stem, moderately persistent; leaf-blade oblong-elliptic or oblanceolate-oblong, 8–13 × (2.5–)4cm, acumen 1cm, base cuneate, with a pair of glands at base on upper surface, lateral nerves 6–8 pairs, forming a looping marginal nerve, lower surface drying orange-brown, minutely granular, upper surface grey-black; petiole 4–5mm. Inundated forest; 700m.
Mefou Proposed National Park: Cheek in MF 54 3/2004; 67 3/2004; 99 3/2004.
Note: restricted to swamp forest. Identified as Chrysobalanaceae by Breteler ix.2010. Distinct in the orange-brown lower surface. Flowering material needed to identify to species. All specimens are from "plot 1" (2004), Cheek ix.2010.

Dactyladenia sp. cf. *floribunda* Welw.

Tree, to 8m; young stems covered with velutinous brown hairs; leaves 13–15 × 7–9cm, elliptic, glabrous upper surface except midrid and veins, scattered hairs on lower surface, more densely so on midrib and veins, glands not obvious; petioles to 7mm, densely velutinous; stipules absent; inflorescences immature, erect cone-like buds. *Raphia* swamp; 710m.
Ndanan 1: Cheek 11753 3/2004.

Magnistipula zenkeri Engl.

Fl. Gabon 24: 93 (1978).
Syn. *Hirtella fleuryana* A.Chev.
Tree, to 8(–35)m; branches lenticellate; leaves elliptic-oblong, 23 × 10.5–11.5cm, glabrous upper and lower surface, apex acute, base cordate, glands few, scattered towards base; petiole 7mm long, glabrous; stipules c. 1 × 0.6cm, persistent; inflorescences branched and congested, to 15cm long; flowers to 33mm long; bracts c. 4 × 3mm; calyx to 2–3mm long, short velutinous ochre-coloured hairs, tube to 15mm long, lobes to 8mm; petals white, 15mm long; stamens 8mm long, on one side. Forest near river; 600m.
Distr.: Sierra Leone, Liberia, Ivory Coast, Equatorial Guinea, Cameroon & Gabon [lower Guinea & Congolian].
IUCN: LC
Mefou Proposed National Park: Etuge 5159 fl., 3/2004.

Parinari hypochrysea Mildbr. ex Letouzey & F.White

Fl. Cameroun 20: 134 (1978); Fl. Gabon 24: 134 (1978); Keay R.W.J., Trees of Nigeria: 184 (1989).
Syn. *Parinari* sp. F.W.T.A. 1: 430 (1958).
Tree, to 30m; buttresses c. 5, 4m high, projecting 2m; bark whitish brown, slash dark red and granular; stems densely tomentose, hairs chestnut brown in immature growth, fawn brown when mature; immature leaves large, 19–28 × 5.3–12cm, apex acuminate, base rounded to cordate, upper surface glabrous except midrib, lower surface tomentose with chestnut hairs; sheathing stipules to 3cm; mature leaves smaller, 5–6 × 3–4cm, broadly elliptic, apex acute, base rounded, upper surface glabrous except for midrib, lower surface densely tomentose with chestnut-coloured hairs; stipule caducous; petiole densely tomentose, 3–4mm in young leaves, to 6mm in mature leaves, 2 yellow glands at midpoint, c. 1mm diam. Forest; 700–710m.
Distr.: Nigeria, Cameroon and Gabon [lower Guinea].
IUCN: LC
Mefou Proposed National Park: Cheek 13104 2/2006;
Ndanan 1: Cheek 11200 10/2002; 11588 3/2004; 11629 3/2004.
Note: young leaves visited by ants (Cheek 11588).

COMBRETACEAE

C. Jongkind (WAG)

Fl. Cameroun 25 (1983).

Combretum comosum G. Don

Syn. *Combretum rhodanthum* Engl. & Diels
Shrub, or liana, to 5m or more, scales absent; stems with densely patent golden-brown hairs both simple and glandular; leaves subopposite, elliptic, c. 14 × 8cm, apex rounded or 5mm, base rounded, then abruptly cordate, lateral nerves 7–8 pairs, upper and lower surfaces glabrous apart from a few hairs on the nerves, finely reticulate; petiole 5mm; panicles axillary, leafy, to 45 × 20cm; flowers 10–14mm excluding stamens, 5-merous, pink/red, sweet-scented; upper receptacle bell-shaped, 5 × 3–4mm; petals 7mm; anthers exserted 10mm; fruit 5 winged, 2.5cm. Semi-deciduous forest; 600m.
Distr.: Guinea (Conakry) to Uganda and Angola [Guineo-Congolian].
IUCN: LC
Mefou Proposed National Park: Etuge 5105 fl., fr., 3/2003.

Combretum mucronatum Schumach. & Thonn.

F.T.A. 2: 426 (1871).
Liana; scales absent, stems black with scattered grey-white patent hairs; leaves sparsely yellow-brown hairy, subopposite, simple, obovate, c. 6–13 × 3–5cm, acumen 3–7mm, base abruptly rounded or subcordate, lateral nerves 6–8 pairs, domatia absent, glands brown, inconspicuous, scalariform; petiole 2–8mm; panicles axillary with several branches 6–20cm, branches with cylindrical heads of flowers 2 × 1cm, white; flowers 4-merous, white, 5mm; upper receptable funnel-shaped, 2 × 2mm, brown hairy; petals 1.5mm; fruit 4 winged, 2 × 2cm. Semi-deciduous forest; 710m.
Distr.: Senegal to Uganda [Guineo-Congolian].
IUCN: LC
Ndanan 1: Cheek 11743 fl., 3/2004; 11817 fr., 3/2004.
Note: F.W.T.A. 1: 272 (1954) considers *Combretum mucronatum* to be a synonym of *C. smeathmannii* G.Don, Harvey x.2010.

Combretum zenkeri Engl. & Diels

Scandent shrub, or liana; scales absent, sparsely puberulent on stems, orange-brown on leaf nerves; leaves subopposite, simple, elliptic, c. 11 × 5.5cm, acumen 0.5cm, base obtuse to truncate, lateral nerves c. 10 pairs, domatia absent, scalariform; petiole c. 12mm, abscissing 2mm from base, developing hooks; panicle c. 20 × 14cm, longest branches c. 6cm; flowers grouped in heads of c. 20, c. 1cm diam., 5-merous, cream, 5mm; upper receptacle cup-shaped, 1.5mm; petals 1.5mm; fruit 5-winged, obovate-elliptic, c. 2cm. Disturbed forest; 700m.
Distr.: Guinea (Conakry) to Cameroon [upper & lower Guinea].
IUCN: LC
Ndanan 1: Harvey 147 fr., 10/2002.

Terminalia mantaly H.Perrier

Fl. Cameroun 25: 79 (1983).

Tree, to 8m; stems spreading, branching in one plane; leaves obovate, 5–6.5 × 2.3–3.3cm, emarginate, base long-cuneate, margin subundulate, glabrous; petiole 4mm; spikes axillary, to 4cm; peduncle 4cm, puberulous; flowers 2mm, white-green; stamens 2.5mm. Villages, ornamental; 700m.

Distr.: Madagascar, widely planted in the tropics [pantropical].

IUCN: LC

Ndanan 1: Darbyshire 256 fl., 3/2004.

Terminalia superba Engl. & Diels

Tree, 35(–50)m with flat crown; stems glabrous; leaves coriaceous, c. 12 × 6cm, lateral nerves 6–7 pairs, glabrous; petiole to 5cm with paired glands; spikes to 13.5cm; samara transversely oblong-elliptic, not fleshy, wings 2 × 2.7cm. Forest; 700m.

Distr.: Guinea (Conakry) to W Congo (Kinshasa) [upper & Lower Guinea].

IUCN: LC

Ndanan 1: Cheek in MF 393 10/2002.

COMPOSITAE

J.-M. Onana (YA), M. Cheek (K), D.J.N. Hind (K) & Seren Thomas (K)

Acmella caulirhiza Delile

Fragm. Flor. Geobot. 36(1) Suppl.1: 228 (1991); Fl. Masc. Compositae: 208 (1993).

Syn. *Spilanthes filicaulis* (Schum. & Thonn.) C.D.Adams

Annual or perennial creeping herb, 5–30cm tall; leaves ovate, up to 5 × 2.5cm, obtuse, margins subentire to dentate; petiole 0.25–1cm; peduncles to 10cm; capitula conical; ray florets present, sometimes tiny, ray limb yellow; pappus usually absent. Forest & stream banks; 700–710m.

Distr.: tropical & subtropical Africa and Madagascar [tropical & subtropical Africa].

IUCN: LC

Ndanan 1: Cheek 11811 3/2004; **Ndanan 2:** Cheek 11154 10/2002.

Adenostemma viscosum J.R.Forst. & G.Forst.

F.T.E.A. Compositae 3: 820 (2005).

Syn. *Adenostemma perrottetii* DC.

Herb, c. 0.75m tall; cauline leaves ovate to trullate, c. 5.5 × 3.5cm, apex acute, base cuneate, margin coarsely serrate; petiole up to 2.5cm long; capitula discoid, c. 6mm wide; involucre of 5 elliptic phyllaries c. 3 × 2mm; corollas with 3-4(-5) lobes, white; achene tuberculate; pappus of 3 pegs <1mm long. Forest gaps; 700–710m.

Distr.: Senegal to Bioko, Cameroon, Sudan, Uganda and Tanzania [tropical Africa (montane)].

IUCN: LC

Mefou Proposed National Park: Cheek in RP 49 3/2004; **Ndanan 1:** Cheek in MF 218 10/2002; Tadjouteu 604 3/2004.

Ageratum conyzoides L.

Annual herb, 0.3–1m tall; cauline leaves ovate, 5 × 2.5cm, apex acute, base acute to truncate, margins serrate; petiole c. 1cm, white pilose; capitula discoid, c. 6mm diam.; numerous in dense terminal aggregations; phyllaries narrowly oblong, c. 4mm long, green, with two white lines; corollas blue; achenes 5-ribbed, black; pappus of 5 awned scales. Farmbush; 700m.

Distr.: Senegal to Tanzania [pantropical].

IUCN: LC

Ndanan 1: Cheek 11022 10/2002.

Centratherum punctatum Cass. subsp. *punctatum*

Dict. Sci. Nat., ed. 2. [F. Cuvier] 7: 384 (1817).

Annual herb, often subshrubby with age, to 50cm tall; stems pubescent; leaves elliptic, 7–7.5 × 3–3.5cm, 8 veins on each side of midrib, serrate, adaxial surface dark green, sparsely pubesecent, abaxial surface paler green, pubescent; capitula 1.5cm diam., usually 2 or 3 together in stalked terminal nodes; phyllaries of 2 types, outer larger and leaf-like, inner small; florets many; corollas purple, central ones shorter than marginal ones; achenes cylindrical to obconical, <3mm long; pappus setae straw-coloured, deciduous. Roadside; 700–720m.

Distr.: India, Phillipines, Australia, Hawaiian Islands, Nigeria, Cameroon, Kenya, Zambia, Zimbabwe, El-Salvador, West Indies, Paraguay, Guyana, Brazil, Bolivia, Argentina [pantropical].

IUCN: LC

Mefou Proposed National Park: Nana 38 3/2004; **Ndanan 1:** Cheek 11711 3/2004; Onana 2834 3/2004.

Note: Onana 2834 is the first record of this taxon in Cameroon.

Chromolaena odorata (L.) R.M.King & H.Rob.

Fragm. Flor. Geobot. 36(1) Suppl.1: 450 (1991); Candollea 47: 645 (1992); Fl. Masc. Compositae: 235 (1993).

Syn. *Eupatorium odoratum* L.

Shrub or liana, to 3.5m tall; leaves opposite, ovate, 3–10 × 2–6cm, dentate; capitula 10mm long, arranged in corymbs; corollas white or mauve; achenes 5-angled; pappus of silky, straw-coloured bristles. Roadsides & villages; 700m.

Distr.: pantropical.

IUCN: LC

Ndanan 1: Cheek 11021 10/2002.

Note: a major weed of farms and roadsides.

Conyza bonariensis (L.) Cronquist

Fragm. Flor. Geobot. 36(1) Suppl.1: 59 (1991).

Syn. *Erigeron floribundus* (Kunth) Sch.Bip.

Syn. *Conyza sumatrensis* (Retz.) E.Walker Kirkia 10: 43 (1975).

Syn. *Erigeron bonariensis* L.

Annual herb, 40cm to 1.5m tall, erect; leaves sessile and narrowly lanceolate, 2–15 × 0.2–3cm, entire or dentate; capitula 4–8mm in lax panicles; florets tiny; corollas whitish with purple tips; pappus of white bristles. Grassland, weedy; 640m.

Distr.: pantropical.

IUCN: LC

Ndanan 2: Darbyshire 318 fr., 3/2004.

Crassocephalum montuosum (S.Moore) Milne-Redh.

Herb, 1.5m tall; cauline leaves elliptic in outline, 8.5 × 6cm, pinnately 2–8-lobed, apical lobe largest, lower lobes decreasing in size, lobes dentate, glabrous; capitula 10–20, in aggregations, discoid, 2–4mm diam.; corollas yellow or brownish yellow to orangish; achenes ribbed; pappus setae white. Forest & forest-grassland transition; 710m.

Distr.: W Cameroon to Congo (Kinshasa) to E Africa & Madagascar [afromontane].

IUCN: LC

Mefou Proposed National Park: Cheek in RP 43 3/2004.

Emilia coccinea (Sims) G.Don

Annual herb, 15–120cm tall; leaves ± subsucculent, 1–20 × 0.4–6cm, often purplish beneath; capitula in terminal corymbs; discoid, phyllaries uniseriate; corollas bright orange; pappus setae white. Forest & farmbush; 600m.

Distr.: tropical Africa [afromontane].

IUCN: LC

Mefou Proposed National Park: Etuge 5267 3/2004.

Melanthera scandens (Schumach. & Thonn.) Roberty

Perennial herb, to 1m tall, or scandent and then to 4m tall; leaves opposite, ovate, 2–13 × 1–7cm, scabrid, strongly 3-veined from base; capitula solitary or in lax corymbs; ray florets limbs orange-yellow; disc florets corollas yellow; pappus of bristles. Roadsides, farmbush; 600–720m.

Distr.: tropical Africa [afromontane].

IUCN: LC

Mefou Proposed National Park: Etuge 5128 fl., 3/2004; **Ndanan 1:** Cheek 11863 3/2004; Onana 2835 3/2004; **Ndanan 2:** Cheek 11169 10/2002.

Mikania chenopodifolia Willd.

Fragm. Flor. Geobot. 36(1) Suppl.1: 460 (1991); Fl. Masc. Compositae: 229 (1993); F.T.E.A. Compositae 3: 835 (2005).

Syn. *Mikania capensis* DC. Fragm. Flor. Geobot. 36(1) Suppl. 1: 458 (1991).

Shrub, 5–9m tall; leaves opposite, ovate, 2–10 × 1–7cm, subsagittate to hastate, glandular-punctate, with 3 main veins; capitula in a dense, leafy corymb; phyllaries 4; florets 4; corollas white, tubular; pappus setae white. Forest & thicket; 700m.

Distr.: tropical Africa & Asia [palaeotropics].

IUCN: LC

Ndanan 1: Harvey 140 10/2002; **Ndanan 2:** Darbyshire 196 fl., fr., 3/2004.

Note: taxon mentioned in F.W.T.A. 2: 286 (1963) as a synonym of *M. cordata* (Burm.f.) B.L.Robinson var. *cordata*.

Struchium sparganophora (L.) Kuntze

Syn. *Sparganophorus sparganophora* (L.) C.Jeffrey Kew Bull. 43: 271 (1988); Fragm. Flor. Geobot. 37: 379 (1992).

Herb, to 1m tall, sparsely pubescent; stems fleshy, rooting at nodes, leafy stems erect; leaves papery, oblanceolate-elliptic, to 10 × 2cm, apex acuminate, base cuneate, 8–9 veins on each side of midrib, crenate-serrate, abaxial surface sparsely pubescent and with minute gland dots, adaxial surface sparsely pubescent; inflorescence sessile in leaf axils, of >3 capitula, yellow-brown; involucre green-brown, 7mm diam.; corollas whitish; achenes 5-angled; pappus a cartilaginous cup. *Raphia* swamp; 600–710m.

Distr.: India, Malaysia, Thailand, Indo-China, Sumatra, Jawa, Borneo, Sarawak, Sabah, Papua New Guinea, Fiji, Senegal, Mali, Guinea (Bissau), Guinea (Conakry), Sierra Leone, Liberia, Ivory Coast, Burkina Faso, Ghana, Togo, Nigeria, Cameroon, Brazil, Panama, Columbia, Venezuela, Ecuador, Peru [pantropical].

IUCN: LC

Mefou Proposed National Park: Cheek in RP 13 3/2004; Etuge 5157 fl., 3/2004; **Ndanan 2:** Darbyshire 212 fl., 3/2004.

Note: C. Jeffrey was incorrect in assigning the new combination under *Sparganophorus*. The name stands under *Struchium*, Hind xi.2010.

Tithonia diversifolia (Hemsl.) A.Gray

Kirkia 6: 56 (1967).

Perennial herb, to 5m tall; leaves large, alternate, ovate to obovate, lobed; capitula solitary, terminal, with swollen stalk; phyllaries in 4 series; receptacle with paleae enclosing disc florets; ray florets sterile, limb large, yellow; achenes 4-angled; pappus of 2 scales and 2 awns. Farms & farmbush; 600m.

Distr.: originally S America, widely naturalised in the tropics [pantropical].

IUCN: LC

Ndanan 2: Etuge 5114 3/2003.

Vernonia amygdalina Delile

Shrub or small tree, 0.5–10m tall; leaves elliptic, 4–15 × 1–4cm, slightly hairy; capitula discoid, in large terminal corymbs; corollas white, scented; pappus of small outer scales and longer inner setae. Cultivated; 663m.

Distr.: Mali to Zimbabwe & Annobón [tropical Africa].

IUCN: LC

Mefou Proposed National Park: Etuge 5303 3/2004.

Local Name: Atet (Ewondo) (Onana pers. comm).

Note: also found in Brazil, S America where it is cultivated, having been described twice as a new species, possibly having been introduced by slaves, Hind xi.2010.

Vernonia stellulifera (Benth.) C.Jeffrey
Fl. Zamb. 6(1): 142 (1992).
Syn. *Triplotaxis stellulifera* (Benth.) Hutch.
Annual herb, 0.1–1.2m tall; leaves ovate or elliptic, 2–10 × 1–3cm, remotely crenate; capitula in corymbose cymes, the corollas mauve to purple; pappus of a rim of scales with a few inner bristles. Forest and farmbush; 600–710m.
Distr.: Guinea (Conakry) to Zambia & Uganda [Guineo-Congolian (montane)].
IUCN: LC
Mefou Proposed National Park: Etuge 5150 fl., 3/2004; **Ndanan 1:** Cheek 11786 3/2004; Harvey 145 10/2002; 146 10/2002; **Ndanan 2:** Cheek 11108 10/2002.

CONNARACEAE

C. Jongkind (WAG), J.-M. Onana (YA) & M. Cheek (K)

Agelaea paradoxa Gilg
Agric. Univ. Wag. Papers 89(6): 140 (1989); Fl. Gabon 33: 31 (1992).
Syn. *Castanola paradoxa* (Gilg) Schellenb. ex Hutch. & Dalziel
Liana, to 10(–40)m; branchlets puberulous; leaves trifoliolate, 13 × 10cm, leaflets ovate-elliptic, terminal leaflet c. 10 × 5.5cm, acumen to 1.4cm, base acute to obtuse, asymmetric in lateral leaflets, glabrous, mucous cells numerous on upper surface; petiole 7–8cm, puberulent; petiolules 0.4cm; panicles axillary, may be numerous on leafless branches, to 11cm, many-flowered, branches puberulent; pedicels 1mm; flowers 3mm, honey-scented; sepals ovate, 1–1.5mm, puberulent; petals oblanceolate, 3mm, white. Forest; 710m.
Distr.: Senegal to Congo (Kinshasa) [Guineo-Congolian].
IUCN: LC
Ndanan 1: Cheek 11720 3/2004.

Agelaea pentagyna (Lam.) Baill.
Agric. Univ. Wag. Papers 89(6): 144 (1989); Fl. Gabon 33: 34 (1992).
Syn. *Agelaea obliqua* (P.Beauv.) Baill.
Syn. *Agelaea floccosa* Schellenb.
Syn. *Agelaea pseudobliqua* Schellenb. F.W.T.A. 1: 746 (1958).
Syn. *Agelaea dewevrei* De Wild. & T.Durand
Syn. *Agelaea preussii* Gilg
Syn. *Agelaea grisea* Schellenb.
Syn. *Agelaea hirsuta* De Wild.
Liana, 6(–25)m; branches furrowed, puberulent; leaves trifoliolate, 14 × 14cm, leaflets ovate, terminal leaflet

10 × 8cm, acumen 1.5cm, base rounded, finely tomentose on lower surface and veins; petiole 11.5cm; petiolule 3mm; panicles to 35cm, glabrous to tomentose; sepals 2.5–5mm, with fringing multi-cellular hair; petals 3–5.5mm. Forest; 700m.
Distr.: Senegal to Mozambique & Madagascar [tropical Africa & Madagascar].
IUCN: LC
Ndanan 1: Cheek in MF 265 10/2002.

Cnestis corniculata Lam.
Syn. *Cnestis grisea* Baker
Syn. *Cnestis sp. A* sensu Hepper
Syn. *Cnestis congolana* De Wild.
Syn. *Cnestis longiflora* Schellenb.
Syn. *Cnestis aurantiaca* Gilg
Liana, to 10m; branchlets yellow-brown pilose to glabrous; leaves imparipinnate, to 45cm, 7–12-jugate, leaflets opposite to alternate, oblong, to 11 × 4.5cm, acumen 1cm, base asymmetric, truncate or subcordate, veins beneath pilose, basal leaflets much reduced, ovate; petiolules 1mm; racemes axillary or cauliflorous, to 10cm, single or in clusters of up to 4, pubescent; flowers 0.5cm; sepals reflexed, puberulent; petals oblong-lanceolate; follicles obliquely ellipsoid with a curved beak at maturity, to 1 × 3cm, golden brown-pilose. Forest & thickets; 700m.
Distr.: Senegal to E Tanzania [tropical Africa].
IUCN: LC
Ndanan 1: Cheek 11032 10/2002.

Cnestis ferruginea Vahl ex DC.
Woody liana; young stems, petioles, lower surface of leaflets, inflorescence axis, densely yellow-brown tomentose; leaves alternate, pinnately compound, 25cm; petiole 1.5–6cm; leaflets c. 5 pairs, drying brown-black above, opposite and alternate, oblong, 8 × 2.5cm, subacuminate, base rounded, lateral nerves 4–5 pairs; petiolule 1mm; inflorescences spike-like, 10–15cm, numerous from stem apex and upper axils, c. 30 forming a mass 20 × 20cm, scented of honey; flowers minute, 2mm, white; fruit fleshy, orange, 4cm, at length splitting. Forest; 700–710m.
Distr.: Gambia to Cameroon, Gabon, Congo (Brazzaville), Congo (Kinshasa), Angola, CAR and Sudan [Guineo-Congolian].
IUCN: LC
Ndanan 1: Cheek 11047 10/2002; 11787 fl., 3/2004.

Rourea minor (Gaertn.) Alston
Agric. Univ. Wag. Papers 89–6: 337 (1989).
Weedy climber, to short or small tree; branches glabrous, furrowed; leafy stems puberulous or glabrous; leaves alternate, estipulate, pinnately compound; petiole 2–9cm; leaflets 2–5-jugate, ovate to oblong-elliptic, 1.5–12 × 0.5–6cm, acumen to 2.5cm, base acute to obtuse; inflorescence to 9cm, glabrous; pedicel 3–7mm; sepals 1.5–4mm; petals 5–8mm, imbricate in bud, ovary hairy on one side; fruit follicles 1, 1–2 × 0.5–1cm,

glabrous, dehiscing in circle above calyx; seed covered in free fleshy layer. Forest; 700m.

Distr.: tropical Africa and Asia [tropica Africa and Asia].

IUCN: LC

Ndanan 1: Cheek in MF 277 10/2002.

Note: plot voucher MF277 is only tentatively identified, flowering material is needed, Cheek x.2010.

Rourea obliquifoliolata Gilg

Agric. Univ. Wag. Papers 89(6): 345 (1989); Fl. Gabon 33: 126 (1992).

Syn. *Roureopsis obliquifolioliata* (Gilg) Schellenb.

Liana; lenticels orange; young stems very shortly grey-white puberulent, extending to petiole and midribs; leaves alternate, pinnately compound, 20–30cm; petiole 0.5–3cm; leaflets c. 10 pairs, basal-most resembling foliaceous stipules, clasping stem, orbicular 2 × 2cm, remaining leaflets rhombic-oblong, c. 5.5 × 2.5cm, apex subacuminate to slightly notched, base obtuse, midrib diagonal; petiolule 1mm; infructescence axillary 20cm, twining, inserted below leaves; fruit sessile, each with 1–2 red follicles 2cm, splitting to show glossy black ellipsoid seeds 1.5cm. Evergreen forest; 700–710m.

Distr.: SE Nigeria to Gabon and Congo (Kinshasa) [lower Guinea & Congolian].

IUCN: LC

Mefou Proposed National Park: Cheek in MF 102 3/2004; **Ndanan 1:** Cheek 11763 3/2004; 11802 3/2004; 11866 3/2004; Mackinder 502 10/2002.

CONVOLVULACEAE

J.-M. Onana (YA), L. Pearce (K) & H. Fortune-Hopkins (K)

Calycobolus sp. of Mefou

Climbing shrub; leaf-blades subcoriaceous, elliptic; flowers small, in axillary cymes; corolla white; bracts not enlarged in fruit, outer 2 sepals markedly accrescent, one much exceeding the other, to 6cm diam., heart-shaped, chartaceous with venation obvious. Roadside weed; 710m.

Ndanan 1: Cheek 11494 3/2004.

Note: Cheek 11494 comprises a fallen fruit, Hopkins x.2010.

Ipomoea alba L.

Climbing herb, glabrous; leaves papery, ovate, c. 15 × 12cm, acumen slender, 1.5cm, cordate; petiole c. 17cm; inflorescence 1–5-flowered, racemose; peduncle as long as rachis; corolla white, night opening, 8cm long, 10cm across; fruit ovoid, 3cm. Farmbush; 700m.

Distr.: native of S America [pantropical].

IUCN: LC

Ndanan 1: Harvey 128 10/2002.

Ipomoea batatas (L.) Lam.

Climber, tubers edible; stems to 15m, glabrous; leaves triangular-ovate, c. 10 × 8cm, long acute, broadly cordate, or pedately 5-lobed, central lobe to 5cm; petiole c. 7cm; inflorescence cymose 5–10-flowered, rachis to 3cm; peduncle 10cm; sepals lanceolate 0.6cm; corolla pale pink, 3cm. Farmbush & farms; 700m.

Distr.: native of S America [pantropical].

IUCN: LC

Ndanan 2: Cheek 11107 10/2002.

Uses: FOOD — Roots — tubers eaten.

Ipomoea mauritiana Jacq.

Large subglabrous liana, with twining hollow stems; leaf-blades palmately 3 or 5-lobed (to entire), to 16 × 22cm, cordate at base, lobes usually acute at apex, both surfaces glabrous, sometime purple below; petioles to 9cm, puberulent; flowers in pedunculate cymes; peduncles to 15cm; flower buds globular with convex sepals tightly clasping corolla tube; corolla pinkish purple, to 6.5cm long, funnel shaped; capsules ovoid, to 1.4cm long. Forest, secondary bush, *Raphia* swamp; 600–710m.

Distr.: pantropical [pantropical].

IUCN: LC

Mefou Proposed National Park: Cheek in RP 14 3/2004; 75 3/2004; 8 3/2004; Etuge 5113 3/2003; **Ndanan 1:** Cheek 11003 10/2002; Tadjouteu 627 4/2004; **Ndanan 2:** Cheek 11153 10/2002.

Note: plot vouchers RP8, RP14 & RP75 have tentatively been placed in this taxon, Pearce viii.2010.

Local Name: Dumsi (Ewondo) (Cheek 11153).

Ipomoea obscura (L.) Ker-Gawl

Twining perennial herb; stems glabrous to pilose (hairs white, to 1.5mm); leaf-blades chartaceous, heart-shaped; leaf-blades 2.7–9 × 0.4–7.5cm, apex acute, base cordate, glabrous; petiole 4–8cm; flowers in small, lax, axillary cymes; peduncles to 4cm; pedicels 1–2cm; sepals pilose; corolla pale yellow, 2.8cm (can be white, yellow, orange, crimson or brown and 1.4–4cm), funnel-shaped; capsules globose, to 1.2cm diam. Roadside; 700–710m.

Distr.: tropical Africa, South Africa, Madagascar, Indian Ocean Is., through tropical Asia to N Australia and the Pacific [pantropical].

IUCN: LC

Ndanan 1: Cheek 11019 10/2002; 11712 3/2004.

Note: according to F.W.T.A. *Ipomoea ochracea* belongs to an aggregate species with *I. obscura* and according to the key in FTEA, both taxa differ only in the size of the corolla (our material is intermediate). Cheek 11019 and 11712 have both been placed under the name *I. obscura*, Hopkins x.2010.

Ipomoea tenuirostris Choisy

Slender climber; stems pilose; leaves lanceolate, c. 4–8 × 1.6–4.5cm, appressed hairy; inflorescence subracemose, 3–12cm, rachis as long as peduncle, 3–10-flowered; calyx lobes c. 0.5cm; corolla 1.5cm, white, purple throat; fruit globose 0.5cm. Forest edge; 710m.

Distr.: Sierra Leone to Cameroon [upper & lower Guinea].

IUCN: LC

Ndanan 1: Cheek 11748 3/2004.
Note: Cheek 11748 is tentatively placed in this taxon, Onana vi.2006.

Lepistemon parviflorum Pilg. ex Büsgen
Climber, glabrous; leaves broadly ovate, c. 14 × 11cm; petiole 12cm; inflorescence capitate, dense, c. 4 × 4cm, c. 50-flowered; peduncle 1cm; bracts absent or inconspicuous; pedicels c. 1–5cm; flowers 0.8cm, 4–numerous. Secondary forest; 640m.
Distr.: Sierra Leone to Cameroon [upper & lower Guinea].
IUCN: LC
Ndanan 2: Darbyshire 307 fr., 3/2004.

Neuropeltis velutina Hallier f.
Climber, or climbing shrub, with dark red hairs on some parts; old stems woody, to 4mm diam., densely hairy (hairs to 1mm); leaf-blades chartaceous, broadly ovate-elliptic, to 17 × 9.5cm, obtuse, rounded or subcordate at base, shortly acuminate and mucronate at apex, upper surface glabrous except for midrib, lower surface adpressed-hairy especially on veins; petiole to 4cm long, densely hairy (hairs 1–2mm long); inflorescence long-spicate; flowers white; fruit with a large papery bract. Forest; 700m.
Distr.: Sierra Leone, Liberia, Nigeria, Cameroon, Gabon, Congo [Guineo-Congolian].
IUCN: LC
Mefou Proposed National Park: Cheek in MF 116 3/2004; **Ndanan 1:** Cheek in MF 210 10/2002; 223 10/2002; 338 10/2002.

Neuropeltis sp. of Mefou
Twining herb; woody towards base, with pale orange hairs; stem 1.5mm diam., shortly hairy (hairs to 0.5mm), glabrescent; leaf-blades chartaceous, broadly ovate-elliptic, to 10.5 × 7.2cm, cordate or subcordate at base, acuminate and mucronate at apex, sparsely hairy above, hairy below, ciliolate; petiole to 1.5cm, densely hairy (hairs 0.3mm); no flowers or fruits present. Forest near swampy area; 700m.
Mefou Proposed National Park: Cheek in MF 20 3/2004.
Note: plot voucher MF20 differs from *N. velutina* in having less-dense indumentum of a different colour, Hopkins x.2010.

CUCURBITACEAE

T. Heller (K), L. Pearce (K), M. Cheek (K), H. Fortune-Hopkins (K) & J.-M. Onana (YA)

Fl. Cameroun 6 (1967).

Bambekea racemosa Cogn.
Herbaceous climber, dioecious; stems ridges, glabrous; tendrils bifid; leaf-blades chartaceous, variable in shape, mostly palmatifid, 3- or 5-lobed to sub-lobed, 10–20 ×

10–24cm, base deeply cordate, lobe-apices acute, margin with mucro-like teeth, upper surface sparsely hairy, lower surface glabrous to minutely felty, sometimes verrucose; petioles to 8cm; flowers in racemes 4–15cm long; corolla bright yellow, c. 1.5cm diam.; fruits ovoid, to 3 × 2.5cm, glabrous, with small projecting annulus around apex. Secondary forest, *Raphia* swamp; 600m.
Distr.: Ivory Coast, Nigeria, Cameroon, Gabon and Congo (Kinshasa) [Guineo-Congolian].
IUCN: LC
Mefou Proposed National Park: Etuge 5219 3/2004.

Coccinia barteri (Hook.f.) Keay
Herbaceous climber, c. 5m; stems ridged, glabrous, angular; leaves palmatifid 3–5 lobed, 12–22 × 13–22cm, base ± deeply cordate, lobe apices acute or acuminate, margins with 1mm deltate teeth; inflorescence a raceme, 2–3cm; fruits ellipsoid or suborbicular, 4–5 × 2.5cm, green, drying blackish, smooth; seeds numerous, c. 5 × 3mm, white, smooth with narrow rim. Roadside; 700m.
Distr.: widespread in (sub-)tropical Africa [tropical Africa].
IUCN: LC
Ndanan 1: Harvey 83 10/2002.

Coccinia cf. *keayana* R.Fern.
Herbaceous climber; stems slender, finely furrowed, villose in places; tendrils simple; leaves palmatisect, 5(–7) lobed, 6–11 × 5–10cm, base cordate (by 1.5cm), lobe apices acute or acuminate, margins gently sinuate-dentate, puberulous; petiole 6–8.5cm; Forest edge; 700m.
Ndanan 2: Darbyshire 188 3/2004.
Notes (Mefou): sterile, juvenile material. Differs from *C. keayana* in having pubescence on stems, petioles and on the nerves and margins of the leaves, Pearce x.2010.

Coccinia cf. *subhastata* Keraudren
Herbaceous climber; stem slender, puberulous then glabrous; leaves entire, narrowly deltate, 7–15 × 4–10cm, base truncate to sub-hastate, apex acute or acuminate, margin with 0.5–1mm deltoid teeth; petiole 1.5–3.5, puberulous then glabrous. *Raphia* swamp; 710m.
Mefou Proposed National Park: Cheek in RP 26 3/2004; 5 3/2004.
Note: plot vouchers RP5 & RP26 are sterile, Heller x.2010.

Cogniauxia podolaena Baill.
Herbaceous climber, to 4m, dioecious; stems ridged, minutely hairy; tendrils bifid, branches subequal; leaf-blades subcoriaceous, ovate-cordate, 9–18 × 10–20cm, base deeply cordate, apex obtuse or acute, margin coarsely dentate to undulate-entire, both surfaces glabrous, 2° veins obvious below and darker than intervenal areas when dry, higher order venation

minutely prominent on both surfaces when dry; petioles to 4.5cm; male flowers in racemes; peduncles to 30cm; pedicels 2cm; female flowers solitary; corolla yellow-orange, 6cm diam.; fruit ellipsoid-cylindrical, to 15 × 8cm, glabrous. Roadsides in secondary forest, semi-deciduous forest; 600–700m.

Distr.: Cameroon, Gabon, Equatorial Guinea, CAR, Congo (Brazzaville), Congo (Kinshasa), Cabinda [Lower Guinea & Congolian].

IUCN: LC

Mefou Proposed National Park: Etuge 5124 fl., 3/2004; Nana 39 3/2004; **Ndanan 1:** Cheek 11056 10/2002; 11482 fl., 3/2004; Harvey 75 10/2002; Tadjouteu 628 4/2004; **Ndanan 2:** Cheek 11165 10/2002.

Local Name: Ambge (Ewondo) (Cheek 11165).

Momordica cabraei (Cogn.) C.Jeffrey

Fl. Cameroun 6: 166 (1967).

Syn. *Dimorphochlamys mannii* Hook.f.

Herbaceous climber, 5m; young stem lightly scabrous; tendrils bifurcate; leaves ovate, subcoriaceous, scabrous, 9.5 × 7–8.5cm, base cordate, apex shortly acuminate, margins finely sinuate-dentate, 5-veined at the base, venation clearly raised below; male flowers white, receptacle rounded funnel-shaped, 0.5–0.7 × 0.7mm (at the throat), winged, wings membranous, to 4mm, sepals rounded, 0.7–0.8 × 0.6mm, petals pedicel 2–5mm, winged for entire length, Forest; 710m.

Distr.: Ivory Coast to Bioko, Cameroon, Gabon, Congo (Kinshasa) and Angola (Cabinda). [Guineo-Congolian].

IUCN: LC

Ndanan 1: Cheek 11823 3/2004.

Momordica camerounensis Keraudren

Fl. Cameroun 6: 170–171 (1966); Adansonia sér. 2, 7: 189. (1967).

Herbaceous climber, to 4m, dioecious?; stems ridged, glabrous; tendrils simple; leaf-blades subcoriaceous, triangular-ovate, 10 × 6.5cm, base truncate, sub-cordate, rounded at 'corners', apex acuminate, margin entire, both surfaces glabrous and verrucose; petiole 2.5cm long; male flowers in racemes; corolla pale yellow, 10cm diam.; female flowers and fruits unknown. Undergrowth, secondary forest, degraded riverine forest; 720m.

Distr.: Cameroon [Cameroon endemic].

IUCN: EN

Ndanan 1: Onana 2869 3/2004.

Note: Onana 2869 is tentatively placed in this taxon, specimen sterile. No material at K. Matched using Fl. Cameroun & JSTOR, Pearce ix.2010.

Momordica charantia L.

Rank-smelling herbaceous climber, monoecious; stems thin, ridged, tomentose; leaves palmatisect c. 7 × 6cm, 5–7 lobed with strongly sinuate margins; fine white hairs at vein margins on both surfaces; flowers solitary;

pedicel 7–12cm, sparsely hairy, with sessile ovate entire bract 1 × 0.7cm approx. midway along; petals yellow, dark veined, 1–2cm; sepal lobes acute, c. 4mm; ovary fusiform, hispid; fruits c. 6.5 × 4cm, irregularly ridged, orange, 3-valved; seeds c. 1 × 0.6 × 0.3cm, coarsely warted. Disturbed vegetation, may be planted; 700–710m.

Distr.: widespread in the tropics and sub-tropics [pantropical].

IUCN: LC

Ndanan 1: Cheek 11710 3/2004; **Ndanan 2:** Darbyshire 190 3/2004.

Momordica cissoides Planch. ex Benth.

Herbaceous climber; stems thin, ridged, glabrous; leaves palmately compound, (3–)5-foliolate c. 8 × 5.5cm, slightly scabrid; leaflets elliptic, base acute, assymetric in lateral leaflets, apex mucronate, margins serrate; dioecious; male umbels and solitary female flowers both enclosed within toothed orbicular bract (2 × 2cm) with cordate base; peduncle 4–5cm; sepals lanceolate c. 3mm; petals elliptic 2.5 × 1cm, white-yellow; fruits c. 3.5 × 2cm, ovate, orange, densely bristled. Forest and farmbush; 700–710m.

Distr.: Guinea (Conakry) to Angola, E Africa [tropical Africa].

IUCN: LC

Ndanan 1: Cheek 11856 3/2004; **Ndanan 2:** Darbyshire 191 3/2004.

Momordica enneaphylla Cogn.

Bull. Acad. Belg. Ser. 3(16): 238 (1888); Fl. Cameroun 6: 156 (1967).

Herbaceous climber, dioecious; stems ridged, glabrous; leaves biternate, to 19 × 12cm, glabrous; leaflets ovate, base obtuse, apex acuminate, mucronate, margin entire except for filiform teeth at vein apices; male inflorescence a c. 4-flowered umbel enclosed within an orbicular sessile bract c. 2 × 2cm; peduncle c. 2cm; female flowers solitary; pedicel c. 1cm, calyx reflexed; petals c. 1.5 × 1cm yellow; ovary ellipsoid, 2 × 0.3cm, ridged; fruits not observed. Forest & forest margins; 710m.

Distr.: Cameroon, Gabon, Congo (Kinshasa) [lower Guinea & Congolian].

IUCN: VU

Mefou Proposed National Park: Cheek in RP 38 3/2004.

Note: plot voucher RP38 is sterile, Pearce iii.2010.

Momordica foetida Schum. & Thonn.

Syn. *Momordica cordata* Cogn.

Herbaceous climber, monoecious; stems deeply ridged, finely pubescent to glabrous; leaves ovate, 13–15 × 10–13cm, base deeply cordate, apex acuminate, margin shallowly dentate, pubescent along veins on upper surface; petioles 7–11cm; male inflorescence semi-umbellate, female solitary; male flowers with peduncle to 6.5cm, finely pubescent, may be kinked; pedicels to

1.5cm subtended by a leafy bract c.1cm; sepals rounded at apex, pubescent, drying black; petals obovate 1.5 × 1cm; stamens 2 with sinuous anthers; fruits pale orange with long, soft, fleshy spines, ovoid to ellipsoid, c. 4 × 2cm. Secondary forest & farmbush; 700m.
Distr.: tropical & S Africa.
IUCN: LC
Ndanan 1: <u>Gosline 429</u> 10/2002.

Zehneria keayana R.Fern. & A.Fern.
Fl. Cameroun 6: 43 (1967).
Herbaceous climber, to 4m, dioecious; stems to 1mm, ridged, glabrous; leaves c. 6.5 × 4.5cm, ovate, base deeply cordate, apex acuminate, margin dentate, irregularly scabrid on upper surface; petioles 2–2.5cm; male flowers on axillary racemes 4–5cm long, 1–2 per axis; pedicels c. 2mm, persistent; flowers c. 1mm with white conical petals; calyx glabrous; female flowers and fruits not observed. Open forest near water; 700m.
Distr.: Liberia, Ghana, Nigeria to Congo (Kinshasa) [Guineo-Congolian].
IUCN: LC
Ndanan 1: <u>Cheek 11182</u> 10/2002; <u>Harvey 209</u> 10/2002; **Ndanan 2:** <u>Cheek 11114</u> 10/2002.
Note: uncommon over most of its range.
Local Name: Wodwog (Cheek 11114).
Uses: MEDICINES (Cheek 11114).

DICHAPETALACEAE

F. Breteler (WAG)

Fl. Cameroun 37 (2001).

Dichapetalum heudelotii (Planch. ex Oliv.) Baill. var. *heudelotii*
Syn. *Dichapetalum johnstonii* Engl.
Syn. *Dichapetalum subauriculatum* (Oliv.) Engl.
Syn. *Dichapetalum ferrugineum* Engl.
Climber, 12m; stem, petiole and midrib above shortly red brown hairy, midrib white hairy below; leaves narrowly elliptic-oblong, c. 17 × 6cm, acute, rounded, nerves c. 9 pairs; stipules caducous; inflorescence axillary, dense, subsessile c. 0.6 × 0.6cm; flowers white, 5mm. Forest; 640m.
Distr.: Guinea (Conakry) to Congo (Kinshasa) [Guineo-Congolian].
IUCN: LC
Ndanan 2: <u>Darbyshire 311</u> fl., 3/2004.

Dichapetalum lujae De Wild. & T.Durand var. *lujae*
Bull. Jard. Bot. Brux. 25: 349 (1955).
Liana; stems densely covered in short bronze-coloured hairs; leaves waxy, elliptic, axile, to 6 × 2.5cm, entire, abaxial surface densely covered in blue-white hairs, adaxial surface glabrous apart from bronze-coloured hairs along the midrib; stipules to 0.5cm, covered in bronze-coloured hairs; pedicel 0.5–1cm, with bronze hairs; flowers white, buds approx. 2–3mm across.

Forest; 679–710m.
Distr.: Cameroon, Gabon, Congo (Kinshasa), Angola [lower Guinea].
IUCN: LC
Ndanan 1: <u>Cheek 11764</u> 3/2004; <u>Cheek in MF 381</u> 10/2002; <u>Onana 2896</u> 3/2004.

Dichapetalum minutiflorum Engl. & Ruhland
Bot. Jahrb. Syst. 33(1): 88 (1902); Fl. Gabon 32: 129 (1991).
Liana, 7–13m, glabrous; leaves oblong-elliptic, 9–15 × 4–5cm, acumen 1cm, base attenuate, 6/7 nerves on each side of the midrib; petiole 2–6(–10)mm; stipules oblong-triangular, 206mm; pedicels c. 3mm; flower buds pale green; fruits subspherical. Evergreen forest; 700m.
Distr.: Cameroon, Gabon [lower Guinea].
IUCN: NT
Ndanan 2: <u>Cheek 11164</u> 10/2002.

DILLENIACEAE

F. Breteler (WAG), C. Couch (K) & M. Cheek (K)

Tetracera rosiflora Gilg
Bot. Jahrb. Syst. 33: 199 (1902).
Woody climber, to 6m; stem glabrous, pale brown; leaves alternate, astipulate, those of fruiting branches elliptic, 10.5 × 5cm, apex abruptly shortly acutely acuminate, base acute, margin shallowly, acutely serrate, lateral nerves 6–11 pairs, lower surface glossy pale brown, midrib brown with sparse appressed hairs, domatia absent; petiole 4–5mm; infructescences terminal on short branches; panicles 7–10 × 6cm, 4–10-fruited; fruits glossy pale brown, 2–3cm, mericarps dehiscing to reveal a glossy seed enveloped in a lacerate aril. Forest; 700–710m.
Distr.: Cameroon, Gabon, Congo (Kinshasa) [Congolian].
IUCN: LC
Ndanan 1: <u>Cheek 11701</u> 3/2004; <u>Harvey 130</u> 10/2002.
Note: Cheek 11701 is tentatively placed in this taxon, the fruits of Harvey 130 are a closer match, Couch ix.2005. Identification follows Kubitzki determinations 1969 replacing former determinations of *T. mayumbensis* Exell and *T. claessenii* De Wild., Cheek x.2010.

Tetracera sp. 1 of Mefou
Sapling, or liana, 40cm; stem hispid; leaves oblanceolate, 20 × 7cm, acumen 1cm, base acute, margin strongly dentate, teeth triangular, spine-tipped, lateral nerves 22 pairs, lower surface densely scabrid; petiole 2cm. Forest; 700m.
Ndanan 1: <u>Cheek in MF 324</u> 10/2002.
Note: plot voucher MF324 has not been matched but possibly a juvenile stage of a known species. Differing greatly from the other Mefou *Tetracera* in the scabrid leaves and hispid stems, Cheek x.2010.

EBENACEAE

Y.B. Harvey (K), H. Fortune-Hopkins (K), E. Fenton (K) & G. Gosline (K)

Fl. Cameroun 11 (1970).

Diospyros bipindensis Gürke

Shrub or tree, 3–20m; bark very hard, shattering when struck; branchlets frequently with ellipsoid swellings c. 2–3cm diam.; leaves glabrous, oblong 10–15(–26) × 4–6(–10)cm, acumen c. 2.5cm long, 5–8 pairs secondary nerves, anastomosing some distance from leaf margin; petioles 5–10(–15)mm; flowers axillary, solitary or in small panicles of 2–3; male flower: calyx c. 4mm, 4 united lobes, petals c. 2cm, united to ¹/₂; female flower: calyx c. 10mm, 4 triangular lobes, corolla 18–20mm with 4 reflexed lobes; fruits with accrescent 4-lobed calyx 2.5cm long, fruit as long as calyx, 7–8 seeded. Forest; 600–700m.
Distr.: Cameroon to Uganda [lower Guinea & Congolian].
IUCN: LC
Mefou Proposed National Park: Cheek in MF 28 3/2004; Etuge 5263 3/2004; **Ndanan 1:** Gosline 407 10/2002; Tadjouteu 600 3/2004.

Diospyros cinnabarina (Gürke) F.White

Syn. *Diospyros simulans* F.White
Tree, to 15m; bark smooth black or dark grey with superficial vertical striations; slash black outside, red and fibrous inside; leaves ovate-elliptic to lanceolate, 7–24 × 3–15cm, (sub-)acuminate, 5–7 pairs lateral nerves, lower surface distinctly pale grey; flowers in crowded cymes, 5–100-flowered, generally on old wood; male flowers 10 × 2mm, sharply conical; female flower 7.5mm with 3 lobes ¹/₃ length of the corolla; fruit globose to ovoid, red, 3–4 × 2–3cm, apex pointed, surface warty when dried, sometimes deeply grooved between the seeds. Forest; 700–710m.
Distr.: E Nigeria to Congo (Brazzaville) [lower Guinea].
IUCN: NT
Ndanan 1: Cheek 11267 10/2002; 11721 3/2004; Gosline 441 10/2002.
Note: uncommon over most of its range. Frank White placed *D. simulans* F. White in synonymy with *D. cinnabarina* in his later work. Senterre (2005) in a Checklist of the Ebenaceae of Equatorial Guinea argues that these are in fact two separate taxa, in which case these specimens become *D. simulans* again, Gosline ix.2010.

Diospyros conocarpa Gürke & K.Schum.

Tree, to 10m, trunk slender; bark black with thin papery scales, slash black outside, pale yellow and fibrous inside, branchlets with bristle-like hairs; leaves harbouring ants, lamina papery, to 25 × 8cm, acuminate, base cordate, bristle-like hairs on midrib beneath, 7–9 pairs lateral nerves; inflorescence axillary; male flowers in lax drooping 3–5-flowered racemes; female flowers solitary; calyx 1–2cm with 4 papery triangular lobes free almost to the base; corolla 2cm, tube 1cm, lobes 4, large; fruits to 3.5 × 1.5cm, ovoid-conical with few bristly hairs, fruiting calyx 1cm with 4 papery lobes. Forest; 700m.
Distr.: Nigeria to Congo (Kinshasa) [Guineo-Congolian].
IUCN: LC
Mefou Proposed National Park: Cheek in MF 9 3/2004; **Ndanan 1:** Cheek 11046 10/2002; Harvey 170 10/2002; Tadjouteu 584 3/2004.

Diospyros crassiflora Hiern

Tree, to 25m; trunk to 30cm dbh; bark dark grey, smooth, very shortly haired; slash black; leaves 11–18 × 5–9cm, ovate-elliptic, apex acuminate, base attenuate, glabrous, or with scattered hairs on the undersurface, particularly the midrib, 6 pairs of lateral veins; petiole (5–)7–13mm long, very shortly pubescent; inflorescence cymose, 1–3-flowered but only one reaching maturity; calyx to 7mm long, pubescent outer surface, tube to 4mm long, 5-lobed, lobes 3 × 3.5mm, ovate; corolla to 31mm long, velutinous outer surface, glabrous within, tube urceolate, to 28mm long, lobes to 3mm long; stamens to 12mm, velutinous; fruit [not seen at Mefou] obovoid, to 10 × 6.5cm. Forest; 679m.
Distr.: Nigeria, Cameroon, Gabon, Congo (Brazzaville), Congo (Kinshasa), Angola and CAR [Guineo-Congolian].
IUCN: EN
Ndanan 1: Onana 2910 3/2004.
Local Name: Mevini (Ewondo) (Onana pers. comm.).
Uses: FOOD. MATERIALS — Wood — timber.
Note: this local name is for the true ebony, but also used for the other species of *Diospyros* (Onana pers. comm.)

Diospyros ferrea (Willd.) Bakh.

Tree or shrub, to 4–5m; bark dark grey; slash black outside, red inside; branchlets with erect hairs; leaves obovate-elliptic, (2.5–)5–7(–8.5) × (1.5–)3–4cm, apex acuminate, base acute, glabrous except for few isolated hairs on undersurface, 5–6 pairs of nerves; petiole 2–3mm, erect hairs; flowers solitary; calyx 4–5mm long, erect hairs, tube to 4mm long, lobes 3, triangular, to 1mm long; corollas in bud, pubescent; fruit orange, 1.8 × 1cm, ellipsoid, glabrous. Secondary forest understorey; 700–710m.
Distr.: tropical Africa, Madagascar, India, Sri Lanka to Australia and Hawaii [pantropical].
IUCN: LC
Ndanan 1: Tadjouteu 570 3/2004; **Ndanan 2:** Cheek 11937 3/2004.

Diospyros gilletii De Wild.

Fl. Cameroun 11: 82 (1970).
Tree, to 9m; bark dark brown, shallowly fissured; branchlets with short, bristle like hairs, glabrous when mature; leaves ovate to broadly elliptic, 12–16.5 × 7–8.5cm, apex acute, base rounded, glabrous except for few isolated hairs at midrib of undersurface, 5–6(–9) pairs lateral nerves; petiole 7–10mm, glabrous; flowers

in crowded cymes, to 16-flowers; pedicel to 1.5mm, dark bristle like hairs; calyx tube to 1.5mm, lobes 5, ovate, to 1.5mm; male corolla 6-lobed, lobes oblong, to 6 × 3mm, tube to 1.5mm; filaments to 3mm long; anthers to 2mm; fruit not seen. Forest overhanging *Raphia* swamp; 640–710m.
Distr.: Cameroon, CAR, Congo (Brazzaville), Congo (Kinshasa) & Gabon [lower Guinea].
IUCN: LC
Ndanan 2: Cheek 11918 3/2004; Darbyshire 303 fl., 3/2004.

Diospyros gracilescens Gürke
Fl. Cameroun 11: 86 (1970); Bull. Jard. Bot. Nat. Belg. 48: 321 (1978); Keay R.W.J., Trees of Nigeria: 385 (1989).
Syn. *Diospyros nigerica* F.White
Tree, to 5m; bark dark brown, smooth, very shortly haired; leaves 5–11 × 2.5–4.5cm, ovate, apex acuminate, base acute to rounded, upper surface glabrous or with scattered hairs, lower surface with short hairs, 4–5 pairs of lateral veins; petioles 1.5–4mm long, short chestnut hairs; flowers solitary, pink in bud; pedicels 2mm, with fine chestnut hairs; calyx to 2mm long, cupular, 4-lobed, lobes triangular, fine chestnut hairs; flower bud to 7mm, narrowly urceolate, with fine chestnut hair; fruits not seen. Forest; 700–710m.
Distr.: Nigeria, Cameroon, Gabon, Congo (Brazzaville) and Congo (Kinshasa) [Guineo-Congolian].
IUCN: LC
Ndanan 1: Cheek 11581 3/2004; Gosline 442 10/2002.

Diospyros hoyleana F.White subsp. *hoyleana*
Tree, 8–15m tall, or shrub to 3m; bark almost smooth, grey-brown to black; slash black and brittle outside, reddish inside; leaves 3–4 × 2cm, obliquely lanceolate, emarginate, lateral veins scarcely visible; flowers axillary and on older twigs, on very short stalks; corolla 6mm long, sausage-shaped with 3 very short lobes; fruits red turning black, 2.5 × 1.5cm, ovoid-cylindric, calyx c. 2mm wide. Forest; 700–720m.
Distr.: SE Nigeria, Cameroon, Gabon, Congo (Brazzaville), Congo (Kinshasa), Angola and Zambia [Guineo-Congolian].
IUCN: LC
Ndanan 1: Cheek 11245 10/2002; Gosline 405 10/2002; Onana 2840 3/2004.

Diospyros kamerunensis Gürke
Tree, to 15m; stems with red-gold pubescence; leaves lanceolate, oblong or elliptic, 15–20 × 5–9cm, acumen to 2.5cm, acute, base rounded, young leaves densely red-gold pubescent, 4–6(–8) pairs lateral nerves, tertiary nerves indistinct; petiole 1–2cm, pubescent; male flowers in axillary panicles of 3–6(–20) flowers, pubescent; calyx 8mm, deeply lobed; corolla 18–20mm; female flowers solitary or few in axils, larger than male; fruit on thick pedicel c. 5mm, subglobose, 5 × 4cm, red-orange, glabrescent. Forest; 693–710m.

Distr.: Liberia to Ghana, Cameroon, Gabon [upper &lower Guinea].
IUCN: NT
Ndanan 1: Cheek 11665 3/2004; 11765 3/2004; Gosline 404 10/2002; Onana 2923 3/2004.
Note: the Mefou specimens have small leaves to c. 7 × 5cm. Fertile material is needed. A variety or subspecies may be appropriate, Gosline ix.2010.
Local Name: Bonj (Gosline 404).
Uses: FOOD — Infructescences — fruits eaten (Gosline 404).

Diospyros polystemon Gürke
Tree, 10–15m tall, or shrub to 5m; young branches and twigs puberulent; leaves 10–18 × 4–7cm elliptic to narrowly ovate, base decurrent, acumen triangular to 2cm long, margin recurved in dried specimens; petioles puberulent 5– 10mm long; inflorescences axillary cymes to 1cm long; male flowers with up to 120 excerted stamens; female flowers with pubescent pedicels to 5mm long; fruits subglobular to 2cm diam.; calyx 1.5cm diam. with 4–5 lobes. Forest and farmbush; 700–710m.
Distr.: Cameroon, Congo (Brazzaville), Congo (Kinshasa) and Angola [Guineo-Congolian].
IUCN: LC
Ndanan 1: Cheek 11906 3/2004; 11955 3/2004; Cheek in MF 241 10/2002; **Ndanan 2:** Gosline 419 10/2002.

Diospyros suaveolens Gürke
Tree, 8–30m; bark scaly, deep purplish-black; slash black outside, deep yellow turning orange inside; leaves lanceolate-elliptic, to 8(–14) × 2(–5)cm, acuminate, base rounded or subcordate, c. 7 pairs of lateral nerves, distinctly hairy; flowers in lax many-flowered cymes on branches and main stems, never axillary; calyx 4mm, hairy; corolla 10mm, glabrous, 5-lobed; fruits broadly ovoid-conical, 35 × 22mm, densely covered with short brown hairs and longer, yellow irritant hairs, calyx 10mm with 4–5 triangular lobes as long as the tube. Forest; 700–710m.
Distr.: Nigeria, Cameroon & Gabon [lower Guinea].
IUCN: NT
Mefou Proposed National Park: Cheek in MF 8 3/2004; 97 3/2004; **Ndanan 1:** Cheek 11490 3/2004; Cheek in MF 239 10/2002; Gosline 406 10/2002.
Note: plot vouchers MF8, MF97 & MF239 have tentatively been placed in this taxon, Hopkins iv.2010.

Diospyros zenkeri (Gürke) F.White
Tree, to 20m; bole black and smooth; slash black outside, pink with a yellow inner edge inside; leaves drying black, elliptic to ovate, to 10 × 5cm, acumen to 2.5cm, 4–5 pairs lateral veins looping far from edge; male flowers in clusters of 1–9 in leaf axils or on older twigs, c. 1cm long, cylindric; calyx with 3 short lobes; female flowers in clusters of 1–3, shorter and wider than males; fruits globose, yellow turning black, 2.5cm, glabrous, calyx not accrescent. Forest; 700m.

Distr.: Nigeria to Congo (Kinshasa) [lower Guinea & Congolian].
IUCN: LC
Mefou Proposed National Park: Cheek in MF 95 3/2004.

EUPHORBIACEAE

G. Challen (K), M. Cheek (K), C. Couch (K),
H. Fortune-Hopkins (K), F. Breteler (WAG),
E. Fenton (K), J.-M. Onana (YA), Seren Thomas (K)
& P. Hoffmann (K)

Acalypha paniculata Miq.
F.T.A. 6(1): 886 (1912).
Herb, 1–1.5m, puberulent; leaves ovate, c. 9 × 4.5cm, apex acuminate, base truncate-rounded, coarsely serrate; petiole c. 3–12cm; male inflorescence axillary, pendulous c. 12cm; female inflorescence a diffuse terminal panicle 18 × 9cm; fruit 2mm, stalked, glandular-hairy. Secondary forest near *Raphia* swamp; 600–710m.
Distr.: [tropical Asia and tropical Africa].
IUCN: LC
Mefou Proposed National Park: Etuge 5245 3/2004; **Ndanan 1:** Cheek 11671 3/2004; **Ndanan 2:** Cheek 11143 10/2002.
Local Name: Djena (Ewondo) (Cheek 11143).
Uses: MEDICINES (Cheek 11143).

Alchornea cordifolia (Schum. & Thonn.) Müll.Arg.
Shrub, to 8m; leaves ovate, c. 17 × 13cm, acute-acuminate, shallowly cordate, margin serrate, crenate, large nectar craters in axils of midrib and lower secondaries, minutely pubescent below; petiole c. 12cm; male inflorescence axillary, pendant, yellow, c. 20cm, highly branched; female unbranched; styles 2, red, to 2cm long; fruit 1cm diam., green. Roadsides & farmbush; 650–710m.
Distr.: widespread in tropical Africa [tropical Africa].
IUCN: LC
Mefou Proposed National Park: Cheek in RP 50 3/2004; Nana 44 3/2004; **Ndanan 2:** Darbyshire 208 fr., 3/2004.

Alchornea floribunda Müll.Arg.
Shrub; leaves often subclustered, oblanceolate, to 35 × 12cm, acumen short, base cuneate then abruptly truncate, nerves 10+ pairs; petiole 1cm; male inflorescence terminal, erect, red; female inflorescence with styles 3, 2cm, red. Farmbush, secondary forest; 600–710m.
Distr.: Senegal to Uganda [Guineo-Congolian].
IUCN: LC
Mefou Proposed National Park: Nana 63 3/2004; 69 3/2004; **Ndanan 1:** Cheek 11028 10/2002; 11738 3/2004; Cheek in MF 382 10/2002.
Local Name: Enguela (Nana 63).

Antidesma laciniatum Müll.Arg. var. *laciniatum*
Tree, 8–10m, puberulent; leaves obovate-oblong, 15(–30) × 8(–15)cm, subacuminate, base abruptly rounded; petiole 0.5cm; stipules ovate, 3–6-lobed, the lobes lanceolate; infructescence non-interrupted, pendulous racemose, c. 15cm; fruits ellipsoid, 7mm, red, fleshy. Forest; 700–720m.
Distr.: Ivory Coast, Nigeria to Congo (Kinshasa) [Guineo-Congolian].
IUCN: LC
Mefou Proposed National Park: Cheek in RP 36 3/2004; **Ndanan 1:** Cheek in MF 297 10/2002; Onana 2866 3/2004.

Antidesma vogelianum Müll.Arg.
Tree or shrub, 3–10m, pubescent; leaves elliptic, rarely slightly obovate, c. 15 × 5.5cm, acumen c. 2cm, base acute; petiole c. 5mm; stipule lanceolate, c. 6 × 2mm, entire; infructescence non-interrupted, pendulous, racemose, c. 15cm; fruits ellipsoid, 7mm, red, fleshy. Forest; 600–710m.
Distr.: Nigeria to Tanzania [lower Guinea & Congolian].
IUCN: LC
Mefou Proposed National Park: Etuge 5171 3/2004; **Ndanan 1:** Cheek 11634 3/2004.
Note: Cheek 11634 is tentatively placed in *Antidesma vogelianum*, Challen v.2010.

Bridelia atroviridis Müll.Arg.
Shrub, 8–12m, inconspicuously puberulent; leaves distichous, elliptic-oblong, rarely oblanceolate-elliptic, drying black above, c. 12 × 5.5cm, acuminate, base rounded, ends of secondary nerves not joining to a marginal nerve; fruit axillary, fleshy, ellipsoid, 7mm. Semi–deciduous forest; 700m.
Distr.: Sierra Leone to Zimbabwe [tropical Africa].
IUCN: LC
Ndanan 1: Gosline 401 10/2002.

Bridelia micrantha (Hochst.) Baill.
Shrub or tree, 2–14m; trunk spiny, glabrescent; leaves oblong-elliptic, 10–15 × 3–5cm, acuminate, drying pale green below (bluish green live), the secondary nerves fawn, joining to form a marginal nerve; inflorescence axillary glomerules; flowers greenish white, 3mm diam.; fruit axillary, sessile, fleshy, ellipsoid, red, 7mm. Forest edge; 658–710m.
Distr.: Senegal to S Africa [tropical Africa].
IUCN: LC
Mefou Proposed National Park: Cheek in RP 34 3/2004; Etuge 5290 3/2004.

Cleistanthus polystachyus Hook.f.
Small tree, 8m; lacking white exudate; stems persistent brown puberulent (extending to petiole and inflorescence), with nearly white streaks of cork; leaves alternate, simple; leaf-blade glabous, oblong or elliptic,

8–12.5 × 3–5.2cm including the 1–1.5cm acumen that arises abruptly from the rounded apex, base acute, sides concave, margin entire, thickened, lateral nerves 5–6 pairs, uniting in a looping marginal nerve, domatia absent, quaternary nerves reticulate, raised; petiole 5–6mm; stipules caducous ovate, toothed 1.5mm; racemes axillary, single, 2cm, c. 10-flowered; pedicels 7mm; flowers orange, 5mm diam.; female flowers at base of spike; ovary superior; stigmas 3, patent, laciniate, stout. Forest; 600m.
Distr.: tropical Africa [tropical Africa].
IUCN: LC
Mefou Proposed National Park: Etuge 5283 3/2004.

Cleistanthus sp. nov.
Tree, 20m; lacking white exudate; stems glabrous, grey, with linear white lenticels, drying with ridges; leaves alternate, simple; leaf-blade elliptic, c. 9 × 4.5cm, acumen 8mm, curved, base acute to rounded, margin revolute, lateral nerves 4 pairs, forming a weak and indistinct marginal nerve, domatia absent, quaternary nerves moderately conspicuous; petiole black, wrinkled, 7mm, sparsely puberulent; stipules triangular, 3mm, brown pubescent; spikes axillary, 1.5–3.5cm, densely white puberulent, c. 20-flowered; flowers in clusters subtended by ovate bracts 3 × 2mm; corollas yellow, 3mm wide. Evergreen forest; 710m.
Ndanan 1: Cheek 11587 3/2004.
Note: Cheek 11587 was determined as *Cleistanthus caudatus* by Breteler ix.2010, but Leonard, Fl. Congo Belge viii (1): 6 (1962) keys that as a shrub with petiole 2–4mm, leaves ovate, glaucous below, secondary nerves forming a 60° angle, buds etc. glabrous, Cheek x.2010.

Codiaeum variegatum (L.) A.Juss cv. '*Weismanii*'
Shrub, 5m; glabrous, no exudate; leaves leathery, narrowly oblong-ligulate, 30 × 4cm, midrib area and secondary nerves orange-red above, lower surface pink, apex subacuminate, base cuneate; petiole 4.5cm; inflorescence pinkish red, racemose, 25cm; pedicels 1.5cm; flowers white, 5mm. Farmbush & farms; 660m.
Distr.: native of New Guinea, cultivated widely in tropics [pantropical].
IUCN: LC
Ndanan 2: Darbyshire 333 fl., 4/2004.

Crotonogyne manniana Müll.Arg.
Shrub, 2–8m; stem and inflorescence indumentum fimbriate-lepidote; leaves oblanceolate-oblong, to 35 × 10cm, apex shortly cuspidate, base acute; petiole to 1.5cm; stipules ovate, to 7mm; racemes to 30cm, simple; male flowers on short spikelets, unbranched; calyx globose, 3-lobed; female flowers, few, solitary; calyx oblong, 5-lobed; petals 5; ovary lepidote; capsule lepidote. Forest; 700–720m.
Distr.: Ghana to Gabon [upper & lower Guinea].
IUCN: NT
Ndanan 1: Cheek 11565 3/2004; 11610 3/2004; 11693

3/2004; Onana 2845 3/2004; 2892 3/2004; Tadjouteu 572 3/2004; **Ndanan 2:** Cheek 11161 10/2002; 11162 10/2002; 11166 10/2002.
Note: after much deliberation and looking at types I think *C. zenkeri* Pax and *C. manniana* Mull.Arg. are one and the same. Have determined these all therefore as *C. manniana* (variations thereof) as this is older name. Onana 2892 has lepidote scales, glands on bracts, but no glands on female calyx though, Challen iii.2010.
Local Name: Abam (Ewondo) (Cheek 11161).
Note: Abam is the name for *Chrysophyllum* spp. and other species in different genera of Sapotaceae (see chapter by Onana and Mezili: Contribution de la langue ewondo à la connaissance des plantes du sud-Cameroun) (Onana xi.2010).

Drypetes angustifolia Pax & K.Hoffm.
Engl. Planzenr. Euphorb.–Phyllanthoid: 261 (1922).
Small tree, to 6m; leafy stems brown puberulous, becoming glabrous; leaf lamina 6.5–11.5 × 3–4.5cm, oblong to lanceolate, chartaceous, glabrous, margin serrulate, drying light green above, light brown/green beneath, apex acuminate, base rounded, often very assymetric, secondary veins 4–6; female flowers solitary in leaf axils; 4 glabrous sepals; ovary glabrous; 2 thickened bifid styles. Swamp forest, river bed; 600–650m.
Distr.: Cameroon, CAR, Equatorial Guinea, Gabon [lower Guinea & Congolian].
IUCN: LC
Mefou Proposed National Park: Etuge 5169 3/2004; **Ndanan 2:** Darbyshire 217 fl., 3/2004.

Drypetes bipindensis (Pax) Hutch.
F.T.A. 6(1): 679 (1912).
Shrub or tree, to 4m; branches slender, minutely puberulous to glabrous; leaf-blade 13.5 × 4.5cm, oblong to oblong-lanceolate, chartaceous, apex acuminate, base rounded, assymetric, secondary veins 6–7 pairs, looped within margin; petioles to 4mm; stipules lanceolate, to 5mm, caducous; male flowers almost sessile in fascicles in leaf axils; sepals 4, pubescent; stamens 7–8 inserted around margin of glabrous disk; female flowers in leaf axils and on older wood; sepals 4, puberulous; 2 flattened stigmas. Forest near *Raphia* swamp; 670–720m.
Distr.: Cameroon to Congo (Kinshasa) [Congolian].
IUCN: LC
Ndanan 1: Cheek 11513 3/2004; 11612 3/2004; 11959 3/2004; Cheek in MF 260 10/2002; Onana 2838 3/2004; **Ndanan 2:** Darbyshire 224 fl., 3/2004.
Note: Cheek 11612 and plot voucher MF260 have been tentatively placed in *D. bipindensis* by Challen x.2010.

Drypetes dinklagei (Pax) Hutch.
F.T.A.. 6(1): 683 (1912).
Shrub-like tree, to 8m; leafy stems minutely brown puberulent when young, becoming glabrous; leaves to 23 × 6.5cm, oblong to oblong-lanceolate, brown

puberulous, margin entire to subentire, drying grey-green on upper surface, brown on lower surface, apex acuminate to 1cm, base acute-rounded, assymetric; stipules ovate-lanceolate, puberulous, soon caducous, lateral nerves 6–7 pairs, slightly impressed above, prominent beneath; flowering in axillary fascicles and on older parts of branchlets from which leaves have fallen. Swampy soil; 700–710m.

Distr.: Cameroon [Cameroon endemic].

IUCN: NT

Mefou Proposed National Park: Cheek in MF 73 3/2004; **Ndanan 1:** Cheek 11719 3/2004; Tadjouteu 605 3/2004.

Note: plot voucher MF73 is tentatively determined as this taxon, Challen x.2010.

Drypetes floribunda (Müll.Arg.) Hutch.

Shrub to small tree, to 6m; leaves drying pale grey-green on upper surface; lamina 5–15 × 3–7.5cm, glabrous, coriaceous, ovate-elliptic to oblong, margin denticulate to spiny dentate, apex acuminate, base acute-rounded, assymetric, secondary veins 6–7 pairs, prominent on leaf underside; petioles to 6mm, thickened; pedicels to 1.2cm; male flowers orange-yellow in many-flowered fascicles on older wood; sepals 5, glabrous externally with ciliate margin; stamens 15–16, inserted in margins of disc; disc sparsely puberulous. Semi-deciduous forest, roadside, river edge; 600–720m.

Distr.: Senegal to Congo (Kinhasa) [Guineo-Congolian].

IUCN: LC

Mefou Proposed National Park: Etuge 5099 3/2003; Nana 27 3/2004; **Ndanan 1:** Cheek 11626 3/2004; 11717 3/2004; Harvey 169 10/2002; Onana 2877 3/2004; Tadjouteu 592 3/2004.

Note: Harvey 169, previously thought to be *Drypetes ugandensis* by Cheek, iv.2006, has subsequently been redetermined by Challen (x.2010) as *D.* cf. *floribunda*. Unlikely to be *D. ugandensis*, most likely to be *D. floribunda*, but female specimens determined as this at Kew have puberulous very young fruit. May be mature version with indumentum fallen off? (as have styles), Challen x.2010.

Erythrococca atrovirens (Pax) Prain

Ann. Bot. 25: 623 (1911).

Shrub or tree, to 4.5m, sparingly to densely yellow puberulent; spines absent; leaves drying green, ovate to elliptic-oblong, 8–13 × 3–7cm, long acuminate, cuneate to rounded, subentire, lateral nerves 4–6 pairs, pubescent below, densest along the nerves; petiole 1–1.5cm; male racemes 2–4cm, long-pedunculate, pubescent; pedicels capillary; female flowers with ovary bilobate; styles smooth; fruit green. Forest; 600–680m.

Distr.: Cameroon, Congo (Kinshasa) to Tanzania [lower Guinea & Congolian].

IUCN: LC

Mefou Proposed National Park: Etuge 5135 3/2004; **Ndanan 2:** Darbyshire 249 fl., 3/2004.

Erythrococca membranacea (Müll.Arg.) Prain

Shrub, 2–3(–5)m; unarmed, pilose, hairs yellowish brown, c. 2mm long; leaves oblong or elliptic, 8–12 × 3.5–5cm, long acuminate, obtuse, margin coarsely and irregularly serrate; inflorescence 6–8-flowered, subumbellate, 3cm; peduncle 2cm; pedicels 5mm; flowers green, 3mm wide; fruit green, bilobed, 1.5cm wide, seeds red. Forest; 679–710m.

Distr.: SE Nigeria & W Cameroon [lower Guinea].

IUCN: NT

Ndanan 1: Cheek 11869 3/2004; Onana 2901 3/2004; 2914 3/2004.

Erythrococca welwitschiana (Müll.Arg.) Prain

Shrub, 1m; branching; stems white, corky, glabrous; leaves alternate, elliptic or ovate-lanceolate, 4–15 × 2.5–8.2cm, apex attenuate or acuminate, base rounded or obtuse, faintly serrate, glabrous on both surfaces, lateral nerves 5–6 pairs; petiole 0.5–1cm; female inflorescence 1.7–3cm long; female flowers with bilobate ovary; male racemes c. 2.5–4cm; long pedunculate, sparsely hairy, peduncle for both male and female inflorescences 1–2cm; flowers green. Forest; 679–700m.

Distr.: Nigeria to Congo (Kinshasa) [lower Guinea & Congolian].

IUCN: LC

Ndanan 1: Cheek 11952 3/2004; Onana 2907 3/2004; **Ndanan 2:** Darbyshire 203 3/2004.

Note: Cheek 11952 & Onana 2907 are tentatively placed in this taxon, Challen x.2010.

Euphorbia cotinifolia L.

Sp. Pl.: 453 (1753).

Shrub, 5m; producing white exudate when cut; stem striate, black, glabrous; leaves purple, opposite, simple, broadly ovate to rounded triangular in outline, 3.5–6.3 × 2.5–5.7cm, glabrous on both surfaces, apex broadly acute, base rounded, margin entire, up to 11 pairs of lateral veins; petiole 3.7–5.8cm. Abandoned village; 710m.

Distr.: ornamental originating in S America.

IUCN: LC

Ndanan 1: Cheek 11885 3/2004.

Uses: ENVIRONMENTAL USES — Ornamentals. ENVIRONMENTAL USES — Boundaries (Cheek pers. comm).

Euphorbia hirta L.

Herb, 30cm, stems densely white puberulent, with long yellow hairs mixed, several stems per plant; leaves opposite, oblong, c. 2 × 1cm; inflorescences dense axillary subumbels c. 7mm diam.; peduncle 2mm; perianth absent; fruit 1mm diam. Farmbush; 700m.

Distr.: pantropical.

IUCN: LC

Ndanan 2: Darbyshire 204 3/2004.

Euphorbia teke Schweinf. ex Pax

F.T.E.A. Euphorbiaceae (pt 2): 481 (1988).

Small succulent spiny tree, 7m; white exudate when cut, glabrous; stems woody green, angled, 2–3cm diam.; leaves alternate, spiralled, dense, succulent-leathery, oblanceolate, 39 × 10cm, apex rounded, base cuneate-decurrent, entire, lateral nerves 3–4 pairs, obscure; petiole stout, 4 × 0.5cm; leaf scars corky, c. 1cm diam. with 2–3mm spine on each side. Swamp forest; 550m.
Distr.: Cameroon, CAR, NE Congo (Kinshasa), S Sudan, W Uganda, Tanzania [Congolian].
IUCN: NT
Mefou Proposed National Park: Etuge 5202 3/2004.
Note: newly recorded here from Cameroon. FTEA (1988): "Clay soil with swamp forest; 900–1200m". Cheek x.2010.

Euphorbia thymifolia L.
Prostrate herb; stem brown hairy; leaves simple, opposite, oblong to obovate, 7–8 × 4mm, rounded to obtuse at apex, base asymmetrical to obtuse, inconspicuously serrulate, pubescent on both surfaces (leaves on inflorescences smaller to 5 × 2mm); petiole to 1mm; inflorescence branches up to 1.5cm, to 13 flowers per inflorescence. Roadsides; 660m.
Distr.: pantropical.
IUCN: LC
Ndanan 2: Darbyshire 329 fr., 4/2004.
Note: a common introduced weed near habitation, Cheek x.2010.

Macaranga angolensis (Müll.Arg.) Müll.Arg.
Whitmore, The Genus Macaranga: 78 (2008).
Liana; stem spiny, leafy stems softly white tomentose, hairs patent 1mm; leaves alternate; leaf-blade 3-lobed, broader than long to 16 × 14.5cm, central lobe c. 7cm, acumen 5mm, base cordate, sinus 90°, upper surface with circular flat gland pair at base, margin shallowly dentate, teeth with nerve connected gland, 3-nerved from base, scalariform, lower surface with inconspicuous minute translucent glands and short white hairs; petiole c. 10 × 0.15cm; stipules triangular, 4mm; Riverine forest; 710m.
Distr.: E Cameroon to Congo (Kinshasa), Angola, Uganda [Congolian].
IUCN: LC
Ndanan 1: Cheek 11826 3/2004.

Macaranga monandra Müll.Arg.
Tree, 8–10m, trunk and stems (below leaves) spiny, rusty stellate, glabrescent; leaves elliptic, c. 13 × 7cm, acuminate, base rounded or obtuse, margin dentate, c. 6 teeth per side; petiole 3–6cm; inflorescence pendulous, c. 3cm; bractlets laciniate; fruit subspherical, 4mm. Secondary forest & farmbush; 700–710m.
Distr.: S Nigeria to Uganda [lower Guinea & Congolian].
IUCN: LC
Mefou Proposed National Park: Cheek in RP 61 3/2004; 76 3/2004; **Ndanan 1:** Cheek 11713 3/2004;

11716 3/2004; Cheek in MF 392 10/2002.
Note: plot voucher RP61 has tentatively been placed in this taxon (very young plant), Challen viii.2010.

Macaranga saccifera Pax
Whitmore, The Genus Macaranga: 231 (2008).
Shrub or climber; stem exudate red when cut, stems red brown, with thinly scattered hispid hairs; leaves alternate; leaf-blade 3-lobed, lobed for 3/4 the radius, c. 20 × 20cm, acumen 1.5cm, base cordate, sinus 3cm, lower surface with minute black glands, base of blade with cupular glands 3mm; petiole 3–16cm, densely white puberulent; stipules red, inflated, hooded, to 10 × 8mm, no ants seen; inflorescence panicles axillary, 3cm. Swamp and inundated forest; 700–710m.
Distr.: Cameroon, Gabon, CAR, Congo (Brazzaville), Congo (Kinshasa), Angola (Cabinda) [Congolian].
IUCN: LC
Ndanan 1: Cheek 11237 10/2002; 11622 3/2004; Tadjouteu 577 3/2004.

Macaranga cf. *schweinfurthii* Pax
Tree, 10–15m; bole spiny; stems crisped and straight hairy, hairs white, 1–2mm; leaf-blade semi-orbicular, to 13 × 18cm, wider than long, apex rounded, base truncate, dentate, teeth 5mm, terminating lateral nerves, base of blade with glands flanking petiole, lateral nerves 4–5 pairs, scalarifom, lower surface with dense black glands and scattered white hairs; petiole 3–16mm; stipules ovate, 6mm, persistent, hairy along midrib only. Riverine forest; 700m.
Ndanan 1: Cheek 11994 3/2004.
Note: Cheek 11994 is possibly a juvenile-leaved variant (leaves not lobed). R Smith sunk *angolensis* thus in FTEA. Cheek i.2006. The specimen has large glands on auricles, persistent stipules etc. Challen viii.2010. Cheek 11994 differing from *M. schweinfurthii* in the entire, not lobed leaves, Cheek x.2010.

Macaranga spinosa Müll.Arg.
Tree, 8m; spiny from trunk to leaf branches (spines c. 1cm), densely pilose-pubescent (hairs 1–2mm); leaves oblong, c. 9 × 5cm, acuminate, base truncate and abruptly cordate, margin sinuous-lobed, c. 4 shallow lobes on each side; petiole 3–5cm; fertile material not seen. Forest; 700–710m.
Distr.: Liberia to Angola [upper & lower Guinea].
IUCN: LC
Mefou Proposed National Park: Cheek in RP 80 3/2004; **Ndanan 1:** Cheek 11700 3/2004; Cheek in MF 320 10/2002.
Note: Cheek 11700 is atypical in the toothed leaves and in this resembling *M. monandra*. It has granular glands black, more likely *M. spinosa* whereas *M. monandra* has resinous glands. Plot voucher RP61 is tentatively placed in this taxon, the leaves are peltate, it is a very young seedling. Challen viii.2010.

Maesobotrya barteri (Baill.) Hutch.

Shrub or treelet, to 3+m, dioecious; twigs pubescent, without latex; leaf-blades elliptic, 5–22 × 2.5–9cm, cuneate to rounded at base, acuminate-cuspidate at apex, margin minutely toothed, with tufts of hairs at ends of veins, both surfaces glabrous except on venation; petioles 1–6cm, pubescent, with pulvinus at either end; stipules small, linear, caducous; male racemes short, mostly on twigs; female racemes 5+cm long in flower, to 10cm in fruit, in dense clusters arising from woody protruberances on trunk; buds red; flowers on short pedicels, green fide Darbyshire 272; fruits ellipsoid, c. 1.2cm long, fleshy, red, glabrous (or sometimes pubescent). Forest; 660–720m.

Distr.: Sierra Leone to Cameroon and Rio Muni [Guineo-Congolian].

IUCN: LC

Mefou Proposed National Park: Cheek in MF 68 3/2004; **Ndanan 1:** Cheek in MF 384 10/2002; Onana 2874 3/2004; **Ndanan 2:** Darbyshire 272 fl., 3/2004. Note: plot voucher MF68 is tentatively placed in *Maesobotrya barteri*, leaves subentire, glabrous on upper surface except for midveins, puberulous beneath. Challen viii.2010.

Maesobotrya klaineana (Pierre) J.Léonard

Fl. Congo B., Rw. & Bur. 8(2): 50 (1995).

Syn. *Maesobotrya dusenii* (Pax) Hutch.

Shrub or small tree, to 4+m; resembling *M. barteri* except leaf-blades 19–27 × 9–11cm, upper surface glabrous, lower surface pubescent on midrib and veins; stipules foliaceous, falcate, 1–2cm long, persistent; male racemes cauliflorous or on twigs; female racemes in dense clusters on trunk or branches; flowers red; fruits glabrous. Evergreen forest; 700m.

Distr.: SE Nigeria to Gabon and Congo (Kinshasa) [lower Guinea & Congolian].

IUCN: LC

Ndanan 1: Cheek 11255 10/2002; Cheek in MF 394 10/2002.

Mallotus oppositifolius (Geiseler) Müll.Arg.

Shrub, 2–10m; puberulent; leaves opposite; leaf-blades ovate to c. 17 × 8cm, apex attenuate, base rounded, obscurely toothed, lower surface with minute red surface glands; inflorescence axillary, erect, spike-like, c. 14cm; flowers 3mm diam., yellow; stamens numerous; fruit 8mm diam., glabrous. Forest edge; 600–700m.

Distr.: tropical Africa & Madagascar [tropical Africa].

IUCN: LC

Mefou Proposed National Park: Etuge 5148 3/2004; **Ndanan 1:** Harvey 133 10/2002; 212 10/2002; 94 10/2002; **Ndanan 2:** Cheek 11088 10/2002.

Manniophyton fulvum Müll.Arg.

Climber; stem and midrib scabrid-spiny; leaves drying dark brown above, brownish white below, ovate, c. 14 × 9cm, acuminate, cordate, margin entire, basal glands dimorphic, 2 ridged, 2 cylindric; petiole 6cm, sparsely pilose. male inflorescence to 30cm, spike-like, flowers clustered; flowers 4mm; capsule 3-lobed, 2.5cm. Forest; 640–710m.

Distr.: Sierra Leone to Congo (Kinshasa) [Guineo-Congolian].

IUCN: LC

Mefou Proposed National Park: Cheek 13110 2/2006; 13123 2/2006; Cheek in MF 80 3/2004; **Ndanan 1:** Cheek 11035 10/2002; 11228 10/2002; 11545 3/2004; Harvey 129 10/2002; **Ndanan 2:** Darbyshire 308 fl., 3/2004.

Local Name: Ecocolot (Cheek 11228).

Margaritaria discoidea (Baill.) G.L.Webster var. ***discoidea***

J. Arnold. Arb. 48: 311 (1967); Meded. Land. Wag. 82(3): 145 (1982); F.T.E.A. Euphorbiaceae (2): 63 (1988); Keay R.W.J., Trees of Nigeria: 167 (1989); Hawthorne W., F.G.F.T. Ghana: 85 (1990); Ann. Missouri Bot. Gard. 77: 217 (1990).

Syn. *Phyllanthus discoideus* (Baill.) Müll.Arg.

Deciduous tree, 10–25m, glabrous; leaves ovate–elliptic, 7–9 × 3.5cm, subacuminate, acute to obtuse, bluish white below, drying green, secondary nerves 10 pairs, arcuate, quaternary nerves reticulate, conspicuous; petiole 8mm; fruit 3–lobed, glossy, 7mm. Farmbush & forest edge; 700m.

Distr.: widespread in tropical Africa [tropical Africa].

IUCN: LC

Ndanan 1: Harvey 141 10/2002.

***Necepsia* sp. 1 of Mefou**

Shrub, to 4.5m; stems pale; leaves alternate to subopposite or clustered at the apex of short shoots; leaf-blade 4.5–14 × 1.8–4.5cm, obovate-elliptic, strongly dentate, subcoriaceous, glabrous, with sparse pubescence on the main vein of the underside, apex mucronate to 2cm, base rounded to abruptly obtuse, lateral veins 4–6 pairs, prominent beneath and tinged red, nerves looping within margin, tertiary venation not prominent, disc-shaped glands present at the base of the upper surface and sparingly scattered over both surfaces; petiole 2–6mm; inflorescence spicate, axillary, immature inflorescences to 10cm, rachis puberulous; flowers mainly in tight buds, but consisting of clusters of flowers subtended by several triangular puberulous bracts to 0.6 × 0.6mm; petals 0; sepals 3, 2 × 1.5mm, glabrous on inner surface, subglabrous on outer; stamens >50m, in a compact ball; filaments free, thecae pendulous; female flowers not seen. *Raphia* swamp; 550m.

Mefou Proposed National Park: Etuge 5189 fl., 3/2004. Note: Etuge 5189 most closely resembles the genus *Necepsia*, however it differs from the generic description in having subglabrous sepals externally and lacking prominent tertiary parallel venation. Further material may be required to confirm the generic identity, Challen x.2010.

Phyllanthus kidna Challen & Petra Hoffm.
Syst. Bot. (in press) (2011).
Tree, 10–15m; monoecious or dioecious, most likely
deciduous; trunk cylindrical, dbh 15–20cm diam.,
dilated at ground level; orthotropic branchlets
(resembling pinnate leaves) terete, simple, with up to 29
leaves per shoot; leaves distichous, glabrous, papery,
green tinged red when young; leaf-blades ovate to
oblong (-orbicular), 0.9–2.3 × 0.5–1.3cm, entire, basally
rounded to obtuse, apically rounded to
obtuse(mucronate), drying discolorous, dull olive
adaxially, pale glaucous green abaxially, veins (5)7–9
pair; petioles 1.0–1.3 × 0.3mm; stipules triangular,
0.5mm; flowers solitary or fasciculate, the fascicles in
the axils of lower leaves of brachyblasts, consisting of 1
pistillate and many staminate flowers, also some solitary
pistillate flowers in the axils of leafy plagiotropic
shoots; staminate flower pedicels 1.3mm long, capillary;
sepals 4, 0.7–1.3 × 0.5–0.8mm, equal, suborbicular,
erect, apically acute, glabrous, white, occasionally
tinged red, margins hyaline; petals 0; mature fruit 2
locular, with 1–2 ovules per locule, indehiscent, 1.3 ×
1.1–1.3cm, subglobose with a fleshy pericarp, drying
mid-brown; pedicels 4.5mm; seeds irregularly ovoid, 0.6
× 0.8cm. Secondary forest; 605–710m.
Distr.: Cameroon endemic.
IUCN: CR
Mefou Proposed National Park: Cheek 13098 2/2006;
13102 2/2006; **Ndanan 1:** Cheek 11531 3/2004;
Harvey 73 10/2002; Tadjouteu 611 3/2004.
Local Name: Kidna (Bety).
Uses: MEDICINES (Victor Ngueben, pers. comm. to
Cheek, 2006).

Phyllanthus muellerianus (Kuntze) Exell
Climber, 8m; stem spines 3-pronged, ultimate branches
c. 15cm, glabrous; leaves ovate, 6 × 2.8cm, acuminate,
truncate, bluish-green below; petiole 3mm;
inflorescences multiflowered, c. 5cm, in fascicles, in
axils of fallen leaves; fruit red, berry, 2.5mm; pedicel
3mm. Farmbush; 600–700m.
Distr.: tropical Africa.
IUCN: LC
Mefou Proposed National Park: Etuge 5107 3/2003;
Ndanan 1: Cheek in MF 409 10/2002; **Ndanan 2:**
Darbyshire 275 fr., 3/2004.

Phyllanthus odontadenius Müll.Arg.
Syn. *Phyllanthus braunii* Pax Bot. Jahrb. Syst. xv: 525
(1893).
Herb to small shrub, 0.1–0.6(–2)m, monoecious,
occasionally dioecious; with reddish stems; branching
phyllanthoid with lateral (plagiotropic) shoots up to
17cm, flattened and narrowly 2-winged; cataphylls
2.5–3mm; leaves of lateral shoots distichous; leaf-
blades oblong, (5–)10–12(–30) × (2–)5–9(–15)mm, red
tinged when young, apex subacute or rounded,
mucronulate, base rounded or truncate, lateral nerves

6–8 pairs; petioles 0.5–1mm; male flowers axillary in
pairs or clusters; pedicels 0.5mm; sepals 6, whitish or
colourless with a central greenish or reddish stripe, disc
glands 6, free; stamens 3, connate into a column;
female flowers with pedicels 1mm; sepals 6, greenish
with darker midrib; disc 6-lobed; ovary smooth with
shortly bifid styles; fruit a dehiscent trilobate
subglobose capsule. Farmbush, roadside; 650–710m.
Distr.: Guinea (Bissau) to Angola [Guineo-Congolian].
IUCN: LC
Ndanan 1: Cheek 11803 3/2004; **Ndanan 2:**
Darbyshire 207 fr., 3/2004.

Pseudagrostistachys africana (Müll.Arg.) Pax &
K.Hoffm. subsp. ***africana***
Tree, 5–20m tall; bark whitish with green spots, slash
brown to red; leaves leathery, elliptic, 30 × 11–18cm,
subacuminate, base rounded then slightly decurrent,
with a pair of flat glands, ± serrate, nerves 21–24 pairs,
venation scalariform; petiole 2.5–5cm; stipule single,
3cm, caducous, scar completely encircling stem.
Forest; 710m.
Distr.: Ghana, Bioko, São Tomé, Cameroon [upper &
lower Guinea].
IUCN: VU
Ndanan 1: Cheek 11874 3/2004.

Ricinodendron heudelotii (Baill.) Pierre ex
Heckel subsp. ***africanum*** (Müll.Arg.) J.Léonard
var. ***africanum***
Deciduous tree, 30m; trunk greyish white, glabrous except
in bud, brown stellate-hairy; leaves alternate, palmately
compound, seedlings lobed, leaflets 5, subelliptic, c. 24
× 10cm, acute at base and apex, denticulate; petiole c.
20cm, gland pair at apex; stipules foliaceous, 2 × 3cm,
deeply dentate, caducous; inflorescence diffuse, c. 45 ×
45cm, paniculate, with new leaves; flowers 2mm wide,
greenish-white; fruit subfleshy, indehiscent, 2-lobed,
grey-green, c. 3 × 3cm; seed yellow, 0.8cm diam. Old
secondary forest; 600–720m.
Distr.: Guinea (Bissau) to Tanzania [Guineo-Congolian].
IUCN: LC
Mefou Proposed National Park: Etuge 5226 3/2004;
Ndanan 1: Onana 2882 3/2004; **Ndanan 2:** Cheek
11121 10/2002.
Local Name: Ezezan (Onana 2882); Njansang (Etuge
5226).
Uses: FOOD — Seeds — seeds edible (Onana 2882).

Sclerocroton cornutus (Pax) Kruijt & Roebers
Biblioth. Bot. 146: 20 (1996).
Syn. *Sapium cornutum* Pax
Shrub, 1–3m, monoecious; twigs glabrous, with white
latex; leaf-blades oblong-elliptic, 10–19 × 4–7.5cm,
rounded to truncate at base, acuminate at apex, margins
toothed, both surfaces glabrous, glossy above, 1–2 large
glands near midrib at base of blade on lower surface,
other glands scattered in blade, and/or around margin

2mm from edge; petiole 0.5–1.5cm; stipules fugaceous; male flowers in clusters on pendulous spikes c. 10cm long; each flower 1mm, yellow-green; female flowers single, at base of spike, apetalous, 4mm diam., green; on pedicel 1.5cm; fruits 3-lobed, 2cm diam., dehiscing into 6 valves, each with a projecting triangular horn 2–6mm long. Farmbush; 640m.

Distr.: Sierra Leone, Liberia, Equatorial Guinea, Cameroon, Congo (Kinshasa), Congo (Brazzaville), Angola [lower Guinea & Congolian].

IUCN: LC

Ndanan 2: <u>Darbyshire 317</u> fl., 3/2004.

Sibangea arborescens Oliv.

Hooker's Icon. Pl. 15: t. 1411 (1883).

Syn. *Drypetes arborescens* (Oliv.) Hutch. F.T.A. 6(1): 680 (1912).

Shrub or small tree, to 7m; branchlets glabrous, ridged, lenticellar; leaves glabrous, drying grey-brown on upper surface; lamina 9–15.5 × 2.5–5cm, oblong to oblong-elliptic, apex acuminate, base sharply acute, occasionally asymmetric, secondary veins 5–6 pairs; petiole to 1cm, dark brown; flowers in leaf axils in many flowered fascicles on older wood; male flowers with pedicels 2–3mm; sepals 5, lanceolate, outer surface puberulous, inner surface glabrous; stamens 3–4. Semi-deciduous forest; 600–710m.

Distr.: Cameroon & Gabon [Congolian].

IUCN: NT

Mefou Proposed National Park: <u>Nana 70</u> 3/2004; **Ndanan 1:** <u>Cheek 11595</u> 3/2004; <u>11632A</u> 3/2004; <u>11742</u> 3/2004; <u>11847</u> 3/2004.

Tragia tenuifolia Benth.

Herbaceous climber, 1m, monoecious (or sometimes dioecious?); lacking latex; ± densely pale pubescent throughout; leaf-blades elliptic-cordate, to 7 × 3.5cm, acuminate at apex, 3(–5)-veined from base, margin sinuous-toothed; petioles to 5cm; stipules ovate, 3mm long; inflorescences to 2cm; female flowers, sepals 6, pinnatifid with apical lobe spathulate, accrescent in fruit, hairs sometimes irritant; fruits deeply 3-lobed, 5 × 9mm, dehiscent; pedicel 1–1.5cm. Semi-deciduous forest; 710m.

Distr.: Guinea (Conakry) and Sierra Leone to Cameroon, São Tomé, Gabon, Congo (Kinshasa), E to Sudan and Uganda, S to Angola (Cabinda) and Zimbabwe [Guineo-Congolian].

IUCN: LC

Ndanan 1: <u>Cheek 11865</u> 3/2004.

Local Name: Sas (Ewondo) (Onana pers. comm).

Tragia volubilis L.

Herbaceous climber, 2m, monoecious; lacking latex, most parts sparingly pubescent-hirsute, some hairs irritant; leaf-blades oblong-triangular, to 5 × 2cm, rounded to cordate at base, acute at apex, 3(–5)-veined from base, margins serrate; petiole 0.5+cm; stipules

lanceolate, 4mm long; racemes c. 2cm long, with numerous persistent bracts; female flower(s) towards base; sepals 6, lanceolate, entire, not accrescent in fruit; fruits dimorphic, either 1-lobed or usually deeply 3-lobed, 3 × 6mm; epicarp densely white-hirsute, dehiscent; pedicel 2–5cm long. *Raphia* swamp; 710m.

Distr.: Sierra Leone, Ivory Coast, Nigeria, Cameroon to Congo (Kinshsasa), E to Sudan and Uganda, S to Angola [tropical America & tropical Africa].

IUCN: LC

Ndanan 1: <u>Cheek 11783</u> 3/2004.

Local Name: as *Tragia tenuifolia* (Onana pers. comm.).

Uapaca acuminata (Hutch.) Pax & K.Hoffm.

Tree, to 20m; with conspicuous arching stilt roots, dioecious; twigs with obvious leaf-scars; lacking latex; leaf-blades obovate, 8–15 × 3.8–6.5cm, cuneate at base, acuminate at apex, glabrous on both surfaces, margin entire or slightly sinuate, lateral nerves 6–8 pairs; petioles 1–3cm, with pulvinus at either end; stipules subulate, fugaceous; inflorescences solitary in axils of leaf-scars; male inflorescences many flowered, the flowers in dense globose capitula, surrounded by yellow glabrous tepaloid bracts, c. 1.2cm diam., honey-scented; peduncle 0.5–1cm long; female inflorescences 1-flowered; peduncle 1–1.5cm in fruit; fruits drupaceous, ellipsoid, c. 2 × 1.7cm, glabrous; 3-seeded. Forest and secondary forest merging into *Raphia* swamp; 600–710m.

Distr.: SE Nigeria, Cameroon and Angola (Cabinda) [lower Guinea].

IUCN: LC

Mefou Proposed National Park: <u>Etuge 5278</u> 3/2004; **Ndanan 1:** <u>Cheek 11069</u> 10/2002; <u>11602</u> 3/2004; <u>Onana 2198</u> 10/2002.

Note: *Etuge* 5278 is tentatively placed in this taxon, it has 6–8 indistinct veins and is glabrous, Challen vi.2010.

Uapaca staudtii Pax

Tree similar to *U. acuminata*; leaf-blades obovate or oblanceolate, 24–30 × 9.5–13.5cm, cuneate or rounded at base, shortly acuminate at apex, glabrous on both surfaces, margin sinuate, lateral nerves 9–11 pairs; petioles 2–4cm; stipules foliaceous, persistent or not; inflorescences as in *U. acuminata*; fruits subglobose, 2cm diam., warty, puberulous. Evergreen forest; 700m.

Distr.: S Nigeria to Gabon [lower Guinea].

IUCN: NT

Ndanan 1: <u>Cheek 11990</u> 3/2004; <u>12001</u> 3/2004.

Uapaca cf. *vanhouttei* De Wild.

Tree similar to *U. acuminata* except young stems, petioles and midribs pilose, hairs to 2mm long; lower surface of blade sparsely hirsute, lateral nerves 9–11 pairs; fruits ellipsoid, 2 × 1.5cm, glabrous. Forest; 720m.

Ndanan 1: <u>Onana 2878</u> 3/2004.

FLACOURTIACEAE

B.J. Pollard & M. Cheek (K)

Fl. Cameroun 34 (1995).

Homalium africanum (Hook.f.) Benth.
Syn. Homalium molle Stapf
Syn. Homalium sarcopetalum Pierre
Tree, 4.5–8m; leaves papery, drying grey-brown above, green below, long-elliptic to oblong, to 20 × 7cm, acuminate, rounded-obtuse, entire to obscurely coarsely serrate, lateral nerves 7 pairs, acute, glabrous; petiole 0.5cm; stipules falcate, to 3cm, persistent; inflorescence terminal, c. 20 × 15cm, with 5–15 branches; petals in fruit c. 2mm. Riverbank in forest; 700–710m.
Distr.: Sierra Leone to Congo (Kinshasa) [Guineo-Congolian].
IUCN: LC
Ndanan 1: Cheek 11524 3/2004; Tadjouteu 571 3/2004.

Oncoba dentata Oliv.
Keay R.W.J., Trees of Nigeria: 62 (1989); Fl. Gabon 34: 52 (1995); Adansonia (Sér. 3) 19: 257 (1997).
Syn. Lindackeria dentata (Oliv.) Gilg Hawthorne W., F.G.F.T. Ghana: 131 (1990).
Tree or shrub, to 15m; leaves oblong-elliptic or ovate, 15–22 × 7–12cm, shortly acuminate, dentate to subentire, lateral nerves 6–10 pairs; petioles 3–16cm; inflorescence 5–10cm; flowers in fascicles of 1–5; sepals 3, 5mm; petals 6–10, 5–10mm; fruit orange, indehiscent, with long bristles, c. 2cm diam. Forest; 600m.
Distr.: Guinea (Conakry) to Sudan [Guineo-Congolian].
IUCN: LC
Mefou Proposed National Park: Etuge 5227 3/2004.

Oncoba flagelliflora (Mildbr.) Hul
Fl. Gabon 34: 60 (1995).
Tree, 4–10m, glabrous; base of trunk with radiating whip-like inflorescences in leaf litter; leaves oblong-elliptic, 25 × 8.5cm, acumen 1cm, acute-obtuse, 5–8 nerves on each side of the midrib, entire; petiole 1.5–2cm; inflorescences 3(–10)cm, internodes 2–4cm; flowers c. 1.5cm; sepals 3; petals white; 5–7; stamens yellow, numerous; fruit subglobular to ellipsoid, to 4.5 × 3.5cm, rugose; Forest, farmbush; 700–720m.
Distr.: Cameroon, Gabon, Congo (Brazzaville), Congo (Kinshasa) [lower Guinea].
IUCN: LC
Ndanan 1: Cheek 11265 10/2002; Gosline 427 10/2002; Onana 2827 3/2004.
Note: very distinctive in the prostrate whip-like inflorescences, from which the flowers arise, appearing as though from the ground. Over 20 specimens are known from Gabon alone (Hul, Fl. Cameroun 1995). Sometimes treated as a *Caloncoba*.

Oncoba welwitschii Oliv.
Keay R.W.J., Trees of Nigeria: 61 (1989); Fl. Cameroun 34: 46 (1995); Adansonia (Sér. 3) 19: 261 (1997).

Syn. Caloncoba welwitschii (Oliv.) Gilg
Tree or shrub, 4–7m; leaves papery, ovate to lanceolate, 18–28 × 7–15cm, long acuminate, obtuse; petiole to 20cm; inflorescence axillary below leaves, or ramiflorous; flowers 1–2; pedicels c. 2cm; flowers 5–10cm diam.; sepals 3, 1.5cm; petals c. 10, 3cm; fruit globose, 4cm diam., woody; spines c. 0.6cm. Forest and forest gaps; 600–700m.
Distr.: SE Nigeria, Cameroon, Gabon, Congo (Brazzaville), Congo (Kinshasa) and Angola [Guineo-Congolian].
IUCN: LC
Mefou Proposed National Park: Etuge 5179 3/2004; **Ndanan 1:** Cheek in MF 244 10/2002; Darbyshire 202 fl., fr., 3/2004.

GUTTIFERAE

M. Cheek (K) & B.J. Pollard

Allanblackia floribunda Oliv.
Tree, 15–20m; 5 fluted to 3m; leaves oblong-elliptic or ovate-elliptic, c. 15 × 4.5cm, glossy, midrib concolorous; flowers pinkish red, petal margins folding to give triangular apex; staminal phalanges with apical flange, anthers only visible on upper surface; central disc deeply honeycombed; fruit cylindrical 25–50cm. Evergreen forest; 700m.
Distr.: Nigeria to Congo (Kinshasa) [lower Guinea & Congolian].
IUCN: LC
Mefou Proposed National Park: Cheek in MF 49 3/2004; **Ndanan 1:** Onana 2833 3/2004; **Ndanan 2:** Cheek 11159 10/2002.
Note: plot voucher MF49 is probably this taxon, Cheek x.2010.
Local Name: Nsangomo (Ewondo) (Cheek 11159, Onana 2833).

Garcinia mannii Oliv.
Tree, c. 30m, bole cylindrical; leaves drying matt yellowish green below, elliptic-oblong c. 13 × 6cm, obtuse to rounded, acumen slender, 2cm, base acute, c. 40 pairs secondary nerves; petiole 8mm; inflorescence subsessile, 3-flowered or terminal on spur shoots; flowers 1.8cm diam.; petals bright glossy deep yellow or red; androecium annular, surrounding ovary. Forest; 640–720m.
Distr.: Nigeria to Gabon [lower Guinea].
IUCN: NT
Mefou Proposed National Park: Cheek in MF 51 3/2004; **Ndanan 1:** Cheek 11239 10/2002; 11593 3/2004; 11659 3/2004; Onana 2824 3/2004; Tadjouteu 589 3/2004; **Ndanan 2:** Darbyshire 291 fl., fr., 3/2004.

Garcinia ovalifolia Oliv.
Tree, to 7m; producing yellow exudate when cut; leafy stems grey, ridged when dry, 1.5–2mm diam.; leaves opposite, simple, estipulate; leaf-blade leathery, elliptic,

to 6.5 × 2.2cm, acumen 4mm, base obtuse-acute, margin revolute, lower surface dull grey-white, lateral nerves c. 20 pairs, bold, with transverse black thread-like resin canals; petioles wrinkled, 6–7mm, with a 2mm cup at base, on upper surface; flowers axillary, 1–2; pedicel 3mm; flower 4mm. Swamp forest; 600m.
Distr.: Guinea (Conakry) to Congo (Kinshasa), Angola and Ethiopia [Guineo-Congolian].
IUCN: LC
Mefou Proposed National Park: Etuge 5174 3/2004.

Garcinia smeathmannii (Planch. & Triana) Oliv.
Syn. *Garcinia polyantha* Oliv.
Tree, 4–15m; leaves thickly leathery, drying pale brown below, narrowly oblong-elliptic, c. 20 × 8cm, obtuse, subacuminate, base obtuse, secondary nerves c. 20 pairs; petiole c. 1.5cm; inflorescence sessile, axillary, on leafy stems, umbellate-fascicled, 15–20-flowered; pedicels c. 15mm; flowers white, 1cm diam.; anthers with free filaments inserted on ligules, as long as petals; fruits globose, 2cm; stigmas 2; pedicels 4cm. Forest; 600–710m.
Distr.: Guinea (Bissau) to Zambia [Guineo-Congolian].
IUCN: LC
Mefou Proposed National Park: Etuge 5272 3/2004; **Ndanan 1:** Cheek 11660 3/2004.

Harungana madagascariensis Lam. ex Poir.
Shrub or small tree, 3–6m, glabrescent; leaves ovate, c. 12 × 5.5cm, acuminate, base rounded, nerves c. 10 pairs; petiole 2cm, producing bright orange exudate when broken; inflorescence a dense terminal panicle, 7–15cm; flowers white, 2mm, petals hairy; berries orange, 3mm. Farmbush; 700–710m.
Distr.: tropical Africa & Madagascar [tropical Africa].
IUCN: LC
Mefou Proposed National Park: Cheek in RP 92 3/2004; **Ndanan 2:** Darbyshire 229 fr., 3/2004.

Mammea africana Sabine
Tree, to 30m; exudate yellow; glabrous; leaves opposite, simple, elliptic to long-elliptic, 10–18 × 4–6cm, sapling leaves 35 × 16cm, subacuminate, acute-obtuse, lateral nerves 15–20 pairs, each separated by 3 inter-secondaries, tertiary nerves visible to naked eye, on lower surface, reticulate, each areole with a raised swelling which in transmitted light appears as a translucent dot, resin canals absent; petiole 1.5cm, base lacking cup; flowers borne just below the leaves, single, axillary, 1–1.5cm diam.; pedicel 1cm; fruit subglobose, 6cm; seeds 2–4. Forest; 600m.
Distr.: Sierra Leone to Congo (Kinshasa), Angola and Uganda [Guineo-Congolian].
IUCN: LC
Mefou Proposed National Park: Etuge 5288 3/2004.

Symphonia globulifera L.f.
Tree, to 30m, glabrous; leaves narrowly elliptic to oblong-elliptic, c. 10.5 × 2.8cm, long acuminate, acute,

lateral nerves c. 30 pairs, resin canals inconspicuous; petiole 0.7cm; inflorescence terminal, c. 10-flowered; pedicels 1.5cm; flowers red, globose, c. 1cm diam.; styles ascending, aculeate, 5; fruit ovoid, 2.5cm, lenticellate; styles persistent. Forest; 700m.
Distr.: tropical America, Africa & Madagascar [amphi-Atlantic].
IUCN: LC
Mefou Proposed National Park: Cheek in MF 41 3/2004; **Ndanan 1:** Cheek in MF 279 10/2002.
Note: sterile plot vouchers MF41 & MF279 are probably this taxon, Cheek x.2010.

HERNANDIACEAE

E. Fenton (K) & X. van der Burgt (K)

Illigera pentaphylla Welw.
Climber, 5m; leaves shortly pubescent, alternate; petioles 12cm, twisting; leaflets 5, elliptic, to 9 × 6cm, subacuminate, sub–5-nerved, nervation scalariform; petiolules 2.5cm; inflorescences axillary, 30cm, 1–5 branches; flowers 7mm; corolla white and purple; fruit 3 × 9cm, 1-seeded, 2-winged. Fallow forest and *Hypselodelphys* thicket; 700–710m.
Distr.: Ivory Coast to Uganda [Guineo-Congolian].
IUCN: LC
Ndanan 1: Cheek 11536 3/2004; Cheek in MF 232 10/2002.
Note: easily confused sterile with *Dioscorea*.

ICACINACEAE

C. Couch (K), X. van der Burgt (K), L. Pearce (K), T. Utteridge (K), M. Cheek (K) & E. Fenton (K)

Fl. Cameroun 15 (1973).

Chlamydocarya thomsoniana Baill.
Woody climber; branchlets pubescent; leaves elliptic to obovate-elliptic, 10–28 × 4–12cm, margin denticulate, pubescent; dioecious; male flowers in catkins 3–6cm long; infructescence a spiked sphere comprising fruits 1.5cm diam., orange, tapering into a 4–8cm tube, covered with bristly hairs. Forest; 600m.
Distr.: Sierra Leone to Congo (Kinshasa) [Guineo-Congolian].
IUCN: LC
Mefou Proposed National Park: Etuge 5182 3/2004.

Icacina mannii Oliv.
Woody climber, or scandent shrub, with a large tuber; leaves elliptic to oblanceolate, 6–25 × 3–14cm, 4–6 pairs lateral nerves, glabrous; flowers highly fragrant, in lax axillary clusters; corolla pentamerous, 8 × 6mm, petals shortly pubescent outside, bearded inside; anthers extending beyond corolla; fruit a drupe, 3–3.5 × 1.5–2cm, red pubescent. Forest; 600–710m.

Distr.: Sierra Leone to Congo (Kinshasa) [Guineo-Congolian].
IUCN: LC
Mefou Proposed National Park: Etuge 5280 3/2004; Onana 2805 3/2004; **Ndanan 1:** Cheek 11782 3/2004; 11884 3/2004.

Pyrenacantha acuminata Engl.
Woody climber, slender; leaves alternate, elliptic to oblanceolate, 3–5 × 6–12cm, cordate-subauriculate base, densely appressed-pubescent beneath, secondary nerves 6–9 pairs; dioecious; male flowers in multi-flowered fascicles on stem just below leaves 1–5cm long, tetramerous, pubescent; fruit a drupe, ovate, transversally flattened, 1 × 1.5cm, pubescent, red orange in colour. Forest; 700–720m.
Distr.: Sierra Leone to Cameroon, Gabon and Congo (Kinshasa) [Guineo-Congolian].
IUCN: LC
Ndanan 1: Harvey 156 10/2002; 167 10/2002; Onana 2818 3/2004.

Pyrenacantha staudtii (Engl.) Engl.
Woody climber; stem striate; leaves elliptic, 11–18 × 7–9cm, base rounded or cordate, apex mucronate, densely pubescent to tomentose beneath, secondary nerves 5–7; petioles 1–2.5cm long; flowers in dense axillary spikes, sometimes grouped into fasicles; fruit a drupe, globose to ellipsoid, rusty tomentose. Forest; 679–700m.
Distr.: Nigeria, Cameroon, Gabon, Congo (Kinshasa), Angola and Uganda [Guineo-Congolian].
IUCN: LC
Ndanan 1: Cheek in MF 322 10/2002; Onana 2895 3/2004.
Note: Onana 2895 is tentatively placed in this taxon, Pearce viii.2009.

Rhaphiostylis sp. 1 of Mefou
Woody, stem-twisting liana; main axis pale grey-brown, slightly grooved, 4mm diam.; lateral leafy branches opposite, each 15–20cm long, horizontal, resembling a pinnately compound leaf; leaves alternate, 9–10 per branch, proximal leaf only half the length of the others, branches and midrib with dense, shaggy curled and patent yellow-brown hairs 1mm; leaf-blade oblong-elliptic (3.8–)7–11 × (2–)3–4.5cm, apex rounded then with an abrupt acumen to 1.5cm, base rounded, lateral nerves 4–6, uniting 5mm from the edge in a looping marginal nerve, intersecondaries conspicuous; petiole 2–3mm; flowers and fruit unknown. Forest; 700m.
Ndanan 1: Cheek in MF 376 10/2002.
Note: plot voucher MF376 is closest to *R. cordifolia* due to long (1mm) indumentum, but that taxon is an upper Guinea endemic, Pearce viii.2010.

Stachyanthus zenkeri Engl.
Syn. *Neostachyanthus zenkeri* (Engl.) Exell & Mendonça

Woody climber, slender, twining, to 8m; young branches and petioles bristly-pilose; leaves elliptic to oblanceolate, 10–23 × 5–10cm, appressed-pubescent beneath, 6–9 lateral nerves; dioecious; flowers on long spikes in fascicles on the stem below the leaves; male spikes to 25cm; flowers 6-merous, c. 6mm; fruit a drupe, ellipsoid with 4 flat surfaces, 3 × 1.8 × 1.2cm, in neat ranks along rachis. Forest; 650m.
Distr.: Nigeria to Congo (Kinshasa) [lower Guinea & Congolian].
IUCN: LC
Ndanan 2: Darbyshire 218 fl., 3/2004.
Note: *Darbyshire* 218 is tentatively placed in this taxon, Couch ix.2005.

IRVINGIACEAE

L. Pearce (K)

Desbordesia glaucescens (Engl.) Tiegh.
Tree, to 55m; basal trunk cylinder sometimes rotting in mature trees to be supported only by buttresses, bark pale brown to greyish-white; stipules 1(–3)cm, those covering terminal buds to 4cm; foliage sometimes flushed pink; leaves elliptic, 6.5–13 × 3–6cm, acumen 1cm, base obtuse to cuneate, glossy above, matt & off-white below, c. 10 lateral nerve pairs; petiole 7–12mm; inflorescence a terminal or subterminal panicle to 11.5cm; flowers 3mm, cream; fruit a samara, flattened, oblong, 8.5–16 × 3–4.5cm; 1(–2)-seeded. Forest; 700–710m.
Distr.: Nigeria, Cameroon, Equatorial Guinea, Gabon, Congo (Brazzaville & Kinshasa) [lower Guinea & Congolian].
IUCN: LC
Ndanan 1: Cheek 11725 3/2004; Cheek in MF 225 10/2002; Gosline 438 10/2002.
Note: both *Cheek* 11725 and *Gosline* 438 are tentative determinations made from sterile material. *Cheek* 11725 is a voucher of a tree seedling. The petioles of this specimen are unusually long (2cm) and the leaves are unusually large (to 20 × 7.5cm). Descriptions of flowers and fruit from D.J. Harris, Irvingiaceae, Species Plantarum: Flora of the World Part 1: 1–25 (1999). Pearce viii.2006. There is a move to replace the long-used name for this well-known species with an obscure name, *D. insignis* Pierre but this is not accepted here, Cheek xi.2010.

Klainedoxa gabonensis Pierre ex Engl.
Bull. Jard. Bot. Nat. Belg. 65: 153 (1996); D.J.Harris, Irvingiaceae, Species Plantarum: Flora of the World Part 1: 1–25 (1999).
Syn. *Klainedoxa gabonensis* Pierre ex Engl. var. *oblongifolia* Engl.
Tree, to 45m; bark dark brown; stipules keeled, 5–15cm; leaves ovate to oblong, 15–17 × 8–8.5cm, apex subacuminate, base obtuse to cordate, lateral nerves 14–22 pairs, areolae 0.2–0.3mm diam.; petiole c. 8mm;

inflorescence a terminal panicle to 13cm; flowers c. 4mm, white tinged pink; fruit a woody drupe with 5 pyrenes, depressed globose , 3.5–6.1 × 4.5–9.2cm. Forest; 710m.
Distr.: Guinea (Bissau), Sierra Leone, Liberia, Ghana, Cameroon, Gabon, CAR, Congo (Brazzaville), Congo (Kinshasa), Sudan, Uganda, Tanzania, Angola and Zambia [Guineo-Congolian].
IUCN: LC
Ndanan 1: Cheek 11535 3/2004.
Note: *Cheek* 11535 is sterile, description of flowers and fruit from D.J.Harris, Irvingiaceae, Species plantarum: Flora of the World. Part 1: 1–25 (1999). Pearce viii.2006.
Local Name: Ngon (Bety) (Cheek 11535).

LABIATAE

J.-M. Onana (YA), B.J. Pollard, Y.B. Harvey (K), C. Couch (K), E. Fenton (K) & M. Cheek (K)

Achyrospermum oblongifolium Baker
Erect herbaceous undershrub, 30–70cm; stems little-branched, tomentose; leaves 10–18 × up to 8cm, narrowly obovate-elliptic, acuminate; inflorescence terminal, 3–5 (–10) × 2cm; bracts broadly ovate, ciliate; calyx teeth as broad as long, margin ciliate; corolla greenish-white. Forest; 660–700m.
Distr.: Guinea (Conakry) to Bioko, Cameroon, São Tomé, Angola (Cabinda) [upper & lower Guinea].
IUCN: LC
Ndanan 1: Harvey 154 10/2002; **Ndanan 2:** Darbyshire 325 fl., 4/2004.

Clerodendrum formicarum Gürke
Woody climber, or climbing shrub; internodes hollow; leaves usually ternate or occasionally opposite, 4–10 × 2–5cm, glabrous, acuminate, shortly cuneate, elliptic; inflorescence terminal; axis short; peduncles long and conspicuously horizontal; flowers small; calyx 2–3mm; corolla tube very short, 4–5mm, yellowish-white. Forest; 700–710m.
Distr.: tropical Africa
IUCN: LC
Ndanan 1: Cheek 11760 3/2004; **Ndanan 2:** Darbyshire 226 fl., 3/2004.

Clerodendrum globuliflorum B.Thomas
Shrub or climber, to 3m; stems hollow; leaves large, elliptic, 15–25 × 6–12cm; inflorescences globose, dense, on older wood; calyx to 2cm long, margin ± fimbriate; corolla less than 12cm, ± glandular. Forest; 700m.
Distr.: Nigeria, Bioko & Cameroon [lower Guinea].
IUCN: NT
Ndanan 2: Cheek 11122 10/2002.

Clerodendrum melanocrater Gürke
Climbing shrub, to 8m; conspicuously drying black or dark purplish-black; leaves opposite, ovate, 5–15cm, cordate; petiole 3cm or more; inflorescence a dense corymb of small flowers; calyx white; corolla tube

glabrous, c. 1 × 0.1cm, brownish-yellow. Forest; 700m.
Distr.: Bioko, W Cameroon, Congo (Kinshasa), Uganda and Tanzania [Guineo-Congolian].
IUCN: LC
Ndanan 1: Harvey 208 10/2002; 84 10/2002.

Clerodendrum silvanum Henriq. var. *buchholzii* (Gürke) Verdc.
Mem. Mus. Natl. Hist. Nat. B. Bot. 25: 1555 (1975); F.T.E.A. Verbenaceae: 110 (1992).
Syn. *Clerodendrum buchholzii* Gürke
Syn. *Clerodendrum thonneri* Gürke
Woody climber, to 10m; stems with petiolar spines; leaves elliptic or ovate, glabrous, 8–20 × 3–10cm, often confined to the canopy; inflorescence an elongate, leafless panicle, frequently cauliflorous; rachis 5–30cm long; calyx enlarged, 8–10mm; corolla white, fragrant; tube (1.5–)1.7–2.5cm; fruits red. Forest; 700m.
Distr.: tropical & subtropical Africa.
IUCN: LC
Ndanan 1: Gosline 436 10/2002.

Clerodendrum splendens G.Don
Climbing shrub, 2–4m; young shoots quadrangular; leaves opposite, sub-sessile, elliptic or ovate, 6–20 × 4–10cm; petioles 0.3–20cm; inflorescences corymbose on young leafy shoots, terminal or axillary; calyx teeth 2–4mm; corolla deep red; tube c. 2.5cm long. Forest; 710m.
Distr.: Senegal to Cameroon, Rio Muni, Gabon, Congo (Kinshasa), Angola and CAR [Guineo-Congolian].
IUCN: LC
Ndanan 1: Cheek 11485 3/2004.

Hoslundia opposita Vahl
Erect or semi-climbing aromatic shrub, to 3–5m; leaves opposite or ternate, ovate-lanceolate, c. 10 × 4cm, gradually acuminate; inflorescences copious, of many-flowered terminal panicles; flowers greenish-cream; calyx very small in flower, campanulate, enlarging to form orange-yellow succulent berry-like fruits, c. 0.5cm. Forest; 700m.
Distr.: tropical and subtropical Africa [afromontane].
IUCN: LC
Ndanan 1: Darbyshire 181 fl., fr., 3/2004.

Leonotis nepetifolia (L.) R.Br. var. *nepetifolia*
Robust herb, to 2.5m; stems tomentellous; leaves long-petiolate, ovate-triangular, broadly cuneate, c. 12 × 6cm, serrate; flowers in large globose whorls at the upper nodes; calyx with significantly longer upper lobe; corolla orange, 20–40mm long, tube base with 3 rings of hairs within. Farmbush, cultivated ground near villages; 700m.
Distr.: pantropical.
IUCN: LC
Ndanan 1: Mackinder 512 10/2002.

Leucas deflexa Hook.f.
Straggling or semi-erect aromatic herb, to 2m; leaves lanceolate, cuneate at base, serrate, with an entire, acute

tip, petiolate; inflorescence a densely globose axillary whorl with numerous linear-subulate bracteoles; corolla white; stamens ascending; anthers often conspicuously hairy, orange. Forest, forest margins, savanna; 600m.
Distr.: Ghana, Bioko, Cameroon, Angola [Guineo-Congolian (montane)].
IUCN: LC
Mefou Proposed National Park: Etuge 5203 3/2004.

Ocimum gratissimum L. subsp. *gratissimum* var. *gratissimum*

A branched erect pubescent shrub, to ± 3m, emanating an aroma similar to that of cloves (*Syzygium aromaticum*); leaves ovate to obovate, 6–12 × 3cm, cuneate, acutely acuminate; inflorescence of several dense spikes, >1cm; calyx dull, densely lanate, horizontal or slightly downward pointing in fruit; corolla small, greenish-white; stamens declinate. Submontane forest, woodland, savanna; 710m.
Distr.: widespread in the tropics from India to W Africa, S to Namibia & S Africa; naturalised in tropical S America [pantropical].
IUCN: LC
Ndanan 1: Cheek 11886 3/2004.

Platostoma africanum P.Beauv.

A slender weedy annual herb, to ± 60cm, often less; leaves long-petiolate, ovate, c. 5 × 2.5cm; inflorescences racemose, numerous, axillary, somewhat lax and many-flowered; bracts persistent; pedicels short; corolla white, speckled mauve. Forest and farmbush; 600–700m.
Distr.: tropical Africa and India.
IUCN: LC
Ndanan 1: Harvey 104 10/2002; **Ndanan 2:** Etuge 5120 3/2003.

Pogostemon micangensis G.Taylor

J. Bot. 69 (Suppl. 2): 166 (1931).
Prostrate annual herb, mint-scented when crushed; glabrescent, stem apex and inflorescence with sparse white hairs; stem square, internodes to 7 × 0.2cm, nodes rooting; stem ascending slightly at apex; leaves opposite; leaf-blade elliptic, to 4.5 × 2cm, acute, base unequally obtuse, crenate-serrate, teeth 1 × 3mm, lateral nerves c. 7 pairs; petiole to 2cm; spike terminal, dense, 3 × 1cm; calyx bell-shaped, 3mm, teeth 1mm, white hairy; corolla pale purple, exserted 1mm; stamens exserted 1.5mm, purple. Sandy river bed in dry season; 600–710m.
Distr.: Cameroon & Angola [Congolian].
IUCN: EN
Mefou Proposed National Park: Etuge 5277 3/2004; **Ndanan 1:** Cheek 11875 3/2004; **Ndanan 2:** Cheek 11921 3/2004.

Vitex grandifolia Gürke

Tree, 10–30m; leaves 5-foliolate; leaflets abruptly short- to long-acuminate, the middle ones 13–40 × 6–20cm, attenuated into a short petiolule, c. 1cm long; inflorescence cymose, at base of leaves; peduncle 0.5–5.0cm; flowers pale-yellowish to yellowish-brown. Forest; 700m.
Distr.: Sierra Leone to Cameroon, Rio Muni and Gabon [Guineo-Congolian].
IUCN: LC
Ndanan 1: Cabral 104 3/2004; 109 3/2004; 110 3/2004; 112 3/2004; Harvey 181 10/2002; 89 10/2002.
Local Name: Evoula (Cabral 104).
Uses: FOOD — Infructescences — fruits edible (Cabral 104).

Vitex phaseolifolia Mildbr. ex W.Piep.

Repert. Spec. Nov. Regni Veg. 26: 161 (1929).
Liana, to 10m, glabrous; stems square, hollow, c. 0.5cm diam.; bark corky, peeling off in long, small, thin flakes; leaves papery, obovate to oblanceolate, to 13 × 7.5cm, acumen 0.5–0.8cm, base acute/rounded, adaxial surface glossy, abaxial surface pale green with red-brown midrib; petiole to 9cm; inflorescence spikes to 15cm, axillary; calyx brownish green; corolla pale yellow to white, 0.5–1cm; fruits obovoid, orange-yellow, smooth, glossy, c. 11mm. Forest edge; 700–720m.
Distr.: Cameroon [Cameroon Endemic].
IUCN: NT
Ndanan 1: Cheek 11184 10/2002; Onana 2864 3/2004.
Note: Cheek 11184 has small red, fiercely-biting ants inhabiting stems and holes.

Vitex sp. of Mefou

Small tree, 2–5m tall, more or less glabrous; stems glossy, dark mahogany brown, with sparse, white, appressed hairs 0.2mm, extending to petioles and inflorescence axis; leaves opposite, digitately compound, 20cm; petiole 5–11 × 0.1cm, black, drying with longitudinal grooves; leaflets (3–)5, dimorphic, outer leaflet pair elliptic, 4.5–5.5 × 2.5–2.8cm, acumen 0.5cm, base acute-obtuse, central leaflets oblanceolate, 12–14 × 3.8–5.5cm, acumen 0.75cm, base acute, upper surface glossy, drying dark grey-brown-green, lower surface pale grey, lateral nerves and midrib purple-brown, 5–7 pairs, arching up to form a looping-uniting marginal nerve, domatia absent, quaternary nerves reticulate, prominent, conspicuous on both surfaces; petiolule 3–5mm, canaliculate; infructescences single in both axils of the second and/or third node from the apex, 8–19 × 2–10cm; peduncle 5.5–9cm, terminating in 3 branches, each bearing a single fruit, or branched again, apex forming a shallow cap, into which the branches are inserted and articulated; branches 7–40mm long, also terminating in an elliptic cup from which the thicker pedicel arises 3–5.5 × 0.6–1mm, dilating slightly towards the apex; calycular cup shallow, 3 × 8mm, lobes shallow, 0.5–1mm, sinus concave, lobe apex obtuse with faint midrib on outer surface; mericarp 1, ellipsoid 5–6 × 4–4.5mm, drying glossy black. Forest; 710m.
Ndanan 1: Cheek 11718 3/2004.
Note: other specimens at K are Manning 2031 & 2094 (both from Central Prov.), *fide* Pollard 2005. Possibly an undescribed species endemic to the Yaoundé area, Cheek x.2010.

LAURACEAE

J.-M. Onana (YA), X. van der Burgt (K) & M. Cheek (K)

Fl. Cameroun 18 (1974).

Beilschmiedia mannii (Meisn.) Benth. & Hook.f.
Tree, 6–10m; stems glabrous, finely red-brown rugulose, soon corky; leaves alternate, estipulate; leaf-blade elliptic, thick, drying bronze below, 15–25 × 6.5–9cm, acumen 1.5cm, base obtuse, lateral nerves 5–8 pairs, arching towards the margin; petiole 1–2cm; inflorescences in upper 3 axils of short branches, 10 × 6–10cm; peduncle 2.5cm, branches 2–3, each 1.5–4mm; flowers broadly cupular, dull red, 2mm wide; calyx lobes enveloping flower, 5, triangular-oblong. Forest; 600–720m.
Distr.: Cameroon, Gabon & Congo (Kinshasa) [lower Guinea].
IUCN: NT
Mefou Proposed National Park: Etuge 5211 3/2004; **Ndanan 1:** Onana 2857 3/2004; **Ndanan 2:** Darbyshire 300 fl., 3/2004.
Note: Onana 2857 is tentatively placed in this taxon, Onana vi.2006.

LEEACEAE

L. Pearce (K)

Fl. Cameroun 13 (1972).

Leea guineensis G.Don
Erect to sub-erect, soft-wooded shrub, to 1.5m; stem lenticellate and swollen at the nodes; leaf bipinnate, c. 50 × 50cm; petiole 13cm, sheathing and tuberculate at base; rachis 10cm, chanelled above, with swellings at the insertions of the pinnae (constrictions in dried material); pinnae 5, c. 28 × 12cm; leaflets 5–7, opposite, oblong-elliptic, to 13 × 5.5cm long, base obtuse to cuneate, apex acuminate; stipules early caducous; inflorescence a cyme; infructescence 13 × 18cm; peduncle thick, tuberculate; fruits globose to oblate, to 1cm, green maturing to red. Forest, swamp; 700–710m.
Distr.: Guinea (Bissau) & (Conakry), Sierra Leone, Ivory Coast, Togo, Ghana, Nigeria, Cameroon, Equatorial Guinea, Gabon, CAR, Congo (Kinshasa) & (Brazzaville), Burundi, Sudan, Uganda, Kenya, Tanzania, Malawi, Zambia, Angola, Madagscar, Reunion, Comoros and Mauritius [tropical Africa].
IUCN: LC
Mefou Proposed National Park: Cheek in RP 22 3/2004; **Ndanan 1:** Harvey 97 10/2002.
Note: fruiting specimen only. Leaf damaged and incomplete.

LEGUMINOSAE-CAESALPINIOIDEAE

B. Mackinder (K), M. Cheek (K) & X. van der Burgt (K)

Fl. Cameroun 9 (1970).

Afzelia bella Harms
Shrub or small tree, to 12m; leaves pinnate; petiolules twisted; leaflets 3–5 pairs, glabrous above and below, up to 15cm long × 7cm wide, midvein sub-central; inflorescence a panicle; flowers with single well-developed petal, white marked red, up to 4.8cm, apex bi-lobed; pod compressed, woody, elastically dehiscent, more or less kidney-shaped, up to 15cm long, brown; seeds black with bright orange-red aril. Primary or secondary forest; 700–710m.
Distr.: Guinea (Conakry) to Congo (Kinshasa) [Guineo-Congolian].
IUCN: LC
Ndanan 1: Cheek 11793 3/2004; 12005 3/2004.

Anthonotha macrophylla P.Beauv.
Tree, to 10m; leaves pinnate; leaflets 2–5 pairs, sparsely hairy or glabrous above, somewhat glossy above, densely but minutely hairy below, terminal pair longest, up to 21 × 7.8cm, lower pairs shorter and relatively broader; inflorescence a raceme, compounded into a panicle, axillary, terminal or cauliflorous; flowers with single conspicuous white petal, up to 9mm, apex bi-lobed, other petals inconspicuous; pod compressed, woody, golden hairy becoming glabrous, up to 26cm; seeds irregular shaped, closely packed, edges flattened at adjoining surfaces. Forest; 550–700m.
Distr.: Guinea (Conakry) to Congo (Kinshasa) [Guineo-Congolian].
IUCN: LC
Mefou Proposed National Park: Etuge 5188 3/2004; **Ndanan 1:** Cheek 11048 10/2002; Tadjouteu 620 3/2004.

Aphanocalyx microphyllus (Harms) Wieringa subsp. **microphyllus** Wieringa
Wageningen Agr. Univ. Pap. 99(4): 140 (1999).
Syn. Monopetalanthus microphyllus Harms Fl. Congo B., Rw. & Bur. 3: 444 (1952).
Tree, 7–45m; leaves paripinnate; leaflets (9–)16–44 pairs; leaflets with distal half greatly reduced, narrowly oblong, largest leaflet 6–20(–33) × 2–4(–7)mm; inflorescence a raceme, either simple or compounded into a panicle; flowers white, 4–10mm long; pods oblong-obovate, 3.5–7 × 1.9–3.3cm, compressed, sparsely villous. Lowland and submontane rainforest; 710m.
Distr.: Cameroon, Gabon, Congo (Brazzaville), Congo (Kinshasa), Angola [lower Guinea & Congolian].
IUCN: LC
Ndanan 1: Cheek 11505 3/2004; 11651 3/2004; 11892 3/2004.
Note: lowland and submontane rainforest habitats in

checklist area, Gabon and parts of Congo (Brazzaville) but in drier habitats in parts of Congo (Brazzaville), Congo (Kinshasa) & Angola.

Baikiaea insignis Benth.

Tree, to 26m; leaves imparipinnate; leaflets 3–8, glabrous, alternate, variable, often elliptic or oblong-elliptic, 9–38 × 3.5–15cm, leathery; inflorescence an axillary or (more commonly) a terminal cluster; flowers white, large, up to 20cm long; pod compressed, woody, brown, velvety, oblong, up to 30cm long, elastically dehiscent, occasionally remaining attached to the tree for some time after dehiscing. Forest; 700m.
Distr.: Sierra Leone to Tanzania [Guineo-Congolian].
IUCN: LC
Ndanan 1: Cheek 11993 3/2004.

Bauhinia monandra Kurz

Shrub, 1m, glabrous; leaves alternate, simple, orbicular in outline, 11–13 × 10–14cm, apex deeply bilobed, by 5cm, base obtusely cordate, lateral nerves palmate, 5 on each side of the midrib; petiole 5–7cm long. River bank; 710m.
Distr.: pantropical, cultivated [pantropical].
IUCN: LC
Ndanan 1: Cheek 11887 3/2004.
Note: Cheek 11887 (tentatively placed in this taxon) found surviving in abandoned village, possibly one of several non-native cultivated species found in W Africa, flowers are needed to identify fully. *Bauhinia monandra* Kurz, *B. acuminata* L., *B. purpurea* L., *B. racemosus* L., *B. tomentosa* L., and *B. variegata* L. other species cultivated (F.W.T.A. 1: 444 (1958)), Cheek x.2010.

Berlinia bruneelii (De Wild.) Torre & Hillc.

Fl. Afr. Cent. 3: 393 (1952); Bol. Soc. Brot. sér. 2, 29: 40 (1955).
Tree, 2–40m; leaves paripinnate; leaflets in (1–)2–4 pairs, opposite or lower pairs sub-opposite, narrowly elliptic to elliptic or obovate, slightly falcate, usually discolorous, 11–35 × 4.1–12cm; inflorescence a panicle or cluster of racemes; bracteoles thick; flowers white; largest petal 48–93 × 40–83mm; pods oblong, woody, compressed, 24–35cm long, gold to reddish brown, short-hairy; seeds 3–7 per pod, c. 3.5–4cm long. Lowland forest, along streams, rivers and at the edge of marshes; 710m.
Distr.: Cameroon, Congo (Kinshasa) and Angola [lower Guinea & Congolian].
IUCN: LC
Ndanan 1: Cheek 11881 3/2004.

Chamaecrista mimosoides (L.) Greene

Lock M., Legumes of Africa: 32 (1989).
Syn. *Cassia mimosoides* L.
Prostrate or more commonly erect herb or subshrub, to 1.5m; leaves paripinnate; leaflets 20–70 pairs, linear to linear oblong, 3–8 × 1–1.5mm; flowers yellow, 4–13mm; pod linear to linear oblong, up to 8cm.

Grassland; 700m.
Distr.: widespread in the palaeotropics [palaeotropics].
IUCN: LC
Mefou Proposed National Park: Nana 42 3/2004.

Dialium bipindense Harms

Fl. Cameroun 9: 39 (1970).
Tree; young stems, petiole and midribs with red-brown hairs patent; leaves alternate pinnate, 18–24cm; petiole 2cm; leaflets alternate, 3–4 on each side of the 8–14cm rachis, leaflets increasing in size towards leaf apex, basal leaflets ovate, 2.5–3cm, distal oblong or elliptic, to 9 × 3.2cm, acumen 1.5cm with a filament like protrusion at apex, base rounded-obtuse, lateral nerves c. 8 pairs, quaternary nerves reticulate, conspicuous, petiolule 2mm, swollen, terminal leaflet with stipels 1mm; stipules linear, 15mm. Evergreen forest; 710m.
Distr.: Cameroon & Gabon [lower Guinea].
IUCN: NT
Mefou Proposed National Park: Cheek in RP 56 3/2004; **Ndanan 1:** Cheek 11890 3/2004.
Note: the specimens cited are saplings, fertile material is needed to confirm identity, Cheek x.2010.

Dialium dinklagei Harms

Tree, to 43m; dbh to 1m; bark pinkish grey; buttresses to 2m, 60cm deep, rounded, c. 15cm thick, up to 10 of them; leaves compound, 11–21 leaflets; rachis 7–15cm, with red-coloured pubescent hairs; leaflets alternate or subopposite, terminal leaflets larger than the lower ones, leaflets oblong, 1.5–7 × 0.6–2.5cm, apex acute or obtuse, rounded at the base, entire, abaxial surface with red hairs, ten arching lateral nerves, pubescent hairs beneath; petiolules 2mm, with red hairs; inflorescence a panicle; flowers to 0.8cm; fruits globular, densely pubescent, hairs red, to 2cm diam. Lowland evergreen wet forest; 700–710m.
Distr.: Guinea (Conakry), Sierra Leone, Liberia, Ivory Coast, Ghana, Nigeria, Cameroon, Gabon, Congo (Kinshasa) [upper & lower Guinea].
IUCN: LC
Ndanan 1: Cheek 11195 10/2002; 11530 3/2004; 11799 3/2004; Mackinder 515 fr., 10/2002.
Note: 6 specimens known from Cameroon alone (Fl. Cameroun).
Uses: FOOD — Infructescences — endocarps eaten as a childrens snack for sugar-acid taste (Cheek 11530); ANIMAL FOOD — Fertile Plant Parts — fruits eaten by parrots (Cheek 11530).

Dialium pachyphyllum Harms

Tree, 10m; leaves with 4–5 alternate leaflets, coriaceous, ovate-elliptic, 10–11.5 × 4–5.5cm, acuminate, base obtuse to rounded, lateral nerves 8–10 pairs, tertiary venation closely reticulate; rachis c. 5cm; petiole 2cm; petiolules 4mm; panicles lax, rachis to 23cm, branches racemose, to 10cm, puberulent; flowers 5mm; calyx grey-puberulent outside; petals white to yellow; stamens with bent filaments; fruits obovoid,

2–3 × 1.5–2cm, brown-black, velvety. Forest; 700m.
Distr.: Nigeria to Congo (Kinshasa) [lower Guinea & Congolian].
IUCN: LC
Ndanan 1: <u>Cheek 11194</u> 10/2002.

Erythrophleum ivorense A.Chev.

Tree, to 35m; leaves bipinnate; leaflets 9–13 per pinna, alternate, ovate or elliptic 4.5–8.5 × 3–5cm long, often drying blackish; inflorescence a panicle; flowers small, reddish, up to 3mm; pods oblong, compressed, up to 10cm, drying black. Forest; 700m.
Distr.: Sierra Leone to Gabon [Guineo-Congolian].
IUCN: LC
Ndanan 1: <u>Cheek 11213</u> fr., 10/2002; <u>Mackinder 514</u> fr., 10/2002.

Erythrophleum suaveolens (Guill. & Perr.) Brenan

Taxon 9: 194 (1960); F.T.E.A. Caesalpinioideae: 18 (1967).
Syn. *Erythrophleum guineense* G.Don
Tree; wood red, very hard; leaves compound, (1–)2–3(–4) pairs of pinnate leaflets, each leaf with 10–13 alternate leaflets; leaflets oval-elliptic, acuminate, rounded at the base, 5–12 × 3–6cm, generally pubescent on the midrib and the lower surface and on the petiole; leaf blade papery, finely reticulate on both surfaces, the secondary nerves more raised than the tertiary nerves; inflorescence a terminal panicle of spikes; flowers approx. 4mm, yellowish-white; pedicel to 1.5mm; pods black, oblong, woody, to 15 × 5cm; seeds 4–10. Semi-deciduous forest; 700–710m.
Distr.: Mali, Guinea (Bissau), Guinea (Conakry), Sierra Leone, Ivory Coast, Burkina Faso, Ghana, Togo, Nigeria, Cameroon, Equatorial Guinea, Gabon, CAR, Congo (Kinshasa), Sudan, Uganda, Kenya, Tanzania, Mozambique, Malawi, Zambia, Zimbabwe [tropical Africa].
IUCN: LC
Ndanan 1: <u>Cheek 11741</u> 3/2004; <u>Mackinder 520</u> fr., 10/2002.
Note: this species is often difficult to distinguish from *E. ivorense* which grows over a large part of the "sudan-Zambezian" region (Fl. Cameroun). 7 specimens known from Cameroon alone (Fl. Cameroun).

Gilbertiodendron brachystegioides (Harms) J.Léonard

Tree; leafy stems densely brown patent puberulent, with 4 rounded longitudinal ridges; leaves alternate, pinnate; petiole c. 1cm, concealed by the orbicular 2.5–3cm, rounded, basally cordate, leathery, persistent, reflexed stipules; rachis c. 20cm, leaflets opposite, basal-most pair broadly elliptic, c. 6 × 4.5cm, apex rounded, other leaflets elliptic-oblong, c. 13 × 6cm, acumen 1cm, base unequally obtuse, margin with scattered glands, lower surface with lateral nerves c. 8 pairs, uniting and forming a looping marginal nerve 5–10mm from margin, midrib orange, quaternary nerves reticulate,

raised, conspicuous. Evergreen forest; 679m.
Distr.: Cameroon & Gabon [lower Guinea].
IUCN: NT
Ndanan 1: <u>Onana 2899</u> 3/2004.

Gilbertiodendron dewevrei (De Wild.) J.Léonard

Tree, to 30m; dbh to 1m; leaves compound, (2–)3(–5)-jugate; leaflets oblong, slightly oblique, apex attenuate, base more or less rounded, asymmetric, pubescent or glabrous, finely papillose underneath, sometimes growing to 40 × 18cm, but often smaller, 15–20 pairs of lateral nerves, very pronounced beneath, arced at the margin, venation particularly pronounced on the abaxial surface; stipules lanceolate, acute, persistent, 2–8cm, 2 auricles, suborbicular or kidney-shaped, from 1–2cm diam., ± persistent; panicles axillary or terminal, velvety brown, rust-coloured hairs; pedicels 2–4cm, hairy as above; bracts 6–7cm, pinkish brown, caducous; sepals 5, lanceolate or almost triangular, 5–6mm, shrunken at the base to form a tube, 2mm; petals red to 1.5cm; pods flattened, obliquely oblong, brownish, 15–30 × 6–9cm, transversely ridged; large seeds, approx. 4–5cm diam. Swamp forest and by streams; 650–710m.
Distr.: Guinea (Conakry), Nigeria, Cameroon, Gabon, Congo (Brazzaville), Congo (Kinshasa) and Angola [Guineo-Congolian].
IUCN: LC
Mefou Proposed National Park: <u>Cheek in MF 1</u> 3/2004; **Ndanan 1:** <u>Cheek 11066</u> fr., 10/2002; <u>11191</u> 10/2002; <u>11198</u> 10/2002; <u>11199</u> 10/2002; <u>11231</u> fr., 10/2002; <u>11497</u> 3/2004; <u>11691</u> 3/2004; <u>Mackinder 498</u> 10/2002; <u>499</u> fr., 10/2002; <u>501</u> fr., 10/2002; <u>507</u> 10/2002; <u>508</u> 10/2002; <u>509</u> fr., 10/2002; <u>511</u> 10/2002; **Ndanan 2:** <u>Cheek 11127</u> 10/2002; <u>Darbyshire 299</u> 3/2004.
Note: Mackinder 498 is tentatively placed in this taxon, Mackinder xi.2005.
Local Name: Abem (Ewongo) (Cheek 11231).

Griffonia physocarpa Baill.

Woody climber; leaves unifoliolate, narrowly ovate or elliptic, 5–11 × 2.2–4.6cm; inflorescence a raceme; flowers with orange-brown calyx and green petals; petals up to 1.5cm; pod inflated, on conspicuous stipe to 2.5cm, deep brown, shiny, leathery, up to 7cm. Forest, secondary forest; 679–710m.
Distr.: Cameroon to Congo (Kinshasa) [lower Guinea & Congolian].
IUCN: LC
Ndanan 1: <u>Cheek 11835</u> 3/2004; <u>Cheek in MF 357</u> 10/2002; <u>362</u> 10/2002; <u>Darbyshire 284</u> fl., 3/2004; <u>Onana 2906</u> 3/2004; **Ndanan 2:** <u>Cheek 11916</u> 3/2004.
Note: Onana 2906 is tentatively placed in this taxon, Mackinder x.2005.

Guibourtia demeusei (Harms) J.Léonard

Tree, to 40m; leaves 2-foliolate; leaflets ovate, falcate, 6.5–20 × 1.5–3cm, translucent glands present; flowers 1–1.2cm (to stamen apices); sepals 4, puberulous

outside; petals absent; fruit broadly elliptic, up to 4cm; seeds lacking an aril. Forest; 700–710m.
Distr.: Cameroon to Congo (Kinshasa) [lower Guinea & Congolian].
IUCN: LC
Ndanan 1: <u>Cheek 11197</u> 10/2002; <u>11861</u> 3/2004; <u>11899</u> 3/2004.

Hylodendron gabunense Taub.
Tree, to 30m; trunk and branches armed with stout spines; leaves imparipinnate; leaflets alternate, oblong-elliptic, 6–13.5 × 1.5–5.4cm; flowers greenish-white, 3–4mm (to stamen apices); sepals 4; petals 0; fruit papery, oblong-elliptic, up to 12cm. Forest; 700–710m.
Distr.: Nigeria to Congo (Kinshasa) [Guineo-Congolian].
IUCN: LC
Ndanan 1: <u>Cheek 11044</u> 10/2002; <u>11532</u> 3/2004; <u>11980</u> 3/2004; <u>Mackinder 516</u> 10/2002; <u>517</u> 10/2002.

Leonardoxa africana (Baill.) Aubrev.
Adansonia (Sér. 2) 8: 178 (1968).
Shrub or small tree, to 15m; leaves pinnate; leaflets (2–)3(–4) pairs, 10–20 × 4–7.5cm, elliptic, slightly falcate, discolorous, acuminate at apex, acuminate to rounded at base; petiole and internodes sometimes swollen to house ants; petiolules twisted; inflorescence a short densely-flowered raceme, usually axillary, sometimes on the old wood; flowers up to 2.5cm; petals 5, subequal, purple; stamens purple; pod woody, compressed, obovate-oblong, up to 10.5 × 4cm, surface finely ridged, 1–2 seeded. Forest; 670–720m.
Distr.: Nigeria, Cameroon, Equatorial Guinea, Gabon [lower Guinea].
IUCN: LC
Ndanan 1: <u>Onana 2836</u> 3/2004; **Ndanan 2:** <u>Darbyshire 223</u> fl., 3/2004.
Note: myrmecophytic. The Mefou material falls in subsp. *letouzeyi* which is in the process of being assessed by B. Mackinder as VU for the Cameroon Red Data book (postscript).

Oxystigma buchholzii Harms
Engl. Jahrb. xxvi: 264 (1899).
Small tree, or shrub, to 15m; wood red-brown; leaflets 3–6, the terminal pair opposite, the others alternate, glabrous; leaf blade ovate-elliptic or lanceolate, 5–15 × 2.5–5cm, falciform, acuminate, acute or rounded at the base, tough, secondary nerves 8–12, impressed lightly on both surfaces; panicle 6–7cm; flowers very small, white, buds c. 1–2mm; sepals orbicular, 2mm, free or almost free, slightly hairy along the edges; no petals; stamens 10; ovary sparsely hairy; pods heart-shaped, flattened, indehiscent, thick, glabrous, smooth, from around 4–4.5cm diam.; only one seed which is very ridged. Evergreen forest near river; 710m.
Distr.: Cameroon, Equatorial Guinea, Gabon, Congo (Brazzaville), Congo (Kinshasa) [lower Guinea & Congolian].

IUCN: LC
Ndanan 1: <u>Cheek 11879</u> 3/2004; **Ndanan 2:** <u>Cheek 11925</u> 3/2004; <u>11935</u> 3/2004.

Pachyelasma tessmannii (Harms) Harms
Tree; cylindric trunk and plank butresses, glabrous; leaves alternate, bipinnate, to 35cm, pinnae 2–5 pairs, opposite; leaflets 9–15 alternate, oblong to oblanceolate, 4–10.5 × 1.5–3.5cm, apex obtuse to notched, base cuneate to obtuse; racemes 18cm, dense-flowered; pedicels 2mm; sepals 5, rounded, 2mm; petals 5, obovate-oblong, 4.5–6mm; stamens free, 10; pods linear-oblong, 15–37 × 3.5–4 × 2cm, very hard, sutures projecting; seeds 10–18. Evergreen forest; 720m.
Distr.: Nigeria, Cameroon, Gabon, Rio Muni, Congo (Kinshasa) [lower Guinea & Congolian].
IUCN: LC
Ndanan 1: <u>Onana 2876</u> fr., 3/2004.

Senna alata (L.) Roxb.
Lock M., Legumes of Africa: 36 (1989).
Syn. *Cassia alata* L.
Shrub, to 5m; leaves pinnate; leaflets in 5–7 pairs elliptic to oblong-elliptic, 5–18.2 × 3.2–11.7cm; inflorescence a raceme; flowers bright yellow, 1.5–2.2cm; pod up to 16cm, winged along the middle of each valve. Villages & farmbush; 700m.
Distr.: pantropical, native to S America.
IUCN: LC
Ndanan 1: <u>Harvey 148</u> 10/2002.
Local Name: Ngom ntangan (Ewondo) (Onana pers. comm.).
Note: probably an escape from cultivation. Ntangan means of the white man, Onana xi.2010.

Senna obtusifolia (L.) H.Irwin & Barneby
Lock M., Legumes of Africa: 38 (1989).
Herb, or subshrub, to 2m; leaves paripinnate; leaflets in 3 pairs, obovate, 1.5–5 × 1–3cm; flowers yellow, 0.9–1.9cm; pod linear, sometimes slightly falcate, up to 23cm. Grassland and a weed of cultivation; 600m.
Distr.: pantropical, introduced in Africa.
IUCN: LC
Mefou Proposed National Park: <u>Etuge 5115</u> 3/2003.

Tessmannia africana Harms
Huge tree, to 50m; dbh 65cm, no buttresses; bark rather smooth, 9mm thick, finely grooved longitudinally, greyish; first branches at 30m above the ground; leaves compound, 7–12 alternate leaflets; rachis 10–12cm, canaliculate; leaflets elliptic or oblong-elliptic, 4.9–5 × 1–4.5cm, obtuse or acuminate and slightly indented at the apex, subsessile, glabrous, tough, secondary nerves arching to form a border at the margin, the network of nerves conspicuous on both surfaces; panicle terminal or axillary, bunches of numerous pink flowers; axis pubescent, hairs yellow-brown; pedicels 5–12mm, covered in pale brown hairs mixed with glandular hairs; sepals 4, lanceolate, 10 × 4mm, slightly imbricate,

covered in hairs & glandular outgrowths, velvety-woolly inside; petals 5, slightly unequal, the 4 larger ones reach 1.7–2.5cm, oblong; pod obliquely suborbicular, up to 5cm diam., hairy or glabrescent, covered in pyramidal projections in the form of spines, which secrete a colourless sticky resin. Lowland wet forest; 700m.
Distr.: Cameroon, Equatorial Guinea, Gabon, Congo (Brazzaville), Congo (Kinshasa) and CAR [Guineo-Congolian].
IUCN: LC
Ndanan 1: Mackinder 503 fr., 10/2002; 504 10/2002; **Ndanan 2:** Cheek 11129 10/2002.
Note: 6 spcms known from Cameroon (Fl. Cameroun).

Zenkerella citrina Taub.

Tree or shrub, to 20m; leaves 1-foliolate; leaflet narrowly-elliptic or elliptic, 8–12 × 4.5–6cm; flowers white and pink, 5–7mm; pod broadly oblong, compressed, somewhat asymmetric, up to 7cm, leathery. Forest; 600–700m.
Distr.: SE Nigeria to Gabon [lower Guinea].
IUCN: NT
Mefou Proposed National Park: Etuge 5165 fl., 3/2004; **Ndanan 1:** Cheek 11196 10/2002.

LEGUMINOSAE-MIMOSOIDEAE

B. Mackinder (K), X. van der Burgt (K), M. Cheek (K) & J.-M. Onana (YA)

Acacia kamerunensis Gand.

Lock M., Legumes of Africa: 69 (1989).
Scandent shrub or climber; stems terete, armed with prickles; leaves bipinnate, gland usually present at the junction of the apical 3–10 pinnae pairs; leaflets numerous, tiny, up to 4 × 1mm; inflorescences capitate; flowers yellowish-white; pods compressed, subcoriaceous, dehiscent, pale to medium brown, up to 14 × 3cm. Forest; 600–700m.
Distr.: Liberia to Congo (Kinshasa) [Guineo-Congolian].
IUCN: LC
Mefou Proposed National Park: Etuge 5106 3/2003; **Ndanan 1:** Cheek 11008 10/2002.

Acacia pentagona (Schum.) Hook.f.

F.T.E.A. Mimosoideae: 100 (1959); Opera Botanica 68: 42 (1983).
Syn. *Acacia pennata* sensu Keay
Climber; stems somewhat angled, armed with prickles; leaves bipinnate, a gland usually present at the junction of the distal 1–4 pinnae pairs; leaflets numerous, up to 7 × 2mm; inflorescences capitate; flowers yellowish-white; pods compressed, coriaceous, indehiscent, dark brown, up to 16 × 3.5cm. Forest; 700m.
Distr.: tropical Africa.
IUCN: LC
Ndanan 1: Cheek in MF 211 10/2002; Mackinder 510 10/2002.

Adenopodia scelerata (A.Chev.) Brenan

Fl. Gabon 31: 83 (1989).
Syn. *Entada scelerata* A.Chev.
Climber, with thorned stems, petioles and leaf rachides; stems subangular, sparsely pubescent, thorns to 2.5mm; leaves bipinnate, 3–6 pairs of pinnae, each pinna to 6cm (lower pinnae shorter), with 7–10 leaflet pairs; leaflets oblong, to 1.4 × 0.5cm, reducing towards the base, terminal leaflet pair obovate; petiole to 10cm, principal rachis to 6.5cm. Forest; 700m.
Distr.: Liberia to Congo (Kinshasa) [Guineo-Congolian].
IUCN: LC
Ndanan 1: Cheek 11007 10/2002; Cheek in MF 217 10/2002.

Albizia zygia (DC.) J.F.Macbr.

Tree, to 25m; leaves bipinnate; leaflets relatively few, asymmetrical, sessile, (1–)2–3(–4) pinnae pairs, with 3–4 pairs of leaflets per pinna, apical pair largest, up to 9 × 5.2cm, lower pairs progressively smaller; inflorescence capitate; calyx and corolla inconspicuous; stamens numerous, showy, red, fused into a tube up to 13mm, the free ends extending a further 4–5mm; pod compressed, coriaceous, glabrous, glossy, up to 20 × 3.8cm. Forest & farmbush; 700–710m.
Distr.: tropical Africa.
IUCN: LC
Ndanan 1: Cheek 11646 3/2004; 12004 3/2004.
Note: Cheek 11646 is tentatively placed in this taxon, Mackinder vi.2005.

Aubrevillea platycarpa Pellegr.

Lock, M., Legumes of Africa: 87 (1989).
Tree, to 25m; leaves bipinnate; pinnae 4–7 pairs, widely spaced, internodes more than 2cm long; leaflets oblong elliptic, somewhat asymmetric, 1.4–3 × 0.6–1.4cm, secondary veins prominent on lower surface; inflorescence a panicle; flowers small, less than 3mm, sessile; pod compressed oblong, narrowing towards base, up to 20cm long, papery. Forest; 710m.
Distr.: Sierra Leone to Congo (Kinshasa) [Guineo-Congolian].
IUCN: LC
Ndanan 1: Cheek 11846 3/2004.

Calpocalyx dinklagei Harms

Tree, to 15m; leaves bipinnate, single pinna pair, pinnules 5–6 pairs, each pair increasing in size from base to apex of the pinna, the apical pair 6–12cm × 3.5–6cm; inflorescence a short spike arranged in terminal clusters or panicles, covered in a dense brown tomentum; pod woody, compressed, asymmetrically kidney-shaped, up to 16cm, valves with single longitudinal ridge, rolling up after dehiscence. Forest; 710m.
Distr.: Nigeria to Congo (Kinshasa) [lower Guinea & Congolian].
IUCN: LC
Ndanan 1: Cheek 11664 3/2004.

Entada gigas (L.) Fawc. & Rendle

Liana, glabrous; leaves alternate, bipinnate, pinnae 2–3-pairs, the end pair usually a tendril; leaflets 2–4 per pinna, 2.5–5 × 1–2.2cm; inflorescence to 30cm; peduncle inserted about 5mm above the leaf axil; flowers small, numerous and dense; pedicels 1–1.5mm; fruits coiled into a single or double spiral, 8–10cm wide, woody, breaking into 1-seeded units. Evergreen forest; 700m.
Distr.: tropica S America and Africa [tropical].
IUCN: LC
Ndanan 1: Cheek 11578 3/2004; 11985 fr., 3/2004.

Entada rheedei Spreng.

Lock M., Legumes of Africa: 92 (1989).
Syn. *Entada pursaetha* DC. F.W.T.A. 1: 490 (1958).
Liana, glabrous; leaves alternate, bipinnate, pinnae 2-pairs, the third pair tendril-like; leaflets elliptic, c. 4.5 × 2cm; spikes 15–22cm, inserted at the axil; peduncle 3.5cm; flowers dense, 2mm; fruit 150 × 10 × 3cm, flat and pendulous, hard, eventually breaking into 1-seeded units. Farmbush; 640m.
Distr.: tropical Africa and Asia.
IUCN: LC
Ndanan 1: Darbyshire 319 3/2004.

Fillaeopsis discophora Harms

Tree, to 40m; leaves bi-pinnate, 1–2 pairs of opposite pinnae; leaflets 4–8, alternate, ovate-elliptic to elliptic; inflorescence a panicle of numerous spikes; flowers numerous, small, 3–4mm long; pods compressed, oblong, 18–60 × 8–20cm; seeds transverse, oblong, 5–14 × 3–5.2cm, winged, attached at the middle of the longer margin by a long funicle. Roadside weed; 710m.
Distr.: Nigeria, Cameroon, Gabon, Congo (Kinshasa), Angola [lower Guinea & Congolian].
IUCN: LC
Ndanan 1: Cheek 11486 3/2004.

Mimosa pudica L.

Herb, to 1m; stems sparsely armed with prickles; leaves bipinnate, closing up when touched, unarmed; (1–)2 pinna pairs; petiole longer than the rachis; inflorescence subcapitate; flowers pink or mauve; pods compressed, bristly on margins only, breaking up into single-seeded segments, margins persistent. Farmbush & roadsides; 600–700m.
Distr.: introduced in the Old World, native of the Neotropics.
IUCN: LC
Mefou Proposed National Park: Etuge 5259 3/2004;
Ndanan 1: Harvey 72 10/2002.

Parkia bicolor A.Chev.

Tree, to 30m; leaves bipinnate; pinnae pairs opposite or sub-opposite, leaflets numerous, asymmetric; inflorescence pendant, claviform, apical part globose, then much narrower, up to 7cm broad; flowers yellow, orange, orangish-red or bluish-red; pods compresssed, up to 40 × 3.3cm; seeds clearly visible through valves, lying diagonally. Forest; 700m.
Distr.: Guinea (Conakry) to Congo (Kinshasa) [Guineo-Congolian].
IUCN: LC
Mefou Proposed National Park: Onana 2812 3/2004;
Ndanan 1: Cheek 11214 fr., 10/2002; Mackinder 500 fr., 10/2002.
Uses: FOOD — Infructescences — fruit pulp edible (Onana 2812).

Pentaclethra macrophylla Benth.

Tree, to 25m; leaves bipinnate, rusty hairy along the axes when young, 12–15 pairs per pinna, middle pinnae longest, up to 12.7cm, others up to 2 × 1.1cm; leaflets sessile, asymmetric, glabrous or nearly so; inflorescence a spike; flowers creamy-yellow or pinkish-white, fragrant; pods compressed, woody, elastically dehiscent, up to 50 × 11cm. Forest; 659–700m.
Distr.: Senegal to Congo (Kinshasa) [Guineo-Congolian].
IUCN: LC
Mefou Proposed National Park: Etuge 5291 fr., 3/2004; **Ndanan 1:** Cheek 11071 fr., 10/2002; 11203 10/2002; Cheek in MF 207 10/2002; 250 10/2002.

Piptadeniastrum africanum (Hook.f.) Brenan

Tree, to 45m, flat crown; leaves bipinnate, 10–12-pairs of alternate (rarely opposite) pinnae; leaflets numeous, up to 7 × 2mm, glabrous; inflorescence a spike, compounded into panicles; flowers yellowish-white; pods compressed, coriaceous, splitting along one margin, up to 36 × 3.2cm; seeds surrounded by a papery brown wing, up to 7.8 × 2.6cm (inc. wing). Forest; 650–710m.
Distr.: Senegal to Uganda [Guineo-Congolian].
IUCN: LC
Mefou Proposed National Park: Cheek in MF 64 3/2004; **Ndanan 1:** Cheek 11202 10/2002; 11540 3/2004; 11703 fr., 3/2004; 11902 3/2004; 11945 3/2004; Cheek in MF 240 10/2002; 326 10/2002; Mackinder 513 fr., 10/2002; Tadjouteu 635 4/2004.
Note: seeds resemble those of *Newtonia* spp. but point of attachment is central (apical in *Newtonia*). Plot voucher MF64 is tentatively placed in this taxon, van der Burgt vi.2009.

LEGUMINOSAE-PAPILIONOIDEAE

B. Mackinder (K), J.-M. Onana (YA), X. van der Burgt (K), Seren Thomas (K) & B. Schrire (K)

Abrus precatorius L.

Liana; leaves alternate, pinnately compound, 7–8cm; petiole 0.9cm; leaflets opposite, c. 14 pairs, oblong, 0.7–1.7 × 0.7cm, margin entire, apex rounded, mucro <1mm, base rounded, glabrous; petiolules 1mm; pods oblong, 2–2.5 × 1.3cm, beak 5mm; seeds glossy black

and ($^2/_3$) red, ellipsoid, 0.6 × 0.5cm. Grassland; 600–720m.
Distr.: pantropical.
IUCN: LC
Mefou Proposed National Park: <u>Etuge 5112</u> 3/2003;
Ndanan 1: <u>Cheek 11644</u> 3/2004; <u>Onana 2841</u> 3/2004.
Uses: MEDICINES (Onana 2841).

Amphimas ferrugineus Pierre ex Pellegr.
Enum. Pl. Afr. Trop. 2: 63 (1992).
Tree, 20 m tall; bark with red sap; branches stout, brittle; leaves in clusters, densely velvety-haired beneath, glabrous above, rachis villous with brown retrorse hairs; inflorescences large and terminal; flowers yellow; fruit as for *A. pterocarpoides*. Secondary forest; 700m.
Distr.: Cameroon, Gabon, Congo (Kinshasa), Angola [Congolian].
IUCN: LC
Ndanan 1: <u>Cheek 11983</u> 3/2004.

Amphimas pterocarpoides Harms
Large tree, c. 35m; trunk triangular; leaves alternate, pinnately compound, 30cm; leaflets alternate, c. 10 pairs, oblong-oblanceolate, to 9.5 × 3.5–5cm, base rounded, apex narrowly retuse, margin entire, glabrous above and beneath, up to 10 pairs lateral veins; petiole 0.5cm, pubescent; panicle terminal, 20 × 20cm; flowers 3mm; fruit flat, papery, wind-dispersed, not dehiscing, oblong, 13 × 4cm. Semi-deciduous forest; 700m.
Distr.: Guinea (Conakry) to Sudan [Guineo-Congolian].
IUCN: LC
Ndanan 1: <u>Cheek in MF 288</u> 10/2002.

Angylocalyx oligophyllus (Baker) Baker f.
Understorey shrub, to 5m; leaves imparipinnate; leaflets (3–)5–7, alternate or subopposite, ovate, 8–25 × 3.8–8.2cm, long acuminate; inflorescence cauliflorus; flowers white with green and red or purple markings, standard petal not reflexed; pod torulose, up to 15cm, beak up to 3cm. Forest and forest gaps; 679m.
Distr.: Liberia to Congo (Kinshasa) [Guineo-Congolian].
IUCN: LC
Ndanan 1: <u>Onana 2916</u> 3/2004.
Note: Onana 2916 is tentatively placed in this taxon, Mackinder v.2005.

Baphia buettneri Harms subsp. *hylophila* (Harms) Soladoye
Shrub or small tree, to 20m; leaves 1-foliolate; leaflet oblong-elliptic or oblanceolate-elliptic, 5–16.8 × 2–7.5cm; flowers white with yellow blotch at base of standard, 11–17mm long; calyx glabrous except for a tuft of hairs at the apex; pod unknown. Lowland rainforest; 600m.
Distr.: Cameroon, Equatorial Guinea, Gabon [lower Guinea].
IUCN: LC
Mefou Proposed National Park: <u>Etuge 5103</u> 3/2003.

Baphia capparidifolia Baker subsp. *multiflora* (Harms) Brummitt
Kew Bull. 40: 324 (1985).
Liane; stem pubescent; leaves alternate, simple, lanceolate-elliptic, 8.3–10.6 × 1.3–4.7cm, apex mucronate, base rounded, hairy, up to 7 pairs of veins, adaxially glabrous, abaxially pubescent, margin entire; petiole to 3.5cm, pubescent; racemes axillary, pubescent, 8–11cm, sometimes branched at the base; pedicel to 0.5cm; calyx pubescent; corolla white, 8mm; stigma pubescent; stamens c. 9; pods crescent shaped, c. 2.8 × 0.4cm, pubescent; pedicel to 1cm. Forest; 600m.
Distr.: Cameroon, Gabon, Congo (Kinshasa), Angola, Rwanda, Burundi, W Tanzania [lower Guinea & Congolian].
IUCN: LC
Mefou Proposed National Park: <u>Etuge 5102</u> 3/2003.

Baphia capparidifolia Baker subsp. *polygalacea* Brummitt
Bol. Soc. Brot (Sér. 2) 39: 163 (1965); Kew Bull. 40: 325 (1985).
Syn. *Baphia polygalacea* (Hook.f.) Baker
Liana, to 5m; stem laxly pubescent; leaves alternate, simple, lanceolate-elliptic, 4.3–7.9 × 2.3–3.4cm, apex narrowly acuminate, base cordate, glabrous midrib and venation, lower surface pubescent, up to 6 pairs of lateral veins; petiole to 2.9cm, pubescent; peduncle up to 9cm; pedicel c. 2mm, pubescent; calyx pubescent; flowers white, standard with yellow basal blotch, keep petals splayed; stigma pubescent; pods crescent shaped. Gallery forest; 710m.
Distr.: Guinea (Conakry) to Cameroon [upper & lower Guinea].
IUCN: LC
Ndanan 1: <u>Cheek 11756</u> 3/2004.

Baphia nitida Lodd.
Shrub or small tree, to 10m; leaves 1-foliolate; leaflets oblong-elliptic to ovate, 7–12 × 4–5cm, glabrous except for nerves on underside; flowers white with central yellow blotch, 1.5–2.1cm, fragrant; pod oblanceolate, compressed, up to 15cm. Forest; 700m.
Distr.: Senegal to Gabon [upper & lower Guinea].
IUCN: LC
Mefou Proposed National Park: <u>Nana 36</u> 3/2004.

Baphia sp. 1 of Mefou
Tree; stem spinulose; leaves simple, alternate; lamina elliptic-lanceolate, 9.8–13.6 × 2.2–5cm, glabrous, margin entire, apex cuspidate, base attenuate, up to 7 pairs veins, secondary veins looking like stars near apex of undersurface; petiole to 1.6cm; pods semi-circular. Forest near river; 710m.
Ndanan 2: <u>Cheek 11934</u> 3/2004.

Calopogonium mucunoides Desv.
Prostrate or scrambling herb; leaves 3-foliolate; leaflets broadly elliptic-rhombic, 2.5–10.5 × 4.6–8.5cm; flowers

blue, 8–11mm; pod narrowly oblong, 31–40 × 5–6mm, golden brown pilose. Forest; 700m.
Distr.: tropical America, cultivated in Africa.
IUCN: LC
Ndanan 1: Harvey 82 10/2002.

Centrosema pubescens Benth.
Scrambling and/or climbing herb; leaves 3-foliolate; leaflets elliptic, 4–8.5 × 1.8–4.5cm; flowers pink and pale purple or white and purple, 1.8–3.2cm; pod linear, up to 12.5cm, margins thickened. Grassland, roadsides, weed of cultivation; 600–700m.
Distr.: native of Neotropics, widely naturalised in tropical Africa.
IUCN: LC
Mefou Proposed National Park: Etuge 5237 3/2004;
Ndanan 1: Harvey 110 10/2002.

Crotalaria doniana Baker
Shrub, 2m; stems pubescent; leaves alternate, trifoliolate, hastate in outline, 3–5 × 4–8cm; petiole to 3cm, pubescent; leaflets elliptic, 1.5–5.2 × 0.9–2.5cm, appressed hairy, margin entire, apex acute, base attenuate, midrib and lateral veins pubescent, to 6 pairs of lateral veins; petiolule c. 1mm; pedicel 0.7–1cm, pubescent; sepals 5, to 1cm, pale green, pubescent; upper petal cream flushed with orange and fawn at throat, keel yellowing at apex, faintly spotted red. Roadside; 700–710m.
Distr.: Sierra Leone to Cameroon, CAR and Congo (Kinshasa) [Guineo-Congolian].
IUCN: LC
Ndanan 1: Cheek 11792 3/2004; Harvey 103 10/2002.

Dalbergia sp. 1 of Mefou
Liane, to 10m; leaves imparipinnate; leaflets 7–11, concolorous, elliptic, 13–47 × 8–21cm, terminal leaflet the largest, lower leaflets becoming progessively smaller, underside pubescent, more densely so along the midvein, ciliate along the margins. Riverbank; 710m.
Ndanan 1: Cheek 11891 3/2004.

Dalbergia sp. 2 of Mefou
Probably a shrub, known only from seedlings, 20cm; stem pubescent; lamina trifoliolate, alternate, deltoid, 2.7–4.8 × 1.6–3.6cm, margin entire and pubescent otherwise leaves glabrous, apex caudate, base rounded-truncate, to 5 pairs of lateral veins; petiole to 2.9cm, pubescent; pod papery, flat, indehiscent, narrowly oblong, constricted at midpoint, to 11.6 × 3.5cm; single central seed. Lowland forest; 700m.
Ndanan 1: Mackinder 506 10/2002.

Desmodium adscendens (Sw.) DC. var. adscendens
Erect or straggling, sometimes prostrate herb, up to 80cm; leaves 3-foliolate; leaflets broadly obovate or broadly elliptic, 1.6–4.5 × 1.2–3.2cm; flowers white, pale purple or pink, 4–6mm; pod narrowly oblong, up to 3.5cm, upper margin straight, lower margin indented to c. 1/2 the pods width. Forest clearings, grassland & farmbush; 650–700m.

Distr.: tropical & subtropical Africa.
IUCN: LC
Ndanan 1: Harvey 106 10/2002; Tadjouteu 633 4/2004.

Desmodium adscendens (Sw.) DC. var. robustum Schubert
Prostrate or straggling herb, to 50cm; leaves 3-foliolate; leaflets broadly obovate or broadly elliptic, terminal leaflet largest, 12–14.8 × 1–3.4cm; flowers white, creamy, pale to deep purple, standards sometimes veined dark purple, 6–10mm long; pod narrowly oblong, 1.8–3.2cm, strongly indented along lower margin, compressed. In light patches of secondary forest, on river banks and streamsides; 700m.
Distr.: Guinea (Bissau), Sudan, Liberia, Togo, Nigeria, Cameroon, Equatorial Guinea, Congo (Kinshasa), Zambia [afromontane].
IUCN: LC
Ndanan 2: Darbyshire 221 fl., 3/2004.

Dioclea reflexa Hook.f.
Robust climber, to 6m long; leaves 3-foliolate, leaflets elliptic or oblong-elliptic, 9–16 × 5.5–11cm, secondary veins 7–8, prominent on lower surface; inflorescence a raceme; bracteoles narrowly lanceolate, up to 20cm long, striate, persistant; flowers reddish-purple with a pale blotch in the middle of the standard petal, 1.6–2.5cm long; pods elliptic (2–3 seeded) or orbicular (1-seeded), up to 15 × 6cm; seeds semi-orbicular, 3–3.6cm at widest point with pale conspicuous hilum around c. 2/3 of the circumferance. Secondary and disturbed forest and along riversides 680m.
Distr.: widespread throughout Old and New World tropics [pantropical].
IUCN: LC
Ndanan 2: Darbyshire 239 3/2004.

Erythrina excelsa Baker
Tree, to 15m; bole armed with stout prickles, axes of young shoots also armed with prickles; leaves 3-foliolate; leaflets with prickles along mid-vein and secondary veins when young, broadly elliptic to rhombic, laterals oblique, 9.5–15 × 4.8–10.5cm; flowers red, 3–3.5cm; pod moniliform (like a beaded necklace). Forest margins, wooded grassland, old farmland; 700m.
Distr.: SW Cameroon to Tanzania [lower Guinea & Congolian].
IUCN: LC
Ndanan 1: Cheek 11049 10/2002.

Indigofera macrophylla Schum. & Thonn.
Scrambling or climbing shrub, to 4m; leaves imparipinnate, 3–5 pairs; leaflets elliptic, 15–36 × 10–14mm; flowers greenish brown except for purple wings, 4–7mm; pod linear; cylindrical, to 3.5cm long. Secondary forest; 600–710m.
Distr.: Guinea (Bissau) to Cameroon and CAR [upper & lower Guinea].
IUCN: LC

Mefou Proposed National Park: Etuge 5145 fr., 3/2004; **Ndanan 1:** Cheek 11915 3/2004; Harvey 80 10/2002.

Leptoderris congolensis (De Wild.) Dunn

Climbing liana, to 25m in trees; leaves imparipinnate, 2 pairs; leaflets narrowly obtriangular or narrowly obovate, 4–21 × 2.2–10.2cm, apex broadly retuse; flowers white to pale yellowish-green, often spotted purple, 7–11mm; pod compressed, papery, winged along one side, 2.2–9.8 × 1–3.6cm; single central seed. Riverine forest and forest at edge of savanna & cultivation; 710m.
Distr.: Nigeria, Cameroon, Gabon, Congo (Kinshasa), Angola [lower Guinea & Congolian].
IUCN: LC
Ndanan 1: Cheek 11896 3/2004.

Mucuna flagellipes Hook.f.

Liana; leaves 3-foliolate; leaflets ovate or elliptic, 3.6–6.5 × 8.3–15.5cm; inflorescence rachis clearly "zig-zag"; flowers greenish-cream, 3.5–5.5cm; pod oblong, up to 16cm, transversely ridged, covered in golden irritant hairs. Secondary forest, edge of old cultivation, along roadsides & overhanging water; 700–710m.
Distr.: Sierra Leone to Uganda [Guineo-Congolian].
IUCN: LC
Mefou Proposed National Park: Cheek in RP 30 3/2004; Etuge 5292 3/2004; **Ndanan 1:** Cheek 11187 10/2002; 11858 3/2004; Harvey 77 10/2002; 79 10/2002.

Mucuna sloanei Fawc. & Rendle

Climbing liana, up to 8m in trees; leaflets obliquely ovate, 5–15 × 3–11cm; flowers cream or yellow, 5–6cm; pods woody, oblong, compressed, up to 15cm long, valves transversely ridged, covered in irritant hairs, bearing 1–3 seeds; seeds suborbicular, 2.3–2.8cm diam., hilum pale, about 4/5 length of the seed circumference. Coastal and secondary forest; 700m.
Distr.: Guinea (Bissau) to Cameroon [upper & lower Guinea].
IUCN: LC
Ndanan 1: Harvey 139 10/2002; **Ndanan 2:** Cheek 11137 10/2002.

Platysepalum violaceum Welw. ex Baker var. *vanhouttei* (De Wild.) Hauman

Shrub or small tree, to 15m; leaves imparipinnate; leaflets 5–9, in pairs plus a terminal leaflet, narrowly ovate to ovate or narrowly elliptic to elliptic; apex acuminate; secondary venation distinct and raised below; 3.2–11.8 × 1.2–5.2cm; inflorescence an axillary or terminal panicle; flowers 9–18mm, standard white with red markings, keel purple with white + red markings; fruits woody, oblong & compressed, 4.5–8.5cm long, valves covered in a dense golden tomentum, valve margins somewhat thickened when mature. Forest; 700m.
Distr.: SE Nigeria, Cameroon, Rio Muni, Gabon,

Congo (Kinshasa) and Angola (Cabinda) [Guineo-Congolian].
IUCN: LC
Ndanan 1: Harvey 138 10/2002.

Pterocarpus soyauxii Taub.

Tree, to 35m; leaves imparipinnate; leaflets 9–13, alternate or subopposite, obovate or elliptic, 2.5–6 × 2–2.7cm; inflorescence an axillary or terminal panicle; flowers deep yellow, 10–13mm; fruit, discoid, 6.5–7.5cm diam.; single central seeds surrounded by a slightly glossy papery wing. Forest; 700m.
Distr.: Nigeria to Congo (Kinshasa) [Guineo-Congolian].
IUCN: LC
Ndanan 1: Mackinder 518 10/2002.

Rhynchosia densiflora (Roth.) DC.

Climbing, procumbent or ascending perennial herb, 0.3–3m long; leaflets 3, elliptic, ovate, rhombic or almost round, the laterals oblique, 1.5–8 × 1–7.2cm, rounded, acute or acuminate apex, rounded to cuneate base, densely covered with small orange or black glands beneath; petiole 2–7cm long; rachis 3–13mm long; petiolules 1–2.5mm; stipules 4–7 × 1–4.5mm, ovate to lanceolate, pubescent and glandular; inflorescences axillary, rachis 2–12.5cm long; peduncle 0.5–1cm long; pedicels 1.5–3mm long; bracts 4–16 × 1–5mm; calyx tube c. 1.5–2mm, lobes 0.5–1.7cm, linear or linear-lanceolate; corolla standard yellow with purple veining, drying orange-brown, 7–14 × 4.5–9mm, elliptic to oblong, glabrous, wings yellow, keel greenish; pods 10–14 × 4.5–5mm, elliptic-oblong or oblong, apiculate, covered with short hairs or dense short tomentum, sparser long hairs and orange-red glands; seeds reddish-brown with black mottling or almost entirely black, 3.5 × 2–3 × 1.5mm, oblong-reniform, compressed. Forest; 700–710m.
Distr.: tropical Africa and India.
IUCN: LC
Ndanan 1: Cheek 11808 3/2004; Gosline 399 10/2002.

Rhynchosia mannii Baker

Climber, to 8m; leaves 3-foliolate; leaflets ovate to broadly ovate, the laterals obliquely so, 8–15 × 5–13.5cm; inflorescence a pseudoraceme or few-branched panicle; lower calyx lobe much longer than the other lobes; flowers pale yellow with purple veins; pods up to 1.5cm. Forest gaps; 700m.
Distr.: Nigeria to Uganda [lower Guinea & Congolian].
IUCN: LC
Ndanan 1: Harvey 109 10/2002.

Rhynchosia pycnostachya (DC.) Meikle

Climbing herb; leaves pinnately 3-foliolate; leaflets broadly ovate to deltoid, often asymmetric at base, 3.5–15.2 × 2.2–10.7cm, 3-nerved from the base; inflorescence a robust axillary raceme; flowers creamy-yellow, streaked red or brown, becoming darker with

age, 12–18mm long; fruit oblong, valves dense grey puberulous; seeds usually 2, metallic blue/black. Primary and disturbed forest, abandoned cultivations and along roadsides; 693m.
Distr.: Senegal to Bioko and SW Cameroon [Guineo-Congolian].
IUCN: LC
Ndanan 1: Onana 2925 3/2004.

LEPIDOBOTRYACEAE

M. Cheek (K)

Fl. Cameroun 14 (1972).

Lepidobotrys staudtii Engl.

Tree, 7m, glabrous; leaves ovate-elliptic, to 17 × 7.5cm, acuminate, obtuse, c. 5 pairs lateral nerves, margin thickened; petiole 1.2cm, joined 0.5cm from the base, the distal 0.4cm swollen; inflorescences axillary, buds resembling a pine-cone, 3mm; fruit ovoid, 3cm, brown tomentose, 2-valved; 1-seeded. Forest; 710m.
Distr.: Nigeria to Congo (Kinshasa) [lower Guinea & Congolian].
IUCN: LC
Ndanan 1: Cheek 11715 3/2004.

LINACEAE

C. Couch (K)

Fl. Cameroun 14 (1972).

Hugonia obtusifolia C.H.Wright

Liana, 0.4–0.6cm diam., glabrous; tendril hooks; leaves elliptic to obovate, to 10 × 4.5cm, 6–7 nerves on each side of the midrib, entire; inflorescences axillary and/or terminal, tomentose; calyx brown 3–4 × 3–5mm; corolla yellow, 14–17 × 8–9mm; fruits globular, slightly square, green, c. 1.5 × 1.5cm; Forest; 700m.
Distr.: Nigeria, Cameroon, Gabon and Congo (Kinshasa) [Guineo-Congolian].
IUCN: LC
Ndanan 1: Cheek 11242 10/2002.
Note: 7 specimens known from Cameroon alone (Fl. Cameroun), (14 specimens in herbarium), 13 specimens known from Congo (Kinshasa) (Fl. Gabon).

Hugonia platysepala Welw. ex Oliv.

Climber, to 20m; young stems with short hairs, soon glabrescent; leaves elliptic to oblanceolate, drying dark green or black, 5.5–23 × 3.5–8cm, acuminate, serrate, lateral nerves c. 22 pairs; petiole 1–2cm; stipules digitate laciniate; inflorescence cymose, 2–14-many flowered; sepals unequal; petals 1.5–2cm, shortly clawed, yellow or white; drupe globose, 1.2cm diam. Forest; 700m.
Distr.: Guinea (Conakry) to Bioko, Congo (Kinshasa), Angola & Uganda [Guineo-Congolian].
IUCN: LC
Ndanan 1: Cheek 11953 3/2004.

LOGANIACEAE

L. Pearce (K), Y.B. Harvey (K), X. van der Burgt (K), J.-M. Onana (YA) & M. Cheek (K)

Fl. Cameroun 12 (1972).

Mostuea brunonis Didr. var. brunonis

Shrub, 0.3–3m; appressed pubescent on stem and midrib; leaves obovate-elliptic, to 10 × 4.5cm, acumen 0.5–1cm, mucronate, obliquely obtuse-acute, nerves 3–7 pairs; petiole 1–6mm; inflorescence sessile, axillary; flowers white with yellow centre; sepals linear, 6mm; fruit bilobed, obovate, flat, 7 × 10mm, brown, sparsely puberulent. Forest; 600–720m.
Distr.: Ghana to Madagascar [tropical Africa & Madagascar].
IUCN: LC
Mefou Proposed National Park: Etuge 5254 3/2004; **Ndanan 1:** Onana 2856 3/2004.

Strychnos boonei De Wild.

Fl. Cameroun Vol. 12: 62 (1972).
Climber; stems terete, lenticelled, pale brown; leaves papery to subcoriaceous, elliptic or ovate, (3–)5–8.5 × (2–)3.5–4.2cm, including acumen (c. 1cm), base rounded, palmately 3-nerved; petiole 4–8mm; tendrils axillary, solitary; inflorescence axillary and/or terminal, solitary, lax; flowers white or cream, yellow in bud, 3–5mm; fruit globose-ellipsoid, 2.5cm, dark purple, 1-seeded. Forests, often on the banks of rivers; 670–720m.
Distr.: Nigeria, Cameroon, CAR, Gabon, Congo (Kinshasa), Uganda [tropical Africa].
IUCN: LC
Mefou Proposed National Park: Cheek in MF 53 3/2004; 70 3/2004; **Ndanan 1:** Cheek 11562 3/2004; Onana 2880 3/2004; **Ndanan 2:** Darbyshire 294 3/2004.
Note: all material sterile/juvenile, inflorescence and fruit characters from Fl. Cameroun 12: 62–63 (1972). Cheek 11562 is a tentative determination (material sterile, leaves narrowly elliptic (to 8 × 3.5cm)), Pearce & Harvey vi.2006.

Strychnos dolichothyrsa Gilg ex Onochie & Hepper

Climber, smelling of cloves; stems terete; leaves thickly leathery, glossy above, narrowly elliptic to narrowly ovate, 9.5 × 3cm, including 1cm acumen, rounded to cuneate, oblique when immature, palmately 3-nerved, tertiary nerves subscalariform; petiole 0.5cm; tendrils axillary, single; inflorescence terminal; flowers 3mm; corolla urceolate, greenish white; fruit globose, 1.5cm. Forest; 700m.
Distr.: Cameroon to Congo (Brazzaville) [lower Guinea].
IUCN: NT
Ndanan 1: Onana 2189 10/2002.

Strychnos icaja Baill.

Liana glabous; tendrils solitary; stems terete, glabrous, green, sometimes with a shallow groove on each side;

leaves opposite, elliptic, 15 × 9cm, subacuminate, base obtuse, 3-nerved at base, junction arising 7mm above the petiole the basal nerves reaching the leaf apex, not looping at junction with other lateral nerves, 1.5cm from margin at midpoint, an additional, weaker although conspicuous, looping nerve inserted 2–3mm from the margin, additional lateral nerves only 1 or 2 connecting with the main basal laterals, intersecondaries numerous, almost as strong as these secondaries, quaternary nerves reticulate, conspicuous, inclusions not seen; petiole 1–1.3cm, articulated with the slightly swollen bases. Forest; 710m.
Distr.: Guinea (Conarkry) to Cameroon, Gabon, Congo (Brazzaville), Congo (Kinshasa), Angola and CAR [Guineo-Congolian].
IUCN: LC
Ndanan 1: Cheek 11673 3/2004; 11833 3/2004.

Strychnos longicaudata Gilg
Meded. Land. Wag. 69(1): 153 (1969); Fl. Cameroun 12: 92 (1972).
Liana; with single tendril (unpaired) 3cm; stems terete, grey-brown, lacking lenticels, minutely, moderately densely, patent puberulent with black hair; leaves opposite, elliptic, to 13 × 7cm, acumen 1.25cm, base obtuse to rounded, drying pale yellow-green below, darker above, 3-nerved from base, junction 3–5mm above the petiole, margin nerves up to 8mm from margin at midpoint, looping only moderately at junction with other secondary nerves (2–3 pairs in distal half of leaf), intersecondaries not prominent, quaternary nerves reticulate, prominent, lower surface with numerous raised rods in transmitted light white; petiole 3–5mm, indumentum as stem, base not very swollen. Forest undergrowth; 700m.
Distr.: Ivory Coast to Congo (Kinshasa) [Guineo-Congolian].
IUCN: LC
Ndanan 1: Onana 2815 3/2004.
Note: originally published as *S. longecaudata* Gilg in Engler Bot. Jahrb. 17: 570 (1893).

Strychnos phaeotricha Gilg
Climber; stems and midribs black-pilose; leaves papery, drying black above, pale green below, oblong-elliptic, 15(–23) × 4(–6)cm, acumen slender, 1.5(–2)cm, truncate-subcordate, weakly palmately 3-nerved, c. 10 pairs secondary nerves; petiole 0.3mm; tendrils paired; inflorescence terminal, lax, 5–8cm; flowers yellowish white, 4–5mm; fruit ellipsoid, 1.8–3cm, orange, soft, 3–7-seeded. Forest; 700–710m.
Distr.: Ghana to Congo (Kinshasa) [Guineo-Congolian].
IUCN: LC
Mefou Proposed National Park: Cheek in MF 17 3/2004; **Ndanan 1:** Cheek 11523 3/2004; **Ndanan 2:** Cheek 11128 10/2002.
Note: Cheek 11128 is sterile.

Strychnos talbotiae S.Moore
Climber; stems slender, terete; leaves (often narrowly) elliptic, 6.5–11 × 3.5–5cm, acumen c. 10mm, base obtuse, palmately 3-nerved; petiole 7–10mm; tendrils paired; inflorescence axillary, 2–3cm; flowers white (green in bud), 3cm; fruit ellipsoid, 3cm; several-seeded. Forest; 600–710m.
Distr.: Nigeria to Congo (Kinshasa) [lower Guinea & Congolian].
IUCN: LC
Mefou Proposed National Park: Etuge 5275 3/2004; **Ndanan 2:** Cheek 11936 3/2004.
Note: specimens in flower only.

Strychnos sp. aff. *staudtii* Gilg
Climber; one pair of tendrils; glabrous, main axis finely ridged, matt pale brown, with white lenticels; leafy-stem glabrous, glossy, slightly 4-angled; leaves opposite, narrow-elliptic, to 16 × 6cm, acumen 1.5cm, base obtuse, drying pale grey-green below, darker above, 3-nerved from base, junction 5mm above the petiole, marginal nerves c. 1cm from margin at midpoint, not looping strongly, other secondary nerves 3–4 pairs, in upper half, intersecondaries prominent, not scalariform, quaternary nerves reticulate, prominent, inclusions absent; petiole 5–6mm, articulated with a bulbous base; tendril 2cm, one pair of branches. Lowland evergreen forest near swamp; 700m.
Mefou Proposed National Park: Cheek in MF 123 3/2004; 18 3/2004; 45 3/2004.

LORANTHACEAE

R.M. Polhill (K)

Fl. Cameroun 23 (1982).

Phragmanthera nigritana (Hook.f. ex Benth.) Balle
Polhill R. & Wiens, D. Mistletoes of Africa: 268 (1998).
Parasitic shrub, to 2m; branchlets tomentose with reddish-brown dendritic hairs; leaves oblong-lanceolate to ovate, blunt, 4–7.5 × 1.5–4cm, brown-pubescent abaxially, with 6–8 pairs of inconspicuous nerves; umbels several, 2–4-flowered; corolla 3.5–4cm, densely hirsute; petals reflexing at anthesis. Forest, riverine forest; 680–700m.
Distr.: Ghana, Nigeria, Cameroon, Bioko, Angola [upper & lower Guinea].
IUCN: NT
Ndanan 1: Harvey 142 10/2002; **Ndanan 2:** Darbyshire 237 3/2004.

MALVACEAE

M. Cheek (K) & C. Couch (K)

Hibiscus surattensis L.
Spiny shrub, 1m; stem with long, fine hairs and prickles; leaves orbicular in outline, 3–5-lobed, divided

by half to three quarters, c. 6cm, serrate; petiole 4cm; stipules ovate, foliaceous, 1.5cm; flowers single, axillary; pedicel c. 6cm; epicalyx 10, each c. 2cm, bifurcate, foliaceous, shorter than calyx; corolla 2.5cm, orange-yellow, centre red. Secondary forest; 700m.
Distr.: palaeotropics [palaeotropics].
IUCN: LC
Ndanan 1: Cheek 11018 10/2002.

Hibiscus vitifolius L. subsp. *vitifolius*

Subshrub, 1.5–2.5m; stems with small spines and sparse patent 1–2mm hairs; leaf-blade 5–10cm diam., palmately 3–5-lobed for half the radius cordate, sparsely hairy; petiole 1.5–6cm; flowers single in upper axils; pedicel 5mm; sepals 5, triangular, 1cm, pubescent; epicalyx bracts 5–6, linear, flat, 6mm; corolla funnel-shaped, 4cm, yellow, with purple centre; staminal column purple-red; fruit with sepals enlarged, 1.5cm; capsule 5 lobed-ridged, 2cm diam.; seeds numerous, reniform, brown, 2mm. Secondary forest; 600–710m.
Distr.: Cameroon to Kenya and South Africa [Congolian E and S African].
IUCN: LC
Mefou Proposed National Park: Etuge 5257 3/2004;
Ndanan 1: Cheek 11809 3/2004.

Sida acuta Burm.

Subshrub, to 1.5m; stems woody at the base, pubescent; leaves lanceolate (occasionally ovate-lanceolate), 3.5–6 × 0.7–1.5cm, serrate; petiole 2–5mm; stipules generally larger than the petiole, linear; flowers solitary or paired; calyx pale green, 6–8mm; corolla orange-yellow. Roadside; 600m.
Distr.: Gambia, Sierra Leone, Liberia, Ivory Coast, Ghana, Togo, Nigeria, Cameroon, Equatorial Guinea, Gabon, CAR, Congo (Kinshasa), Burundi, Sudan, Uganda, Kenya, Tanzania, Mozambique, Malawi, Zambia, Zimbabwe, South Africa [tropical Africa].
IUCN: LC
Mefou Proposed National Park: Etuge 5100 3/2003.

Sida javanensis Cav.

Blumea 14: 178–184 (1966).
Syn. *Sida veronicifolia* Lam.
Syn. *Sida pilosa* Retz. Bol. Soc. Brot (Ser. 2) 54: 11 (1980).
Prostrate herb, rooting at nodes; stem puberulent; leaves ovate, 4cm, acute, cordate, crenate; petiole 3cm; pedicel 1cm; corolla orange, c. 0.7cm. Forest edge; 700–710m.
Distr.: pantropical.
IUCN: LC
Ndanan 1: Cheek 11221 10/2002; 11805 3/2004.
Note: Cheek 11221 & 11805 determined as *Sida veronicifolia* by Couch x.2005 (synonym of *S. javanensis*), Harvey x.2010.

Sida rhombifolia L. var. *a*

Subshrub, 0.5–1m, stellate-puberulent; leaves lanceolate-rhombic, to 6.5 × 2cm, rounded, acute,

serrate, lower surface completely obscured by minute white stellate hairs; flowers axillary, single; pedicels extending to 2.5cm in fruit; corolla yellow; mericarps c. 8, sides rugose-reticulate. Roadsides; 700m.
Distr.: pantropical.
IUCN: LC
Ndanan 1: Cheek 11020 10/2002.
Note: Cheek 11020 is tentatively placed in this taxon, Couch x.2005.

Urena lobata L.

Subshrub, 0.6m, stellate hairy; leaves elliptic, c. 6cm, slightly 3-lobed or entire, base rounded, teeth glandular, densely grey hairy below; petiole 2cm; flowers subsessile; epicalyx 5-lobed in upper half, 7mm; corolla pink, centre purple, 1cm; mericarps 5, spines with grapnel ends. Farmbush; 600–700m.
Distr.: pantropical.
IUCN: LC
Mefou Proposed National Park: Etuge 5250 3/2004;
Ndanan 1: Cheek 11016 10/2002.

Wissadula amplissima (L.) R.E.Fr. var. *rostrata* (Schum. & Thonn.) R.E.Fr.

Woody herb, to 2m; stems densely stellate-pubescent; leaves broadly ovate-cordate to suborbicular-cordate, up to 10 × 7.5cm, palmately nerved, abaxial surface densely stellately hairy to tomentose, adaxial surface minutely stellately hairy; petiole to 5cm, tomentose; terminal raceme to 20cm; pedicels 1–3cm; calyx c. 3mm, stellate pubescent; corolla yellow, c. 5mm. Roadside; 710m.
Distr.: Cape Verde Islands, Senegal, Mali, Guinea (Bissau), Guinea (Conakry), Sierra Leone, Liberia, Ivory Coast, Ghana, Togo, Niger, Nigeria, Cameroon, São Tomé, Congo (Kinshasa), CAR, Burundi, Sudan, Ethiopia, Eritrea, Uganda, Kenya, Tanzania, Mozambique, Malawi, Zambia, Botswana, Angola, South Africa [tropical Africa].
IUCN: LC
Ndanan 1: Cheek 11705 3/2004.

MELASTOMATACEAE

E. Woodgyer (K), E. Lucas (K), X. van der Burgt (K), M. Cheek (K) & S. Yasuda

Fl. Cameroun 24 (1983).

Dicellandra barteri Hook.f. var. *barteri*

Epiphyte-climber, to 4m; stems appressed to tree bark, adventitious roots numerous, densely red-brown hairy, hairs patent, apex curled, 1mm; leaves ovate-lanceolate, 11 × 5.5cm, subacuminate, rounded-abruptly cordate; petiole to 1.5cm; inflorescence cylindrical, to 6cm long; flowers purple. Forest; 700–710m.
Distr.: Liberia to Bioko & Cameroon [upper & lower Guinea].
IUCN: NT
Mefou Proposed National Park: Cheek in RP 17

3/2004; 29 3/2004; **Ndanan 1:** Cheek 11266 10/2002; 11623 3/2004.
Local Name: Abelbonga (Ewondo) (Cheek 11266).
Uses: MEDICINES (Cheek 11266).

Dichaetanthera africana (Hook.f.) Jacq.-Fél.
Syn. *Sakersia africana* Hook.f.
Shrub to small tree, c. 6–12m tall; leaves elliptic, strigose, apex acuminate; petiole c. 2cm long; inflorescence terminal panicle; flowers 4-merous; hypanthium and calyx lobes glabrous; buds c. 0.5cm long, petals pink; fruit a capsule. Secondary bush, lowland rainforest; 650–710m.
Distr.: Sierra Leone to Angola [upper & lower Guinea].
IUCN: LC
Mefou Proposed National Park: Cheek 13109 2/2006; **Ndanan 1:** Cheek 11647 3/2004; Tadjouteu 634 4/2004.

Dinophora spenneroides Benth.
Shrubby herb, 1.5m, glabrous; leaves ovate-oblong, 5–14 × 3–6cm, membranous, acuminate, cordate, 5–7-nerved at base; inflorescence lax, terminal panicles; calyx dentate, white; petals 1cm, pink; berries white. Forest edge; 700m.
Distr.: Guinea (Conakry) to Congo (Kinshasa) & Angola [Guineo-Congolian].
IUCN: LC
Ndanan 1: Cheek 11188 10/2002; 11229 10/2002; Harvey 189 10/2002.

Dissotis multiflora (Sm.) Triana
Erect herb, 0.6–1m; stems appressed hairy; leaves opposite; leaf-blade lancolate or elliptic, 6–9 × 3–4cm, apex acute, base rounded, abruptly decurrent, margin entire, 3 basal nerves, upper and lower surface with scattered appressed hairs; petiole 0.6–2cm; inflorescence terminal, 3–10-flowered, subcymose, 3–8cm; flower 1.5–2.5cm wide, pink; sepals broadly triangular with apical awn 2–3mm long, crowned with a ring of hairs, margins with bristles; capsule bell-shaped, broadest at base, 8mm, apex truncate, grey-white, hairs appressed. Semi-deciduous forest; 700–710m.
Distr.: Guinea (Conakry), Sierra Leone, Liberia, Ivory Coast, Nigeria, Cameroon, Congo (Brazzaville), Congo (Kinshasa) [lower & upper Guinea].
IUCN: LC
Mefou Proposed National Park: Cheek in RP 10 3/2004; **Ndanan 1:** Cheek 11005 10/2002; 11529 3/2004.

Guyonia ciliata Hook.f.
Prostrate herb, sparsely ciliate on stems, leaves and fruit; leaves ovate, 1.5cm, finely serrate; petiole to 2cm; flowers solitary, pink, 1cm diam.; stamens equal; fruit 5mm with leafy calyx. Forest edge; 710m.
Distr.: Sierra Leone to Tanzania [Guineo-Congolian (montane)].
IUCN: LC
Mefou Proposed National Park: Cheek in RP 55 3/2004.

Heterotis buettneriana (Cogn. ex Büttner) Jacq.-Fél.
Fl. Cameroun 24: 40 (1983).
Subprostrate herb; stem 4-angled, with stiff hairs on the angles; leaves opposite, ovate, 2.5–4 × 1.5–2.5cm, pale whitish green below, apex acute, margin denticulate, upper surface with scattered appressed hairs, lower surface with scattered hairs, 3 basal veins; petiole variable, to 2.8cm; inflorescence terminal, 1–2 flowered, not inclosed in bracts; flowers pale pink, <1cm diam.; calyx 7mm, bearing plumed linear appendages, calyx lobes awn-like, 3mm, hairy; stamens yellow, 10, all equal; fuits 9 × 6mm. Evergreen forest; 700m.
Distr.: S Cameroon, Gabon to W Congo (Kinshasa) [lower Guinea].
IUCN: LC
Ndanan 2: Cheek 11111 10/2002; Darbyshire 185 fl., fr., 3/2004.
Note: *Heterotis* is usually sunk into *Dissotis*, but here we follow Fl. Cameroun, Cheek x.2010.

Medinilla mirabilis (Gilg) Jacq.-Fél.
Fl. Cameroun 24: 115 (1983).
Syn. *Myrianthemum mirabile* Gilg
Stem twining liane; leaves in whorls of 3, oblanceolate or elliptic, to 22 × 9cm, subacuminate, cuneate-obtuse, 3-nerved; petiole to 3cm; inflorescence cauliflorous, near ground, fascicles of c. 40 flowers; peduncles to 2cm, much branched; pedicels 1.5cm; flowers 1cm, blue; berries 1cm. Forest; 600–700m.
Distr.: SE Nigeria, Bioko, Cameroon & Gabon [Lower Guinea].
IUCN: NT
Mefou Proposed National Park: Etuge 5212 3/2004; **Ndanan 1:** Onana 2199 10/2002.

Memecylon englerianum Cogn.
Syn. *Memecylon obanense* Baker f.
Shrub, 2–3m; stems rounded; leaves opposite, simple, broadly elliptic, 6–11 × 15–22cm, abruptly or gradually acuminate, 1–1.5cm, base obtuse-rounded then abruptly cuneate, lateral nerves 8–10 pairs, uniting in a looping marginal nerve 2–4mm from the edge, conspicuous; petiole 5–6mm; cymes axillary, 1–2 per node, lax, 4–7cm; peduncle 2–4cm, first branches 1–3cm; flowers umbellate; pedicels 2–4mm; hypanthium broadly campanulate, 3 × 2mm; calyx truncate - 4-sinuate; petals 2mm; anther connectives conical, with gland; fruit ellipsoid, 13 × 6mm. Evergreen forest; 720m.
Distr.: Guinea (Conakry) to Cameroon [upper & lower Guinea].
IUCN: LC
Ndanan 1: Onana 2851 3/2004; 2855 3/2004.

Tristemma leiocalyx Cogn.
Fl. Cameroun 24: 64 (1983).
Syn. *Tetraphyllaster rosaceum* Gilg
Herb, 20cm; leaves opposite, ovate, c. 4–5 × 2–3cm, upper and lower surface appressed hairy, 5 basal nerves; flowers terminal, 1–3 per head, pink, c. 2.5cm wide;

calyx ellipsoid, 8 × 7mm, glabrous, without rings of hairs; sepals 5, triangular, 3 × 2mm; stamens yellow. *Raphia* swamp; 710m.
Distr.: SW Cameroon, Congo (Kinshasa) and Uganda [Guineo-Congolian (montane)].
IUCN: LC
Ndanan 1: Cheek 11751 3/2004.

Tristemma mauritianum A.Juss.
Fl. Cameroun 24: 64 (1983).
Syn. *Tristemma incompletum* sensu Keay
Erect weedy shrub, 0.5–2m; stems 4-winged, c. 1cm, hairs fine, soft, grey; leaves ovate, c. 17 × 8cm, puberulent below; inflorescence capitate, sessile, c. 6-flowered; involucral bracts numerous; hypanthium with 2 rings of hairs; flowers pink, 2cm; stamens yellow, equal, without connective; berries red. Farms & farmbush; 600–710m.
Distr.: Nigeria to Congo (Kinshasa) [lower Guinea & Congolian].
IUCN: LC
Mefou Proposed National Park: Etuge 5268 3/2004; **Ndanan 1:** Cheek 11582 3/2004.

MELIACEAE

H. Fortune-Hopkins (K), M. Cheek (K), J.-M. Onana (K) & E. Koenen (WAG)

Carapa sp. of Mefou
Tree sapling, 7m, glabrous; lenticels raised, white, orbicular; leaves alternate, estipulate, pinnately compound, 80cm; petiole 24cm, terete; leaflets opposite, 4-jugate, terminal leaflet absent, narrowly elliptic or oblong, to 17 × 5cm, acumen slender, 6–7mm, base cuneate, lateral nerves 8 pairs, quaternary nerves reticulate, conspicuous, lower surface pale brown, midrib dark red-brown; petiolules 8mm. Lowland evergreen forest, secondary forest; 600–700m.
Mefou Proposed National Park: Cheek in MF 29 3/2004; Etuge 5285 3/2004; 5286 3/2004.
Note: a revision of *Carapa* by Kenfack is expected soon, enabling improved species delimitation. Fertile meterial is needed from Mefou is needed to enable identification to specie. This may be *C. parviflora* Harms. Cheek x.2010.

Entandrophragma sp.
Large timber tree; outer slash marbled red-pink and white, scented of Meliaceae; leaves lacking stipules, pinnately compound c. 40cm, with numerous elliptic-oblong leaflets, terminal leaflet usually absent; flowers white, 4mm with conspicuous staminal tube; fruit cigar-shaped-cylindrical 15–30 × 3–4cm, woody, with 5 falling valves revealing an angular woody central column with large winged seeds. Semi-deciduous forest.
Distr.: tropical Africa
IUCN: VU

Mefou Proposed National Park: Essama sight record 3/2004.
Note: this is likely to be one or several of *E. angolense*, *E. cylindricum* and *E. utile* all of which are IUCN VU listed widespread timber species, Cheek xi.2010.

Guarea cedrata (A.Chev.) Pellegr.
Large tree, to 45m; bark grey with concentric marking; slash scented of Meliaceae; young growth densely grey-brown softly tomentose; leaves (non-reproductive stems: dimensions larger) alternate, estipulate, pinnately compound, 10-40(-75)cm, 4-6-jugate with terminal leaflet; petiole 4-11cm, adaxially grooved, dilated at base; leaflets opposite or subopposite, oblong, to 30 × 6.5cm, acumen 1–1.5cm, base unequally obtuse-acute, lateral nerves 9–15 pairs, quaternary nervation finely reticulate, prominent; flowers cream coloured, in densely clustered panicles; fruit dull orange/brown, globose, 5–6cm, dehiscing to reveal bright orange, fleshy kidney-shaped seeds 2–3cm. Forest; 700m.
Distr.: Sierra Leone to Congo (Kinshasa) and Uganda [Guineo-Congolian].
IUCN: LC
Mefou Proposed National Park: Cheek in MF 23 3/2004; **Ndanan 1:** Cheek in MF 294 10/2002; 340 10/2002.

Guarea glomerulata Harms
Shrub or small tree, 1–5(–8)m; leaves c. 9–17-foliolate, rachis 20–38cm; petioles 6–17cm, terete, puberulent; leaflets elliptic or oblanceolate-elliptic, c. 5–20 × 3–6cm, long acuminate, acute, lateral nerves c. 15–30 pairs; inflorescence a pendulous spike, c. 0.5–1.5m; flowers pink, scattered on distal part; fruit vermilion/deep red, subglobose, rostrate, with 4 distinct lobes, c. 1.5cm, densely brown-hairy. Forest; 700m.
Distr.: Nigeria to Congo (Kinshasa) [lower Guinea and Congolian].
IUCN: LC
Mefou Proposed National Park: Onana 2801 3/2004; **Ndanan 1:** Harvey 160 10/2002; 164 10/2002; Tadjouteu 610 3/2004; **Ndanan 2:** Cheek 11152 10/2002.
Local Name: Ebege bemvah (Ewondo) (Cheek 11152).
Uses: MEDICINES (Cheek 11152).

Heckeldora ledermannii (Harms) J.J. de Wilde
Blumea 52(1): 184 (2007).
Shrub or small tree, to 5m; leaves 5–13-foliolate; petiole 4.5–15cm long; rachis (4–)16–29cm long; leaflets c. 10–20 × 2–7cm, acuminate; inflorescences raceme-like, to c. 70cm long; pedicel c. 1.5mm long; calyx tiny, <1 × 2mm, puberulous; flower 6–7 × 2–2.5mm, ovary glabrous; fruit (immature) obovoid, 2–3 × 1–2cm, dull grey-green, puberulous. Transition zone of evergreen tropical rainforest of low and medium altitudes towards submontane forest; 680–710m.
Distr.: Cameroon (SW province) [Cameroon endemic].
IUCN: EN

Mefou Proposed National Park: <u>Nana 47</u> 3/2004; **Ndanan 1:** <u>Cheek 11631</u> 3/2004; **Ndanan 2:** <u>Cheek 11091</u> 10/2002; <u>Darbyshire 201</u> 3/2004; <u>244</u> 3/2004.
Note: Darbyshire 201 is tentatively placed in this taxon (*H.* cf. *ledermannii*), by Koenen, v.2010. Darbyshire 244, Nana 47 and Cheek 11091 have a stout infructescence, ± perpendicular to stem, Koenen v.2010.

Heckeldora zenkeri (Harms) Staner
Blumea 52(1): 195 (2007).
Shrub or treelet, to 4m; leaves 3–13-foliolate; petiolules 1–5mm long, subglabrous; leaflets narrowly elliptic to elliptic to narrowly obovate, 4–21 × 2.5–7cm, acuminate; inflorescence a raceme to 80cm long; calyx 2–3.5 × 3–4mm, subglabrous; corolla 8–11 × 2.5–4mm, greenish-white to pink; ovary sericeous to sparsely hirsute; fruit ovoid to obovoid but irregularly shaped and knobbed, 3–7 × 2–5cm, yellowish, velutinous. In shrub layer or evergreen, semi-deciduous and secondary forest; 600m.
Distr.: Nigeria, Cameroon, Equatorial Guinea, Gabon, Congo (Brazzaville) [lower Guinea & Congolian].
IUCN: LC
Mefou Proposed National Park: <u>Etuge 5229</u> 3/2004.

Khaya ivorensis A.Chev.
Tree, 30–45m; buttresses blunt to 3–4m from ground; slash red, bark dark, scaly; leaves paripinnate, 5–6-jugate, rachis c 10cm; petiole 3cm; leaflets oblong c. 10 × 5cm, bluntly acuminate, base rounded, lateral nerves 5 pairs; fruit globose, 4–7cm, 5-valved. Forest; 710m.
Distr.: Ivory Coast & Ghana, Nigeria to Gabon [upper & lower Guinea].
IUCN: VU
Mefou Proposed National Park: <u>Cheek sight record 1</u> 3/2004.
Note: known only from a sight record in the habitat notes of Cheek 11829.

Lovoa trichilioides Harms
Tree, 20m; bark dark, lenticellate; leaves paripinnate, 6–jugate; leaflets elliptic-oblong to 20 × 9cm, acuminate, base obtuse to rounded; inflorescence subcorymbose, 10–15cm; flowers white, c 4mm; fruit sharply cylindrical, c. 4–7cm, thinly woody; seeds winged. Forest; 700–710m.
Distr.: Sierra Leone to Ghana, Nigeria to Congo (Kinshasa) [Guineo-Congolian].
IUCN: VU
Ndanan 1: <u>Cheek 11533</u> 3/2004; <u>Cheek in MF 272</u> 10/2002.
Uses: MATERIALS — Wood — timber (Cheek 11533).

Trichilia gilgiana Harms
Tree, 12m; 1m girth, slash red; densely puberulous; leaves imparipinnate, c. 5-jugate, rachis 14cm; petiole 10cm; leaflets with dots and dashes in transmitted light, oblanceolate to oblong, c. 13 × 3.5cm, acuminate, acute, lateral nerves, c. 12 pairs; fruits globose 2cm, yellow-brown puberulent. Forest; 700m.

Distr.: Nigeria to Congo (Kinshasa) [lower Guinea & Congolian].
IUCN: LC
Ndanan 1: <u>Cheek in MF 213</u> 10/2002; <u>387</u> 10/2002.
Note: plot voucher series MF213 & MF387 (MF387 could be *T. gilgiana* or *T. monadelpha*) are possibly this taxon, Hopkins iii.2010.

Trichilia monadelpha (Thonn.) J.J.de Wilde
Meded. Land. Wag. 68(2): 108 (1968); Keay R.W.J., Trees of Nigeria: 351 (1989); Hawthorne W., F.G.F.T. Ghana: 178 (1990).
Syn. *Trichilia heudelotii* Planch. ex Oliv.
Tree, to 15–18m, densely puberulous; leaves imparipinnate, c. 5-jugate, rachis 14cm; petiole 10cm; leaflets without dots and dashes, oblanceolate to oblong, c. 13 × 3.5cm, acuminate, acute, lateral nerves c. 12 pairs; fruits globose, 2cm, yellow-brown puberulous, stipe c. 3mm. Forest; 700m.
Distr.: Guinea (Bissau) to Congo (Kinshasa) [Guineo-Congolian].
IUCN: LC
Ndanan 1: <u>Cheek 11058</u> 10/2002.

Trichilia rubescens Oliv.
Tree, (4–)5–12(–18)m; twigs densely lenticellate, hollow; leaves c. 5-jugate, rachis 18cm; petiole 5cm; leaflets drying red-brown below, oblong-elliptic, c. 16 × 5cm, acuminate, acute, lateral nerves c. 10 pairs; inflorescences terminal c. 25cm; flowers yellow-white, c. 4mm wide; fruit 2.5cm, glabrous, glossy dark brown. Forest; 700–710m.
Distr.: Nigeria to Tanzania [lower Guinea & Congolian].
IUCN: LC
Ndanan 1: <u>Cheek 11729</u> 3/2004; <u>11848</u> 3/2004; <u>Cheek in MF 299</u> 10/2002; <u>Tadjouteu 576</u> 3/2004; <u>619</u> 3/2004.

Trichilia tessmannii Harms
Meded. Landbou. Wag. 68–2: 171 (1968).
Syn. *Trichilia lanata* A.Chev.
Tree, 12m, densely yellow-brown patent-tomentose; leaves 8-jugate, rachis 18cm; petiole 10cm; leaflets narrowly oblong or oblanceolate, subsessile, 12 × 3.5cm, acuminate, obtuse, lateral nerves c. 19 pairs, highly prominent below, impressed above; fruit subglobose 3–3.5cm, sparsely long-hairy. Forest; 700m.
Distr.: Sierra Leone to Ghana, Nigeria to Congo (Kinshasa) [Guineo-Congolian].
IUCN: LC
Ndanan 1: <u>Cheek 11031</u> 10/2002; <u>Cheek in MF 304</u> 10/2002; <u>378</u> 10/2002.
Note: plot voucher series MF304 and MF378 are sterile. If they do prove to be Meliaceae, they are probably this taxon, *Trichilia tessmannii*, Hopkins iv.2010.

Trichilia zewaldae De Wild.
Meded. Land. Wag. 68–2: 190 (1968).
Tree, to 25m; dbh to 60cm, glabrous; leaves compound,

imparipinnate, leaflets 4–5 pairs; leaflets elliptic to oblanceolate, to 20 × 7cm, acumen 1cm, base attenuate, 12–13 nerves on each side of the midrib, entire; inflorescence a panicle, terminal, to 20cm; flowers very fragrant; calyx 2mm, pale green; corolla creamy yellow, 0.5–0.8cm; fruit a hard, brown capsule, approx. 3cm diam.; seeds black and orange-red. Forest; 600m.
Distr.: Cameroon [Cameroon Engdemic].
IUCN: EN
Mefou Proposed National Park: Etuge 5220 3/2004; **Ndanan 1:** Cheek in MF 254 10/2002.
Note: plot voucher series MF254 is possibly *Trichilia zewaldae*, Hopkins iii.2010.

Turraea vogelii Hook.f. ex Benth.
Woody climber, to 12m, rarely a shrub to 2m, glabrous; leaves simple, elliptic, c. 11 × 5cm, subacuminate, obtuse; petiole 0.5cm; inflorescence subumbellate, axillary; peduncle 3.5cm; pedicels 1cm; flowers white, 3–10; staminal tube 1.5cm; fruit globose, 2.5cm. Forest; 600–700m.
Distr.: Ghana to Uganda [Guineo-Congolian].
IUCN: LC
Mefou Proposed National Park: Etuge 5252 3/2004; **Ndanan 1:** Gosline 425 10/2002; Harvey 71 10/2002.

MENISPERMACEAE

C. Couch (K), L. Pearce (K), X. van der Burgt (K) & M. Cheek (K)

Jateorhiza macrantha (Hook.f.) Exell & Mendonça
Woody climber, to 10m; stems, petioles and leaf-margins with stiff brown hairs; leaves to 20cm diam., 3–5-lobed, each lobe acutely acuminate, chartaceous, venation reticulate, the nerves setose; petiole to 20cm; ♂ flowers in lax panicles, to 30cm long, sessile on lateral branchlets; ♀ flowers in simple racemes, with conspicuous laciniate, deciduous bracts; fruit ovoid, c. 2.5cm long, conspicuously echinate. Forest; 700m.
Distr.: S Nigeria, Cameroon, Bioko, Gabon, Congo (Kinshasa), Angola (Cabinda) [Guineo-Congolian].
IUCN: LC
Mefou Proposed National Park: Nana 40 3/2004; **Ndanan 1:** Cheek in MF 379 10/2002; Harvey 95 10/2002.

Kolobopetalum auriculatum Engl.
Woody climber, to c. 20m; outer layer of stems pale brown-yellow, sloughing off, fruiting stems c. 5mm diam.; leaves drying ± greyish beneath, elliptic to ovate, acutely or obtusely auriculate, to widely cordate, caudate-acuminate, often remotely unevenly-dentate, 5–15 × 3.5–9cm; petioles twining, 2–9cm; ♂ inflorescence to 60cm; lateral branches short; ♀ flowers green; fruiting on woody parts; fruit 1–1.5cm long. Forest; 700–710m.
Distr.: Ghana, Nigeria, Cameroon, Gabon [Guineo-Congolian].
IUCN: LC

Ndanan 1: Cheek 11901 3/2004; Cheek in MF 386 10/2002; Onana 2188 10/2002.
Uses: FOOD — Infructescences — fruits edible (Cheek 2188, 11901).

Penianthus zenkeri (Engl.) Diels
Shrubbery or small treelet, 1.5–6m; leaves drying bright green, elliptic to oblanceolate, 24–41 × 8–15cm, cuneate to shortly rounded; midrib very abruptly prominent above; petioles up to 30cm; perianth spreading to reflexed (not ± closed); fruits 2–4 × 1.1–1.8(–2.0)cm. Forest; 600–710m.
Distr.: SE Nigeria, Cameroon, E Congo (Kinshasa) [Guineo-Congolian].
IUCN: LC
Mefou Proposed National Park: Nana 65 3/2004; **Ndanan 1:** Cheek 11543 3/2004; 11662 3/2004.

Platytinospora buchholzii (Engl.) Diels var. *macrophylla* Diels
Pflanzenr. 46: 170 (1910).
Twining herb, to 8–10m; stems and branches spindly; leaves peltate, 10–13 × 5.5–7.5(–9.5)cm, submembranaceous, ovate to ovate-oblong, rounded to very shallowly cordate, acuminate (acumen linear, 6–8mm), entire or very rarely denticulate towards base; petiole 4–5(–9)cm, geniculate at base, ± twisted; ♂ inflorescences in axillary or supra-axillary racemes, 4–6cm; pedicels c. 1.5mm; bracteoles c. 1mm; ♀ inflorescences and flowers unknown; infructescence racemose, 10–20cm; drupes ovoid, c. 2 × 1.5cm. Forest, riverine forest, farmbush; 693–700m.
Distr.: Cameroon, Nigeria [Guineo-Congolian].
IUCN: EN
Ndanan 1: Cheek in MF 310 10/2002; Onana 2926 3/2004.

Stephania dinklagei (Engl.) Diels
Woody climber, to c. 10m, glabrous; leaves ovate to suborbicular, 7–15cm diam., margin undulate; petiole 5–12cm; inflorescence to 50cm, the main axis bearing numerous lateral, long-pedunculate racemes or pseudo-umbels; flowers greenish. Forest; 700–710m.
Distr.: Guinea to Cameroon, Congo (Brazzaville), Angola [Guineo-Congolian].
IUCN: LC
Ndanan 1: Cheek 11807 3/2004; 11871 3/2004; Cheek in MF 235 10/2002.

Synclisia scabrida Miers
Woody climber; leaves ovate-cordate, 7.5–10 × 5–6.5cm, hirsute; petioles 2.5–3cm; flowers unknown; infructescence axillary; fruit a drupe, 1.5–2 × 1cm, hirsute. Forest; 700m.
Distr.: SE Nigeria, Cameroon, Gabon, Congo (Kinshasa) and Angola [Guineo-Congolian].
IUCN: LC
Ndanan 2: Cheek 11079 10/2002.

Syntriandrium preussii Engl.

Climber, to c. 10m; stems glabrous, striate, ± herbaceous; leaves coriaceous, usually conspicuously 3-foliolate, sometimes lobed or entire, often obliquely auriculate at base, cordate, drying pale green, nerves yellow; ♂ inflorescence a very long and slender panicle to 150cm; flowers greenish-yellow, minute, early caducous; ♀ inflorescence racemose, c. 10–20cm, much < than ♂ fruits c. 1cm diam. Forest; 650–693m.
Distr.: S Nigeria, Cameroon, Gabon, Congo (Kinshasa) [Guineo-Congolian].
IUCN: LC
Ndanan 1: Onana 2928 3/2004; Tadjouteu 626 4/2004.

Triclisia sp. 1 of Mefou

Liana seedling, to 15cm; stem striate, ferruginous-tomentose; lamina to 11 × 4cm, lanceolate, base subcordate, apex mucronate, densely tomentose on primary and secondary veins otherwise sparsely pubescent; petiole to 3.4cm, densely ferruginous-tomentose. Forest, near swamp-edged river; 700m.
Mefou Proposed National Park: Cheek in MF 50 3/2004.
Note: plot voucher MF50, material insufficient for further determination, Pearce iii.2010.

MORACEAE

N. Rønsted, H. Fortune-Hopkins (K), B.J. Pollard, J.-M. Onana (YA) & X. van der Burgt (K)

Fl. Cameroun 28 (1985).

Antiaris toxicaria Lesch. subsp. *welwitschii* var. *welwitschii* (Engl.) C.C.Berg

Bull. Jard. Bot. Nat. Belg. 48: 466 (1978); Fl. Cameroun 28: 106 (1985); Keay R.W.J., Trees of Nigeria: 298 (1989).
Syn. *Antiaris welwitschii* Engl.
Syn. *Antiaris toxicaria* Lesch. subsp. *africana* (Engl.) C.C.Berg var. *welwitschii* (Engl.) Corner Bull. Jard. Bot. Nat. Belg. 47: 316 (1977).
Tree, to 40+m high with self-pruning lateral branches, monoecious or dioecious; leaves distichous; leaf-blades subcoriaceous to coriaceous (chartaceous in juveniles), oblong-elliptic, 6–15 × 3–12cm, obtuse to subcordate at base, obtuse to acuminate at apex, margin subentire (often dentate in juveniles), both surfaces with stiff adpressed hairs; petioles 3–10mm long; stipules semi-amplexicaul, 3–10mm long; inflorescences unisexual, on minute spurs; ♂ 0.6–1.2cm diam., with numerous flowers; on peduncles 0.5–1.5cm long; ♀ uniflorous; fruit drupaceous with enlarged red or orange fleshy receptacle, ellipsoid, to 1.5 × 1cm. Forest; 700–710m.
Distr.: Sierra Leone to Congo (Kinshasa), Angola, Uganda, Tanzania and Zambia [Guineo-Congolian].
IUCN: LC
Mefou Proposed National Park: Cheek in MF 44 3/2004; **Ndanan 1:** Cheek 11670 3/2004; Cheek in MF 257 10/2002; 291 10/2002.

Dorstenia kameruniana Engl.

Fl. Gabon 26: 34 (1984); Fl. Cameroun 28: 32 (1985).
Syn. *Craterogyne kameruniana* (Engl.) Lanjouw
Shrub or undershrub, 0.5–3(–6)m high, branched; leaves distichous; leaf-blades chartaceous to subcoriaceous, variable in shape but often elliptic, 7–16 × 2.5–9cm, cuneate to round and asymmetric at base, distally often truncate and coarsely dentate with acuminate apex, margin coarsely dentate or lobed (to ± entire), both surfaces sparsely hairy; petioles 3–10mm long; stipules semi-amplexicaul, 3–10mm long; inflorescences 1(–3) per axil; peduncle 2–9mm long; receptacle broadly turbinate to discoid, upper surface 3–8mm diam., puberulous; bracts in 2 rows around margin. Forest, including regrowth, often near streams; 710m.
Distr.: Guinea (Conakry to Tanzania and SW to Angola [tropical Africa].
IUCN: LC
Ndanan 1: Cheek 11842 3/2004.

Dorstenia poinsettiifolia Engl. var. *poinsettiifolia*

Rhizomatous herb with upright stems, 0.3–1m high; woody near base, young parts with dense pale hooked hairs; leaves spirally arranged, blades chartaceous, elliptic-obovate and attenuate towards base, 4–20 × 2–8cm, subcordate at base, acute or acuminate at apex, margin subentire, upper surface glabrous, lower surface sparsely hairy on venation; petioles to 2cm; stipules subulate, to 6mm long; inflorescences often paired; peduncle to 1cm; receptacle discoid in flower and turbinate in fruit, upper surface elliptic, 6–15 × 2–10mm, margin lobed, 1–5mm wide, with a projection at either end. Forest, often near rivers or in swamps; 700m.
Distr.: Cameroon and Gabon [lower Guinea].
IUCN: NT
Ndanan 1: Cheek 11063 10/2002.

Dorstenia psilurus Welw. var. *psilurus*

Fl. Gabon 26: 77 (1984); Fl. Cameroun 28: 74 (1985).
Syn. *Dorstenia scabra* (Bureau) Engl.
Syn. *Dorstenia tenuifolia* Engl.
Herb, to 60cm; stem ascending to erect, often branched, densely puberulous; leaves crowded apically; lamina glabrous, lanceolate to elliptic, (2–)5–19 × (1–)2–8cm, often papyraceous, margin usually denticulate to coarsely dentate; inflorescences solitary or paired; receptacle vertical, naviculate; fringe almost lacking, appendages terminal, 2, filiform, the upper one 2–7(–10)cm, the lower 0.3–3.0cm. Forest, particularly along rivers; 700m.
Distr.: Nigeria, Cameroon to Sudan, Uganda, S to Angola, Mozambique and Zimbabwe [tropical and subtropical Africa].
IUCN: LC
Ndanan 2: Darbyshire 186 fl., 3/2004.

Dorstenia turbinata Engl.

Syn. *Dorstenia ledermannii* Engl.
Syn. *Dorstenia buesgenii* Engl.

Shrub, 0.5–1.2m high; leaves ± distichous; leaf-blades chartaceous, elliptic-obovate, 3.5–5 × 9–13cm, cuneate at base, acuminate at apex, margin entire, glabrous above, sparsely and minutely hairy below; petioles 0.5–1cm long; stipules semi-amplexicaul, to 5mm long; inflorescences solitary or in pairs; peduncle 1.5cm long; receptacle turbinate, upper surface star-shaped, 5mm diam., margin bearing (2–)4(–6) appendices in a single row, each narrowly spatulate, to 2.5cm long. Forest, primary and secondary; 700–710m.
Distr.: Sierra Leone to Cameroon and Gabon [Guineo-Congolian].
IUCN: LC
Mefou Proposed National Park: <u>Nana 58</u> 3/2004; **Ndanan 1:** <u>Cheek 11722</u> 3/2004; <u>Harvey 197</u> 10/2002; **Ndanan 2:** <u>Cheek 11086</u> 10/2002.

Ficus asperifolia Miq.
Syn. *Ficus warburgii* Winkl.
Climber, to 10m; stem twining anti-clockwise; leaves elliptic, 8–22 × 3.5–7.5cm, long-acuminate, cuneate, sinuate-denticulate, secondary and tertiary nerves white-margined; petiole 1.75cm; stipules caducous; figs on naked stems c. 2mm diam., 0–2m above ground, 1–8 from brachyblasts, spherical, 0.8(–2)cm, glabrous, orange; pedicel 4mm; bracts 3, alternate. Forest; 645m.
Distr.: Senegal to Tanzania [Guineo-Congolian].
IUCN: LC
Mefou Proposed National Park: <u>Rønsted 215</u> 3/2004.

Ficus cyathistipula Warb.
Fl. Gabon 26: 232 (1984); Fl. Cameroun 28: 246 (1985); Kew Bull. 43: 82 (1988); Hawthorne W., F.G.F.T. Ghana: 122 (1990).
Epiphytic strangling shrub; leaves oblong-oblanceolate, to 18 × 6cm, prominently veined below; stipules triangular, 1 × 1cm, glabrescent; figs c. 1.8cm diam., glabrous. Forest; 650–710m.
Distr.: Ivory Coast to Angola, Uganda, Kenya, Tanzania, N Zambia [afromontane].
IUCN: LC
Mefou Proposed National Park: <u>Rønsted 204</u> 3/2004; <u>219</u> 4/2004; **Ndanan 1:** <u>Cheek 11638</u> 3/2004.

Ficus dryepondtiana Gentil ex De Wild.
Tree, to 30m or shrub, glabrous; leaves drying a characteristic dark brown, oblong, 22–33 × 10–13cm, apex acuminate, base cordate, glabrous except midrib and lateral nerves below; petiole 4–8cm; stipules persistent, 2.5–3.5cm; figs globose, 3.8cm diam., wrinkled; bracts 2, 4mm. Secondary forest and farmbush on cinder cone; 675m.
Distr.: Cameroon to Congo (Kinshasa) & CAR [lower Guinea and Congolian].
IUCN: LC
Mefou Proposed National Park: <u>Rønsted 209</u> 3/2004.

Ficus exasperata Vahl
Tree, c. 15m tall, beginning as a strangler; leaves

elliptic (lobed and longer when juvenile), 10 × 6.5cm, obtuse at base and apex, obscurely crenate-dentate, scabrid above and below, lower 2 nerve pairs with waxy axils; petiole c. 2.5cm; stipules not amplexicaul; figs c. 1.5cm on leafy branches, single; bracts 3, scattered on pedicel; ostiole circular. Farmbush and secondary, deciduous forest; 600–700m.
Distr.: tropical Africa to S India & Sri Lanka [paleaotropics].
IUCN: LC
Mefou Proposed National Park: <u>Etuge 5244</u> 3/2004; <u>Rønsted 200</u> 3/2004; <u>217</u> 3/2004; **Ndanan 2:** <u>Cheek 11172</u> 10/2002.
Note: distinctive in the combination of non-amplexicaul stipules, leaf indumentum and fig position.
Local Name: Sandpaper leaf (Rønsted 200).
Uses: MATERIALS — Unspecified Materials — for cleaning, as sponge (Rønsted 200).

Ficus lingua Warb. ex De Wild. & T.Durand subsp. *lingua*
Tree, to 30m, hemi-epiphyte or terrestrial; branchlets 1–4mm diam., with brownish puberulous hairs, periderm often flaking; leaf-blades coriaceous, oblanceolate to narrowly obtriangular, 0.5–5 × 0.3–2cm, cuneate to obtuse at base, obtuse, truncate or emarginate at apex, both surfaces glabrous, secondary veins 5–8 pairs; petioles 2–8mm long; stipules 2–5mm long, usually persistent; figs in pairs in leaf-axils or below leaves; on glabrous peduncles 2–5mm long; receptacles ± 5mm diam. when fresh, thin-walled, with caducous basal bracts. Forest; 675–710m.
Distr.: Ivory Coast, Cameroon and Gabon to Uganda and Kenya, and S to Mozambique [Guineo-Congolian].
IUCN: LC
Mefou Proposed National Park: <u>Rønsted 208</u> 3/2004; **Ndanan 1:** <u>Cheek 11676</u> 3/2004.

Ficus lyrata Warb.
Shrub or tree, to 15m, or an epiphyte; branchlets stout, glabrous, longitudinally grooved when dry; leaf-blades coriaceous, obovate to violin-shaped, very large, 20–45 × 13–26cm, cordate-auriculate at base, round or truncate-apiculate at apex, glabrous on both surfaces, secondary veins 5–6 pairs; petioles to 3cm long; stipules large, triangular, persistent; figs axillary, sessile, globose, c. 4cm diam., with 3 adpressed basal bracts. Semi-deciduous forest; 710m.
Distr.: Sierra Leone to Cameroon [upper & lower Guinea].
IUCN: LC
Ndanan 1: <u>Cheek 11929</u> 3/2004.

Ficus mucuso Welw. ex Ficalho
Tree, to 40m; trunk to 1m, smooth, grey and brown; leaves broadly ovate c. 10 × 9cm, apex obtuse to rounded, sometimes subacuminate, base deeply cordate, scabrid above, subvelvety below; petiole 3cm, long hairy; stipules ovate, c. 12 × 6mm, acuminate, long hairy; figs

in groups on main branches, 3–4cm diam.; bracts 3, whorled. Farmbush and secondary forest; 700m.
Distr.: Guinea (Bissau) to Tanzania, Ethiopia [Guineo-Congolian].
IUCN: LC
Mefou Proposed National Park: Rønsted 207 3/2004.

Ficus ottoniifolia (Miq.) Miq. subsp. *ottoniifolia*
Shrub or tree, to 17m, usually an epiphytic strangler at first; branchlets smooth pale; leaves elliptic-oblong, to 12 × 6cm, base obtuse or acute, secondary nerves bifurcating c. 1.5cm from margin; petiole c. 4cm; receptacles ellipsoid, 1.3–2.3cm diam.; figs long-stalked, in pairs or clusters on older branches below leaves, globose, 1.5–2.5cm in diam., bracts caducous; peduncle 1cm. Forest; 700m.
Distr.: Sierra Leone to Uganda [Guineo-Congolian].
IUCN: LC
Ndanan 1: Cheek 11960 3/2004.

Ficus ovata Vahl
Epiphytic shrub, glabrous; leaves thickly coriaceous, ovate to 28 × 17cm, acuminate, obtuse, entire, lateral nerves c. 10 pairs; petiole c. 6cm; stipules caducous; figs axillary, single, sessile, elliptic, 3.8 × 2.5cm, rarely globular, dark yellow puberulent; basal bracts forming a sinuate-margined puberulent cup, 1cm diam. Farmbush; 650m.
Distr.: Senegal to Mozambique [tropical Africa].
IUCN: LC
Mefou Proposed National Park: Rønsted 205 3/2004.

Ficus preussii Warb.
Strangler, 15m; leaves coriaceous, elliptic, c. 23 × 8cm, sharply acuminate, obtuse to rounded entire, lateral nerves c. 5 pairs, prominent, white, tertiary nerves conspicuous; petiole c. 4cm; stipules c. 35 × 10mm, persistent; figs axillary, single, sessile, globose, 4cm diam., densely pubescent; basal bracts 2, 5mm. Forest; 675–700m.
Distr.: Nigeria to Uganda [lower Guinea & Congolian].
IUCN: LC
Mefou Proposed National Park: Rønsted 211 3/2004; 220 4/2004; **Ndanan 1:** Cheek 11962 3/2004.

Ficus sansibarica Warb. subsp. *macrosperma* (Mildbr. & Burret) C.C.Berg
Kew Bull. 43: 94 (1988); Kirkia 13(2): 276 (1990).
Syn. *Ficus macrosperma* Mildbr. & Burret
Tree, to 25m high, hemi-epiphytic (and strangling) or terrestrial; branchlets 2–5mm diam., ± glabrous, periderm often flaking; leaf-blades coriaceous or subcoriaceous, elliptic or narrowly elliptic, 4.5–10 × 2–6cm, rounded to cordate or subacute at base, acuminate to subacute at apex, both surfaces glabrous, secondary veins 8–12 pairs; petioles c. 2cm long; stipules to 1.5cm long, glabrous; figs 2–4 on a spur; peduncles to 2.5cm long; receptacles 2–6(–10)cm diam., thick-walled, wrinkled when dry. Forest, including riverine and lake-side; 675–700m.

Distr.: Sierra Leone to Angola, Uganda and into N Zambia [tropical & subtropical Africa].
IUCN: LC
Mefou Proposed National Park: Rønsted 212 3/2004; 218a 3/2004; **Ndanan 1:** Cheek 11963 3/2004.
Note: Cheek 11963 and Rønsted 218a are tentatively placed in this taxon, Rønsted ix.2004.

Ficus sur Forssk.
Syn. *Ficus capensis* Thunb.
Tree, 5–20(–30)m, but fruiting at only 5m; stem to 60cm dbh; leaves elliptic-oblong, c. 14 × 7cm, shortly acuminate, rounded to obtuse, margin with c. 5 well-marked serrations on each side, subglabrous; petiole c. 4cm; stipules caducous; figs on branches c. 15cm long on main branches or trunk apex, (lowest c. 6m from ground), c. 2cm diam.; peduncle 7mm; bracts 3, c. 1.5mm, whorled. Farmbush and secondary forest; 700m.
Distr.: Senegal to South Africa [tropical Africa].
IUCN: LC
Mefou Proposed National Park: Nana 26 3/2004.

Ficus vogeliana (Miq.) Miq.
Tree, 6–20m, closely resembling *F. sur*, but usually specific to riverbanks and swamp forest; leaves usually conspicuously hairy and mottled white on suckers and juvenile plants; stipules subpersistent; figs at ground level on leafless branches several metres long with bracts c. 5mm. River edges; 600–700m.
Distr.: Guinea (Bissau) to Uganda [Guineo-Congolian].
IUCN: LC
Mefou Proposed National Park: Etuge 5132 3/2004; Rønsted 201 3/2004; 202 3/2004; 213 3/2004; 214 3/2004; **Ndanan 2:** Gosline 417 10/2002.
Local Name: Tol (Etuge 5132).

Milicia excelsa (Welw.) C.C.Berg
Bull. Jard. Bot. Nat. Belg. 52: 227 (1982); Fl. Gabon 26: 6 (1984); Fl. Cameroun 28: 9 (1985); Keay R.W.J., Trees of Nigeria: 298 (1989); Hawthorne W., F.G.F.T. Ghana: 118 (1990); Enum. Pl. Afr. Trop. 2: 141 (1992).
Syn. *Chlorophora excelsa* (Welw.) Benth.
Tree, 40m; leaves oblong-elliptic, c. 13 × 7.5cm, apex rounded or slightly acuminate, base unequally truncate to subcordate, sinuate, lateral nerves c. 15 pairs; petiole 6cm. Semi-deciduous forest; 600–720m.
Distr.: Ivory Coast to Mozambique [tropical Africa].
IUCN: LC
Mefou Proposed National Park: Etuge 5151 3/2004; **Ndanan 1:** Cheek 11903 3/2004; Onana 2885 3/2004.
Uses: MATERIALS — Wood — timber (Etuge 5151).

Streblus usambarensis (Engl.) C.C.Berg
Kon. Ned. Akad. Wetensch. (Ser. C) 91: 357 (1988).
Syn. *Neosloetiopsis kamerunensis* Engl.
Syn. *Sloetiopsis usambarensis* Engl. Fl. Gabon 26: 10 (1984); Fl. Cameroun 28: 12 (1985).
Shrub or small tree, to 5(–10)m, mostly dioecious; leaves distichous; blades subcoriaceous, elliptic,

3–11(–16) × 1–3.8(–6)cm, cuneate at base, acuminate-caudate at apex, margin entire or sometimes toothed, both surfaces glabrous or subglabrous; petioles 2–7mm long; stipules semi-amplexicaul, lanceolate, 2–8mm long; inflorescences axillary; ♂ catkin-like, to 5cm long; ♀ uniflorous with bracts around base and far-exserted; style forked; fruit a dehiscent drupe, ellipsoid, partly enclosed by accrescent tepals. Forest, often riverine, also in secondary regrowth; 700–710m.
Distr.: Guinea (Conakry) to Mozambique [tropical Africa].
IUCN: LC
Ndanan 1: <u>Cheek 11204</u> 10/2002; <u>11561</u> 3/2004.

Treculia acuminata Baill.
Bull. Jard. Bot. Nat. Belg. 47: 396 (1977); Fl. Gabon 26: 17 (1984); Fl. Cameroun 28: 19 (1985).
Syn. *Treculia zenkeri* Engl. F.W.T.A. 1: 613 (1958).
Shrub, to 3m, dioecious; leaves distichous; leaf-blades chartaceous, elliptic, 6–24 × 2.5–8cm, broadly cuneate at base, acuminate at apex, margin entire, glabrous above and sparsely puberulent on venation below; petioles 3–9cm; stipules amplexicaul, subpersistent, 3–8mm long, ciliolate; inflorescences globose; × axillary, 1 × 0.4cm, on short shoots with bracts; × c. 4mm diam.; infructescences to 3.5cm diam., covered by numerous bracts, each 5mm long with a hooked tip. Forest; 550m.
Distr.: SE Nigeria, Cameroon and Gabon [lower Guinea].
IUCN: LC
Mefou Proposed National Park: <u>Etuge 5196</u> 3/2004.

Treculia africana Decne. subsp. *africana* var. *africana*
Bull. Jard. Bot. Nat. Belg. 47: 384 (1977); Fl. Cameroun 28: 16 (1985); Keay R.W.J., Trees of Nigeria: 300 (1989); F.T.E.A. Moraceae: 10 (1989).
Syn. *Treculia africana* Decne. var. *africana*
Tree, to 30+m, dioecious; leaves distichous; leaf-blades coriaceous, elliptic, somewhat asymmetric, 10–25(–45) × 4–12(–20)cm, obtuse to slightly cordate at base, acuminate at apex, margin entire, glabrous above and sparsely hairy on venation below; petioles c. 1cm long; stipules amplexicaul, to 18mm long; inflorescences solitary or in small groups, globose; ♂ axillary; ♀ proximal to leaves on twigs and on branches and trunk, 2.5–10cm diam.; flowers numerous; infructescences to 30cm diam., covered by numerous accrescent, peltate bracts. Forest, including riverine; 650–682m.
Distr.: Senegal to Mozambique [tropical Africa].
IUCN: LC
Mefou Proposed National Park: <u>Rønsted 203</u> 3/2004; <u>210</u> 3/2004; <u>216</u> 3/2004.
Uses: FOOD — Seeds — harvested for edible seeds.

Trilepisium madagascariense DC.
Bull. Jard. Bot. Nat. Belg. 47: 299 (1977); Fl. Cameroun 28: 103 (1985); Keay R.W.J., Trees of Nigeria: 302 (1989).
Syn. *Bosqueia angolensis* Ficalho

Tree, 20m, glabrous; leaves elliptic c. 10 × 4.5cm, acuminate, obtuse, entire, lateral nerves c. 5 pairs, basal pair acute; petiole 1cm; stipules caducous; inflorescence axillary at leafless nodes, ellipsoid, 1.5 × 0.8cm, glabrous; peduncle 1.5cm; flowers emerging from apical aperture in inflorescence, in cluster c. 0.5 × 0.5cm. Forest; 600–700m.
Distr.: Guinea (Conakry) to Congo (Kinshasa) [Guineo-Congolian].
IUCN: LC
Mefou Proposed National Park: <u>Etuge 5294</u> 3/2004; <u>Rønsted 206</u> 3/2004; 218b 3/2004; **Ndanan 1:** <u>Cheek in MF 358</u> 10/2002; <u>389</u> 10/2002; <u>397</u> 10/2002.
Note: *Rønsted* 218b has been tentatively placed in this taxon, Rønsted ix.2004.

MYRISTICACEAE

X. van der Burgt (K), C. Couch (K), M. Cheek (K) & H. Fortune-Hopkins (K)

Fl. Cameroun 18 (1974).

Coelocaryon preussii Warb.
Tree, to 30m; bole long, clear, without buttresses; red exudate when cut; leaves 15–25 × 6–8cm, veins prominent beneath; lamina lanceolate, apex shortly acuminate; petiole 10–12mm; inflorescence branched twice; fruits rounded at the extremities, 25–40 × 15–30mm. Forest; 680–710m.
Distr.: Nigeria, Cameroon, CAR, Gabon, Congo (Brazzaville), Congo (Kinshasa) [Guineo-Congolian].
IUCN: LC
Ndanan 1: <u>Cheek 11038</u> 10/2002; <u>11733</u> 3/2004; <u>Cheek in MF 276</u> 10/2002; <u>375</u> 10/2002; **Ndanan 2:** <u>Darbyshire 241</u> 3/2004.

Pycnanthus angolensis (Welw.) Warb.
Syn. *Pycnanthus microcephalus* (Benth.) Warb.
Forest tree, to 40m; specimens drying reddish-brown; bole clear for 10–15(–20)m, to 5m diam.; crown of horizontal, whorled branches; red exudate when cut; leaves with 15–60 pairs of lateral nerves; lamina often with conspicuous insect damage, indumentum of rusty brown, branched hairs; inflorescences axillary, paniculate; male flower heads c. 5mm across; fruits oblong, 3–4cm long, dehiscent; seed solitary, covered by a bright red laciniate aril. Forest; 700m.
Distr.: Guinea (Conakry) to Angola and Uganda [Guineo-Congolian].
IUCN: LC
Ndanan 1: <u>Cheek 11073</u> 10/2002; <u>Cheek in MF 270</u> 10/2002; <u>292</u> 10/2002; <u>Harvey 193</u> 10/2002.
Local Name: Eten (Ewondo) (Cheek, M. 11073).
Uses: MEDICINES (Cheek 11073).

Staudtia kamerunensis Warb. var. *gabonensis* (Warb.) Fouilloy
Fl. Cameroun 18: 104 (1974); Keay R.W.J., Trees of Nigeria: 38 (1989); Fl. Zamb. 9(2): 40 (1997).

Syn. *Staudtia stipitata* Warb.
Syn. *Staudtia gabonensis* Warb. Fl. Gabon 10: 99 (1965).
Tree; exudate red; stems terete, young growth with inconspicuous dark brown indumentum; leaves alternate, estipulate, simple; blade lanceolate or oblong, rhombi-oblanceolate, 8–11 × 3–4.5cm, apex gradually tapering into an acumen, base otuse-rounded, both surfaces glabrous, drying mid-brown, secondary nerve pairs c. 10, very inconspicuous; petiole 1cm; fruits 20–30 × 20mm. Evergreen forest; 700m.
Distr.: Nigeria, Cameroon, Gabon, Congo (Kinshasa) and Angola (Cabinda) [Guineo-Congolian].
IUCN: LC
Mefou Proposed National Park: Cheek in MF 39 3/2004; **Ndanan 1:** Cheek in MF 314 10/2002; 348 10/2002.

MYRSINACEAE

Y.B. Harvey (K)

Ardisia dolichocalyx Taton
Bull. Jard. Bot. Nat. Belg. 49: 98 (1979).
Monopodial shrub, 1–2.5m; glabrous throughout with the exception of very young growth; leaves elliptic to obovate, 14–22 × 6–10cm, orange pellucid gland dots and streaks, margin subentire; petiole 15–25mm, occasionally sparsely hairy; flowers in axillary fascicles, 4–6 flowers per fascicle; peduncles to 5mm; pedicel 11–12mm, extending to 22mm in fruit; calyx c. 2.5mm wide; flowers to 6mm, red with glandular streaks; fruit globose, 8–10mm diam., red with purplish gland dots. Forest; 679m.
Distr.: Cameroon [lower Guinea].
IUCN: NT
Ndanan 1: Onana 2911 3/2004.
Note: known from c. 13 localities, mainly in S Region and SW Region, Cameroon.

Ardisia staudtii Gilg
Bull. Jard. Bot. Nat. Belg. 49: 112 (1979).
Syn. *Afrardisia cymosa* (Bak.) Mez
Syn. *Afrardisia staudtii* (Gilg) Mez
Shrub, (0.5–)1.5–4(–5)m tall, glabrous; leaves elliptic to ovate, 90–180 × 30–70mm, glandular dots present on lower surface, very shallowly crenate; petioles 7–15mm; flowers in axillary fascicles; peduncles 2–5mm; 6–12 flowers per fascicle; pedicel 6–10mm; calyx c. 2.5mm wide, fimbriate margin; flowers white or pink, to 4mm with glandular spots/streaks; fruits globose, 3–6.5mm, red with red gland-dots. Lowland & submontane forest; 700m.
Distr.: Nigeria to Congo (Kinshasa) & CAR [Guineo-Congolian (montane)].
IUCN: LC
Ndanan 1: Gosline 435 10/2002.

Maesa lanceolata Forssk.
Tree or shrub, 6–8m tall; stem glabrous, dark brown with

paler lenticels; leaves elliptic, 9–16 × 3.5–6cm, serrulate, glabrous; petioles 2–2.5cm, glabrous; inflorescence many branched, 5–7mm, profusely covered with minute hairs (<0.5mm); flowers pale green, to 1.5mm, subsessile; fruits globose, 4–5mm; pedicel to 3.5mm. Forest-grassland transition, montane forest; 710m.
Distr.: Guinea (Conakry) to Madagascar [afromontane].
IUCN: LC
Ndanan 1: Cheek 11907 3/2004.

NYMPHAEACEAE

C. Couch (K)

Nymphaea lotus L.
Kew Bull. 44: 484 (1989); F.T.E.A. Nymphaeaceae: 3 (1989).
Aquatic herb, rhizome bottom-rooting; leaves floating on surface of water, orbicular, c. 20cm diam., margin sharply toothed, primary to tertiary nerves strongly produced below; petioles to c. 2m; flowers white, c. 8cm diam., floating on surface. Lake edge; 600m.
Distr.: SE Europe, Africa, India, Malaysia [palaeotropics & subtropics].
IUCN: LC
Mefou Proposed National Park: Etuge 5160 fl., 3/2004.

OCHNACEAE

I. Darbyshire (K), J.-M. Onana (K) & M. Cheek (K)

Campylospermum elongatum (Oliv.) Tiegh.
Bull. Jard. Bot. Brux. 35: 396 (1965); Fl. Congo, Rw. & Bur. Ochnaceae: 35 (1967); Bot. Helv. 95: 67 (1985).
Syn. *Ouratea elongata* (Oliv.) Engl.
Tree, to 6m; stems sparingly branched; stipules triangular, persistent; leaves crowded towards apex, oblanceolate, 44–50 × 9.5cm, subsessile, base sub-cordate or truncate, apex acute, margins serrulate, midrib raised below; flowers on pendulous racemes; peduncles flattened, 0.5cm wide, c. 25cm long; rachis c. 18cm; flowers single or in clusters of 2–3; corolla yellow; calyx lobes in fruit 0.8 × 0.3cm, apex acute; torus fleshy, 0.8cm diam.; fruits red. Forest; 700–710m.
Distr.: Nigeria to Congo (Kinshasa) [lower Guinea & Congolian].
IUCN: LC
Mefou Proposed National Park: Nana 43 3/2004; **Ndanan 1:** Cheek 11014 10/2002; 11620 3/2004; 11841 3/2004; Onana 2890 3/2004; **Ndanan 2:** Darbyshire 200 3/2004.

Campylospermum flavum (Schum. & Thonn.) Farron
Bull. Jard. Bot. Brux. 35: 397 (1965).
Syn. *Ouratea flava* (Schum. & Thonn.) Hutch. & Dalziel
Shrub, 1.5–4m tall, evergreen, glabrous; leaves glossy, leathery, alternate, elliptic-oblong, c. 23 × 6cm, acute at

base and apex, margin acutely toothed, 2 teeth per cm; lateral nerves c. 15 pairs, quaternary nervelets raised; petiole 2mm; inflorescence terminal, c. 15 × 10cm, with 2–3 lateral branches; flowers clustered in 4–5s; pedicels persistent, 2mm; flowers 1.5cm wide; petals 5, yellow, stalked; sepals 5, red, free; fruiting calyx persistent, 5mm diam.; fruitlets glossy black, torus 5mm. Forest; 710m.
Distr.: Guinea to Bioko, Cameroon and CAR [Guineo-Congolian].
IUCN: LC
Ndanan 1: Cheek 11636 3/2004.

Campylospermum glaucum (Tiegh.) Farron
Bull. Jard. Bot. État Bruxelles 35: 397 (1965); Lebrun J.-P. & Stork A., Trop. Afr. Fl. Plants 1: 492 (2003).
Shrub, 1–2m, glabrous; leafy stems with epidermis grey, flaking, scale-leaves 3–4m, sparse; leaves leathery, drying dull grey-green, alternate, elliptic, 11.5(–22) × 4(–9)cm, acumen 0.5cm, base acute, margin finely, acutely serrate, lateral nerves 9–12 pairs, arching, tertiary nerves patent to midrib; petiole 0.5–0.7cm, winged; inflorescence dense, broad, terminal spike, c. 4.5 × 2.5cm; flowers yellow, c. 2cm diam. Evergreen semi-deciduous forest; 600–720m.
Distr.: Cameroon & Gabon [lower Guinea].
IUCN: NT
Mefou Proposed National Park: Etuge 5109 3/2003; Nana 62 3/2004; **Ndanan 1:** Cheek 11551 3/2004; 11652 3/2004; 11663 3/2004; 11766 3/2004; 11767 3/2004; 11829 3/2004; Cheek in MF 317 10/2002; Darbyshire 183 fl., fr., 3/2004; Onana 2814 3/2004; 2821 3/2004; Tadjouteu 612 3/2004; **Ndanan 2:** Darbyshire 198 fl., 3/2004; 199 fl., fr., 3/2004; 255 fl., fr., 3/2004.
Note: plot voucher MF317 and Etuge 5109 are tentatively placed in this taxon, Darbyshire vii.2006 & Cheek x.2010.

Lophira alata Banks ex Gaertn.f.
Emergent tree, 40–60m with long narrow crown; leaves flushing vivid red, obovate-oblanceolate, 18–28 × 7.5–9cm, coriaceous, base cuneate, apex emarginate (immature) to acute (mature), midrib prominent, raised, secondary veins very numerous, parallel; inflorescence paniculate; pedicels not clustered, articulated to 1.5cm from base; corolla white, 2–3cm diam.; stamens numerous; anthers linear, c. 4mm long; fruit an ellipsoid nut with persistant style; sepal lobes expand irregularly to form asymmetric wings, the longest c. 11 × 2cm. Forest; 600–710m.
Distr.: Guinea (Conakry) to Congo (Kinshasa) [Guineo-Congolian].
IUCN: VU
Mefou Proposed National Park: Etuge 5231 3/2004; **Ndanan 1:** Cheek 11821 3/2004.

Rhabdophyllum affine (Hook.f.) Tiegh.
Bull. Jard. Bot. Brux. 35: 390 (1965).
Syn. *Ouratea stenorrhachis* Gilg
Syn. *Ouratea myrioneura* Gilg

Shrub, 1.5–3m, branching, glabrous; stem with two ranks of scale-like leaves, 3mm long; leaves alternate, narrowly elliptic, 6.5–12 × 2.5–5cm, subacuminate, base cuneate, margin entire, thickened by nerve, lateral nerves numerous, parallel, separated by 0.5mm; petiole 1–2mm; flowers yellow; fruiting racemes axillary from scale-leaves, 3–5-flowered, 10cm, pendulous; peduncle thread-like, 5cm; pedicel 1.5mm, mericarps globose, peduncle, pedicel, calyx and torus bright red. Forest; 640–710m.
Distr.: Sierra Leone to Cameroon [Guineo-Congolian].
IUCN: LC
Ndanan 1: Cheek 11894 3/2004; Tadjouteu 573 3/2004; **Ndanan 2:** Darbyshire 210 fl., fr., 3/2004; 310 3/2004.

OLACACEAE

J.-M. Onana (YA), X. van der Burgt (K), M. Cheek (K) & G. Gosline

Fl. Cameroun 15 (1973).

Coula edulis Baill.
Tree, to 40m; branches pale grey, fissured; leafy stems, apical bud and young leaves with very short dense, scurfy red-orange indumentum; leaves alternate, estipulate; blade oblong, 20–30 × 7.5–11cm, acumen 1.5–3cm, base obtuse, lateral nerves 10–13 pairs, tertiary nerves finely scalariform, domatia absent, lower surface glabrous but with raised round or star-shaped swellings, appearing brown in transmitted light; petiole 2–3cm, swollen at apex; fruits ellipsoid 4 × 3cm, with a single large edible seed. Forest; 700–710m.
Distr.: Sierra Leone to Cameroon, Gabon, Congo (Kinshasa) and Angola (Cabinda) [Guineo-Congolian].
IUCN: LC
Ndanan 1: Cheek 11243 10/2002; 11827 3/2004; Onana 2187 10/2002.
Local Name: Kome (Ewondo) (Onana 2187).
Uses: FOOD — Seeds — seed edible (Cheek 11827).

Heisteria parvifolia Sm.
Tree or shrub, to 10m; branchlets narrowly winged; leaves oblong, 8–12 × 3–6cm, lateral nerves c. 6 pairs; inflorescence axillary with 2–3 very small flowers; fruits with enlarged star-shaped calyx, bright red, 3–5cm diam.; fruit ellipsoid; single seeded. Forest; 700–710m.
Distr.: Senegal to Uganda [Guineo-Congolian].
IUCN: LC
Ndanan 1: Cheek 11836 3/2004; Cheek in MF 331 10/2002; Harvey 120 10/2002; 123 10/2002.

Octoknema genovefae Villiers
Bull. Inst. Fondam. Afrique Noire, Sér. A., Sci. Nat. 34(1): 5 (1972).
Shrub, 0.8–1.5m by waterside; densely brown stellate hairy on stems, petioles and nerves; leaves alternate, estipulate; blade oblong, 16–21 × 4–7cm, acumen to

2cm, base rounded and peltate (petiole inserted on lower surface of blade, the blade margins uniting with each other), lateral nerves c. 10 pairs; petiole 14–20mm, swollen at apex; infructescence 1–2cm, terminal, with c. 3 fruits; fruit ellipsoid, bright orange, 1.5–1.25cm. Riverine; 650m.
Distr.: Cameroon endemic.
IUCN: NT
Ndanan 2: Darbyshire 263 3/2004.

Olax gambecola Baill.
Shrub, to 3m; branches 4-winged, covered with whitish dots; leaves ovate-elliptic to lanceolate, 7–15 × 3–6cm; petiole 0–2mm; inflorescence an axillary raceme 1–2.5cm long; flowers 2.5–3mm long, white; stamens 3; staminodes 5–6; fruits red; calyx a disc at base, not accrescent enveloping fruit. Forest, especially by streams; 660–700m.
Distr.: Guinea (Conakry) to Uganda [Guineo-Congolian].
IUCN: LC
Ndanan 1: Harvey 114 10/2002; 187 10/2002; **Ndanan 2:** Darbyshire 270 fl., 3/2004; Gosline 414 10/2002.

Olax latifolia Engl.
Shrub, to 3m; branches 4-sided; leaves variable, elliptic to oblong, 7–22 × 3–10cm, 5–12 lateral nerves; petiole 0–4mm long; flowers pink, c. 4mm long; stamens 6; staminodes 3, in axillary racemes; fruits globose, 2cm diam., completely enveloped by accrescent calyx. Forest; 700–710m.
Distr.: Cameroon to Congo (Kinshasa) [lower Guinea & Congolian].
IUCN: LC
Mefou Proposed National Park: Cheek 11065A 10/2002; Nana 64 3/2004; **Ndanan 1:** Cheek 11556 3/2004; 11624 3/2004; Harvey 101 10/2002; 115 10/2002; 183 10/2002; 196 10/2002; 87 10/2002.

Olax mannii Oliv.
Shrub, to 3m; branches 4-winged; leaves lanceolate to elliptic, 6–15 × 2–7cm; petiole 0–4mm, 5–6 lateral nerves, brownish when dried; flowers greenish-white, 6–7mm long; stamens 3; staminodes 4–5; fruit orange, globose 2–3cm diam.; calyx accrescent, enveloping fruit, drying light brown. Forest; 600–710m.
Distr.: Sierra Leone to Congo (Kinshasa) [Guineo-Congolian].
IUCN: LC
Mefou Proposed National Park: Etuge 5142 fr., 3/2004; **Ndanan 1:** Cheek 11065B 10/2002; 11910 3/2004; Harvey 186 10/2002; **Ndanan 2:** Gosline 439 10/2002.

Olax triplinerva Oliv.
Fl. Cameroun 15: 120 (1973).
Shrub, to 2.5m; branches 4-winged; leaves oblanceolate to oblong, 8–20 × 3–8cm, lateral nerves 4–6 pairs, the lowest pair originating at same point, joining the next pair 1/2 to 3/4 up the length of the leaf; petiole 0–2mm,

flowers with 5 epipetalous stamens; staminodes 3; fruiting calyx accrescent, enveloping fruit, yellowish to reddish. Forest and farmbush; 700m.
Distr.: Cameroon, Rio Muni, Gabon and CAR [Guineo-Congolian].
IUCN: LC
Ndanan 1: Gosline 409 10/2002.

Ptychopetalum petiolatum Oliv.
Tree or shrub, to 10m; twigs with 2 wings; leaves lanceolate, oblong, elliptic or oblanceolate 7–19 × 3–7cm, lateral nerves 4–8(–10) pairs, ascendant, anastomizing 5mm or more from edge of leaf; inflorescence few-flowered, axillary, rachis 5–12mm; bud 4.5 × 1.5–2mm; flower 5-merous; pedicel 2–3mm; calyx absent; petals c. 7mm long, reflexed at top; fruit a red ellipsoid drupe, 17–21 × 12–14mm, often with 6 longitudinal grooves; single seeded. Forest; 700–720m.
Distr.: SE Nigeria, Cameroon, Rio Muni & Gabon [lower Guinea].
IUCN: LC
Ndanan 1: Cheek 11594 3/2004; 11679 3/2004; Harvey 203 10/2002; Onana 2842 3/2004.
Note: fairly common in lowland Cameroon, less so in the rest of its range.

Strombosia pustulata Oliv.
Syn. *Strombosia glaucescens* Engl.
Tree, to 20m; branchlets subterete; leaves oblong to elliptic-oblong 4–10 × 2–5cm, 4–6 main lateral nerves, underside of leaf with minute pustules; petiole 1–1.5cm; flowers in axillary clusters; pedicels with small bracteoles, 5-merous; petals 4–5mm long. Forest; 700m.
Distr.: Senegal to Congo (Kinshasa) [Guineo-Congolian].
IUCN: LC
Mefou Proposed National Park: Cheek in MF 26 3/2004; **Ndanan 1:** Cheek in MF 327 10/2002.

Strombosiopsis tetrandra Engl.
Tree, 4–30m tall, red exudate; leaves oblong to lanceolate 5–25 × 3–10cm, 6–10 pairs lateral nerves, tertiary nerves parallel; inflorescence a short congested raceme, rachis 2–5mm; flowers cream-coloured, 4-merous; fruits ellipsoid 2.5–3.5 × 2–3cm; calyx accrescent, enveloping fruit, with 4 teeth at summit. Forest; 700–710m.
Distr.: SE Nigeria to Uganda [Guineo-Congolian].
IUCN: LC
Mefou Proposed National Park: Cheek in MF 101 3/2004; 63 3/2004; **Ndanan 1:** Cheek 11563 3/2004; 11592 3/2004.

OLEACEAE

M. Cheek (K)

Jasminum sp. of Mefou
Climber; stem, petiole and midrib densely puberulent with grey hairs curved at apex; leaves opposite, simple,

180

lacking stipules; blade elliptic or oblong, 2.5–5.5 × 1.5–2.75cm, acumen to 0.5cm, base rounded, 3-nerved from base, basal laterals glabrous, yellow, extending to apex with hairy domatia, other laterals c. 5 pairs, lacking domatia, arising nearly 90° from midrib and joining the basal lateral; petiole 2–4mm. Forest; 700m.
Ndanan 1: <u>Cheek in MF 302</u> 10/2002.
Note: plot voucher MF302 not matched with any other *Jasminum* in our area due to the combination of characters shown. Fertile material needed to place with more certainty, Cheek x.2010.

ONAGRACEAE

M. Cheek (K)

Ludwigia abyssinica A.Rich.
Webbia 27: 496 (1972).
Syn. *Jussiaea abyssinica* (A.Rich.) Dandy & Brenan
Shrubby herb, 0.7–2m; stems terete, glabrous; leaves narrowly elliptic or (larger leaves) oblong-elliptic, 3–7.5 × 0.5–1.5cm, apex rounded, base cuneate-decurrent, larger leaves with 1–2 teeth; petiole 0–0.5cm; flowers 1–2 per axil, sessile, inferior ovary 7mm; sepals 4, triangular, 1mm; petals 4, yellow-orange subequal to sepals; fruit 1cm, pale brown; seeds in 1 row. Lake or river edges; 600–660m.
Distr.: Guinea (Conakry) to South Africa [tropical & subtropical Africa].
IUCN: LC
Mefou Proposed National Park: <u>Etuge 5156</u> fl., 3/2004; **Ndanan 2:** <u>Darbyshire 327</u> 4/2004.

Ludwigia africana (Brenan) Hara
J. Jap. Bot. 28: 291 (1953).
Syn. *Jussiaea africana* Brenan
Herb, 0.9m, erect; resembling *L. abyssinica*, but stems, leaves and inflorescence with thin spreading white hairs; leaves elliptic, to c. 6 × 3.5cm, acute; petiole c. 1.5cm; inflorescence c. 5-flowered; peduncles 1.5cm; ovary 10mm; sepals 5mm; fruit not seen. River banks; 680–710m.
Distr.: Guinea (Conakry) to Congo (Kinshasa) [upper & lower Guinea].
IUCN: LC
Ndanan 1: <u>Cheek 11584</u> 3/2004; <u>11752</u> 3/2004; **Ndanan 2:** <u>Cheek 11920</u> 3/2004; <u>Darbyshire 232</u> 3/2004.

Ludwigia decurrens Walt.
Reinwardtia 6: 327 (1963).
Syn. *Jussiaea decurrens* (Walt.) DC.
Herb, 0.2–1.3m; resembling *L. abyssinica*, but stems 3-winged, glabrous; largest leaves lanceolate to 14 × 2.5cm; sepals to 7mm; fruits 1.5cm, 4-winged. Roadside; 600m.
Distr.: S America, introduced to W Africa.
IUCN: LC
Mefou Proposed National Park: <u>Etuge 5276</u> 3/2004.

OXALIDACEAE

J.-M. Onana (K) & M. Cheek (K)

Oxalis barrelieri L.
Willdenowia 8: 17 (1977).
Erect-stemmed annual herb, c. 30cm; leaves subopposite, c. 9cm; leaflets ovate, 3–4 × 2–3cm, apex entire, central leaflet with petiolule 1cm, lateral leaflets sessile; inflorescences axillary, many-flowered, dichasial; peduncle 4cm; flowers 1cm, pink, centre yellow. Farms & farmbush; 600–700m.
Distr.: pantropical.
IUCN: LC
Mefou Proposed National Park: <u>Etuge 5101</u> 3/2003; **Ndanan 1:** <u>Cheek 11077</u> 10/2002; <u>Harvey 86</u> 10/2002.

Oxalis corniculata L.
Prostrate, rooting stemmed herb; stems to c. 30cm, borne horizontally, densely white-hairy when young; leaflets transversely elliptic, 1cm long, apical sinus 0.5cm, base obtuse, sessile; inflorescence 1-flowered; flower 0.7cm, yellow. Grassland & forest edge; 660m.
Distr.: cosmopolitan.
IUCN: LC
Ndanan 2: <u>Darbyshire 328</u> fl., fr., 4/2004.

PANDACEAE

J.-M. Onana (YA) & M. Cheek (K)

Fl. Cameroun 19 (1975).

Microdesmis haumaniana J.Léonard
Fl. Gabon 22: 26 (1973); Fl. Cameroun 19: 54 (1975).
Shrub, 2m; young growth sparsely puberulent, soon glabrescent; apical bud claw-like; leaves alternate, estipulate, simple, oblong-elliptic, 6–10 × 1.7–4cm, acumen 0.7cm, base unequally acute, lower surface glossy, lateral nerves 5–6 pairs, not uniting, quaternary nerves spreading, raised, conspicuous; petiole 6–10mm; pedicel 3–6mm; sepals 1mm; flowers axillary, single, 3–5mm diam., pink; styles fimbriate, persistent; fruits globose, <1cm diam. Forest; 700–710m.
Distr.: Cameroon to Congo (Kinshasa) [lower Guinea & Congolian].
IUCN: NT
Ndanan 1: <u>Cheek 11263</u> 10/2002; <u>11632B</u> 3/2004; <u>11731</u> 3/2004; **Ndanan 2:** <u>Gosline 416</u> 10/2002.

Microdesmis klainei Leonard
Fl. Cameroun 19: 47 (1975).
Shrub, 2–3m; densely softly, persistently moderately grey tomentose; apical but claw-like; leaves alternate, estipulate, simple; blade oblanceolate or elliptic, thickly papery, drying pale green below, to 16 × 6cm, acumen 1cm, with 1mm mucro, base unequally acute, lateral nerves c. 5 pairs, uniting strongly in a marginal nerve, quaternary nerves finely reticulate, prominent; petiole

6–9mm; flowers in axillary fascicles of 3–7, orange or pink, c. 3mm diam, densely puberulous on pedicel and outer petal surfaces, inner surface glabrous; petals 5; fruit globose, orange, 1–2cm diam. Secondary forest, near *Raphia* swamp; 600–700m.
Distr.: Cameroon & Gabon [lower Guinea].
IUCN: NT
Mefou Proposed National Park: Etuge 5216 3/2004; **Ndanan 1:** Cheek 11012 10/2002; Harvey 134 10/2002; 182 10/2002.

Microdesmis puberula Hook.f. ex Planch.
**Syn. *Microdesmis zenkeri* Pax F.T.A. 6: 742 (1897).
Shrub, or tree, 1.5–8m; stems, petioles and blade nerves densely and persistent grey-brown hairy; leaves alternate, estipulate, simple; blades oblong or elliptic, 9–16 × 3.5–7.5cm, acumen 0.5cm, base unequally rounded to obtuse, lateral nerves 4–6 pairs, drying white above, not uniting in a looping marginal nerve; petiole c. 5cm; fruits <1cm, orange. Forest 600–710m.
Distr.: Nigeria, Cameroon, Bioko, Gabon, Congo (Kinshasa), CAR and Uganda [Guineo-Congolian].
IUCN: LC
Mefou Proposed National Park: Cheek in MF 104 3/2004; **Ndanan 1:** Cheek 11862 3/2004; Cheek in MF 293 10/2002; Etuge 5111 3/2003; Tadjouteu 580 3/2004; **Ndanan 2:** Darbyshire 240 fr., 3/2004.

PASSIFLORACEAE

M. Cheek (K)

Adenia bequaertii Robyns & Lawalree subsp. *occidentalis* De Wild.
Meded. Land. Wag. 71–18: 246 (1971); Fl. Afr. Cent. Passifloraceae: 42 (1995).
Climber, to 5m; stems glabrous; leaves alternate, estipulate; blade reniform to 6 × 9cm, apex rounded sometimes abruptly obtuse, base cordate, sinus 80–100° wide, 1cm deep, upper surface drying brown, lower surface white, black punctate, lateral nerves brown, 2 main pairs in distal third of the leaf; petiole 1–4mm, apex with a single ellipsoid gland, 3 × 1.5mm on upper surface; panicles axillary, 3cm, c. 10-flowered; peduncle 2cm; flowers white, 5mm. Forest, farmbush; 700m.
Distr.: Cameroon, CAR, Congo (Brazzaville), Congo (Kinshasa), Angola [Congolian].
IUCN: LC
Ndanan 1: Gosline 428 10/2002.

Adenia cissampeloides (Planch. ex Benth.) Harms
Climber, 12m, glabrous; leaves trilobed, broadly ovate, 5 × 5.5cm, lobes 1.5cm, pinnately nerved; petiole gland single; peduncle 2.5cm, c. 8-flowered; flowers white, 1.5cm; fruit ellipsoid 1.5–2cm. Forest; 679–710m.
Distr.: Guinea (Conakry) to Kenya [tropical Africa].
IUCN: LC
Ndanan 1: Cheek 11813 3/2004; Onana 2908 3/2004.

Adenia gracilis Harms subsp. *gracilis*
Climber, closely resembling *A. cissampeloides* but peduncle 0.5–1cm; fruit ellipsoid, 1cm, smooth. Forest edge; 710m.
Distr.: Nigeria to Uganda [Guineo-Congolian].
IUCN: LC
Mefou Proposed National Park: Cheek in RP 91 3/2004.
Note: de Wilde notes that leaves of *A. gracilis* average 4cm wide, *A. cissampeloides* 7cm.

Barteria fistulosa Mast.
**Syn. *Barteria nigritana* Mast. subsp. *fistulosa* (Mast.) Sleumer Blumea 22: 14 (1974).
Tree, 9m, glabrous; lateral stems horizontal, dilated (c. 1cm), hollow, inhabited by large, fierce, black, biting ants; leaves elliptic-obovate, simple to c. 45 × 19cm, shortly acuminate, unequally obtuse; petiole 1.5cm. Forest; 700m.
Distr.: Nigeria to Congo (Kinshasa) [lower Guinea & Congolian].
IUCN: LC
Ndanan 1: Cheek in MF 354 10/2002.

Passiflora foetida L.
Herbaceous climber, 0.1–2m, foetid, softly hispid; leaves suborbicular, 5–10cm, 3-lobed, acute, base cordate, ciliate; tendrils axillary; flowers axillary, solitary, to 5cm diam.; bracts pinnatisect, stalked-glandular; petals white, corona striped, blue-purple; fruit ovoid, 4cm, enclosed by bracts and calyx. Forest & farmbush; 600–720m.
Distr.: tropical Africa, introduced from tropical America.
IUCN: LC
Mefou Proposed National Park: Etuge 5123 3/2004; **Ndanan 1:** Darbyshire 205 fr., 3/2004; Onana 2884 3/2004.
Uses: FOOD — Infructescences — fruits eaten (Onana 2884).

PENTADIPLANDRACEAE

M. Cheek (K), C. Couch (K) & E. Fenton (K)

Fl. Cameroun 15 (1973).

Pentadiplandra brazzeana Baill.
Tree, or climber, 20m; leaves alternate, elliptic 12.5 × 4cm, acuminate, 1.5cm; petiole 1cm; inflorescence axillary, 0.7cm, few-flowered; flowers 3cm, white mottled purple; calyx, corolla and androecium free; fruit ovoid c. 8 × 7cm, hard, mottled-grey, pointed, flesh-orange; seeds 2cm, white. Forest; 600–710m.
Distr.: Cameroon to Congo (Kinshasa) [lower Guinea & Congolian].
IUCN: LC
Mefou Proposed National Park: Nana 30 3/4; **Ndanan 1:** Cheek 11527 3/2004; **Ndanan 2:** Etuge 5119 3/2003.

PHYTOLACCACEAE

Y.B. Harvey (K) & X. van der Burgt (K)

Hilleria latifolia (Lam.) H.Walt.

Herb, 0.3–1m, glabrous; leaves alternate, elliptic, c. 8 × 3.5cm, acuminate, acute; petiole 3cm; stipules absent; inflorescence a raceme, terminal, 1-sided, c. 10cm; pedicels 0.4cm; flowers pink and white, 2mm; calyx lobes persistent, unequally united at base, outermost longest; fruit a capsule, 2mm. Forest edge; 600–710m.
Distr.: tropical Africa, introduced from tropical America.
IUCN: LC
Mefou Proposed National Park: Etuge 5134 fl., 3/2004; **Ndanan 1:** Cheek 11061 10/2002; 11812 3/2004.
Local Name: Secaa (Etuge 5134).
Uses: FOOD — Leaves — green vegetables (Etuge 5134).

Phytolacca dodecandra L'Hér.

Dioecious shrub, sometimes scandent, to 4m, glabrous, slightly fleshy; leaves alternate, ovate-elliptic, c. 8 × 3cm, acute, base rounded; petiole c. 2cm; inflorescence racemose, c. 20cm; pedicels 0.6cm; sepals 4, petals absent; stamens c. 15; berries red, c. 0.8cm. Forest; 600m.
Distr.: tropical & subtropical Africa.
IUCN: LC
Mefou Proposed National Park: Etuge 5180 3/2004.

PIPERACEAE

E. Fenton (K) & X. van der Burgt (K)

Piper guineense Schum. & Thonn.

Hemiepiphyte-climber, reaching 20m above ground; peppery when crushed; stem twining and rooting adventitiously; leaves ovate-elliptic, to 19 × 10cm, obliquely-obtuse at base; inflorescence single, leaf-opposed, 3cm. Forest; 600–710m.
Distr.: Guinea (Bissau) to Uganda [Guineo-Congolian].
IUCN: LC
Mefou Proposed National Park: Cheek in RP 33 3/2004; Etuge 5152 3/2004; **Ndanan 2:** Cheek 11106 10/2002.
Uses: FOOD — Unspecified Parts — vegetable (Etuge 5152); FOOD ADDITIVES — Unspecified Parts — spice (Etuge 5152).

Piper umbellatum L.

Pithy shrub, 1–2m; peppery when crushed; leaves orbicular, c. 24 × 24cm, deeply cordate, white below; inflorescences 2–5, clustered on peduncle, c. 3.5cm. Forest edge; 600–710m.
Distr.: Guinea (Conakry) to Cameroon [upper & lower Guinea].
IUCN: LC
Mefou Proposed National Park: Cheek in RP 71 3/2004; Etuge 5251 3/2004; **Ndanan 1:** Harvey 93 10/2002.

POLYGALACEAE

C. Couch (K) & M. Cheek (K)

Carpolobia alba G.Don

Syn. *Carpolobia glabrescens* Hutch. & Dalziel
Shrub or tree, to 3(–6)m; stems puberulent towards apex; leaves papery, elliptic, oblong or obovate, 7.5–11.5 × 3.2–4.2cm, apex acuminate, base acute to obtuse, lateral nerves 6–9 pairs, midrib sparsely puberulent; petiole 0.3cm, puberulent; inflorescence axillary, 2–4-flowered, rachis <1cm, puberulent; pedicel c. 5mm; sepals ovate, largest 0.7cm, others 0.4–0.5cm; corolla white, turning yellow; petals 5, keel petal to 1.5cm, limb as long as claw, 2–3mm wide when folded; fruit ovoid, 3-lobed, 1.7–2cm, rostrate, smooth, orange, with paired white rings at articulation with pedicel. Forest; 600–710m.
Distr.: Guinea (Conakry) to Liberia & Nigeria to Angola [Guineo-Congolian].
IUCN: LC
Mefou Proposed National Park: Etuge 5133 3/2004; **Ndanan 1:** Cheek 11548 3/2004; Cheek in MF 399 10/2002; **Ndanan 2:** Darbyshire 279 fr., 3/2004.
Note: plot voucher MF399 is probably this taxon, Cheek x.2010.
Local Name: Tombo (Etuge 5133).

POLYGONACEAE

I. Darbyshire (K)

Persicaria setosula (A.Rich.) K.L. Wilson
Kew. Bull. 45(4): 632 (1990); Fl. Zamb. 9(3): 13 (2006).
Syn. *Polygonum nyikense* Baker
Syn. *Polygonum setosulum* A.Rich.
Decumbent herb, to 60cm tall; ochrea to 3cm, glabrescent or sparsely pubescent, apical cilia 7–8mm; leaves ovate-lanceolate, 14.5–20 × 3–4.2cm, base acute, margin entire-ciliate, midrib and secondary nerves of lower surface with short coarse hairs; petiole to 2cm; spikes dense to 9cm long, 2–several together; peduncles glabrous; bracts c. 4mm with an apical fringe of short bristles; flowers 2.5–3mm long, shortly pedicellate, perianth white; nut lenticular or with one side swollen, 2.5mm long, shining dark brown. Lake margins, damp disturbed areas; 640m.
Distr.: tropical & S Africa.
IUCN: LC
Ndanan 2: Darbyshire 316 fl., 3/2004.

PORTULACACEAE

J.-M. Onana (K)

Talinum fruticosum (L.) Juss.
Gen. Pl. [Jussieu] 312 (1789).
Succulent herb, to 20cm; stems pale green, round; leaves shiny, pale green above, beneath pale green;

peduncle triangular; flowers violet; fruits yellow. Villages; 600m.

Distr.: pantropical, cultivated herb.

IUCN: LC

Mefou Proposed National Park: Etuge 5236 3/2004.

Uses: FOOD — Unspecified Parts — vegetable (Etuge 5236).

RHAMNACEAE

M. Cheek (K) & H. Sook Choung

Fl. Cameroun 33 (19991).

Gouania longipetala Hemsl.

Shrubby tendrillate climber, to 4m; stems striate, red-brown pubescent to glabrescent, hairs to 0.5mm; leaves ovate, to 9 × 4.7cm, base rounded to subcordate, apex subacuminate, margin crenate-serrate, lateral nerves 3–4 pairs, sparsely hairy only on nerves of lower surface; petiole to 2cm, red-brown pubescent; stipules rounded to ovate, early caducous, 1mm; leaves ovate, to 9 × 4.7cm, base rounded to subcordate, apex subacuminate, margin crenate-serrate, lateral nerves 3–4 pairs, sparsely hairy only on nerves; petiole to 2cm; stipules 1mm; inflorescence an axillary, spike-like thyrse, to 8cm; flower buds in fascicles, white-puberulent, to 1mm; fruits 3-winged, 0.8 × 1cm. Forest; 700m.

Distr.: Guinea (Conakry), Sierra Leone, Liberia, Ivory Coast, Ghana, Togo, Nigeria, Cameroon, CAR, Equatorial Guinea, Gabon, Congo (Brazzaville) and Congo (Kinshasa) [Guineo-Congolian].

IUCN: LC

Ndanan 1: Cheek 11481a 3/2004; Harvey 149 10/2002.

Lasiodiscus marmoratus C.H.Wright

Kew Bull. 50 (3): 495–526 (1995).

Tree, 8m; young branchlets tomentose to villose with cream lenticels showing through, hairs brown, 1.5–2.5mm; leaves ovate to oblong, 11.5–22.5 × 4.5–13.5cm, base symmetric and rounded, apex obtuse, margin irregularly serrulate, upper surface glabrescent, lower surface with appressed to spreading hairs to 2mm; petioles to 1.4cm, villose; stipules deltate, united on both sides of the branchlets, villose when young, later tomentose along the centre and glabrous at the margin; inflorescences partial compound dichasia (variable), to 10.5cm; peduncle puberulous and with longer spreading hairs, c. 100-flowered; bracts ovate, 4 × 1.5mm, velutinous to villous; flower buds to 2mm, with short dense cream indumentum and occasional tufts of long brown hairs; sepals deltoid, 1.5 × 1.5mm, recurved; petals cucullate 0.75 × 0.5mm, white; stamens erect to incurved at anthesis; ovary hairy at the apex. Forest and forest edge; 710m.

Distr.: Nigeria, Cameroon, Gabon, Congo (Brazzaville), Congo (Kinshasa) [lower Guinea & Congolian].

IUCN: LC

Ndanan 1: Cheek 11586 3/2004.

Note: leaf apices all damaged but should be mucronulate. See Figueiredo, E. 1995. A revision of *Lasiodiscus* (Rhamnaceae). Kew Bull. 50. (3): 495–526 (1995), Hopkins x.2010.

Maesopsis eminii Engl.

Tree seedling, 3m; young stems smooth, brown, finely puberulent, becoming glabrescent with age; leaves subopposite or alternate, papery, elliptic-oblong, c. 7.5–13 × 2.5–5.5cm, apex acuminate, base rounded to subcordate, margin serrate, with teeth more numerous towards the apex and round glands on teeth, lateral nerves c. 6, those of younger leaves covered in a fine puberulence; petiole 0.3–0.7cm; stipules subulate, c. 2mm, puberulent, caducous. Forest & grassland; 700m.

Distr.: Sierra Leone to Uganda [Guineo-Congolian].

IUCN: LC

Ndanan 1: Cheek 11180 10/2002.

Note: Cheek 11180 is sterile. Introduced and invasive in E Africa, Hopkins x.2010.

RUBIACEAE

M. Cheek (K), S. Dawson (K), B. Sonke (YA), R. Mbom (K), J.-M. Onana (YA) & O. Lachenaud (BR)

Aoranthe cladantha (K.Schum.) Somers

Bull. Jard. Bot. Nat. Belg. 58: 47 (1988).

Syn. *Porterandia cladantha* (K.Schum.) Keay

Tree, 8m, glabrescent; stipule oblong, 22 × 7cm; leaves papery, obovate, 20 × 10cm, rounded, acute; petiole 1.5cm; cauliflorous; corolla pinkish-white, tube 20–25mm, lobes 15mm; fruit axillary, few-fasciculate, ridged, ellipsoid, 1.5 × 0.7cm; calyx 4mm; pedicel 1–2cm. Evergreen forest; 700m.

Distr.: Nigeria to Cabinda [lower Guinea].

IUCN: LC

Ndanan 1: Cheek 11186 10/2002.

Argocoffeopsis subcordata (Hiern) Lebrun

Bull. Jard. Bot. Nat. Belg. 51: 366 (1981).

Syn. *Coffea subcordata* Hiern

Climber; side-branches reflexing; bark mid-brown, falling in strips; young stems patent pubescent, pale brown; stipules triangular, aristate, 2–3mm; leaf-blade obovate or elliptic, 3–5 × 1.5–2.5cm, acumen rounded, 4mm, base subcordate, lateral nerves 4 pairs, domatia white hairy, nerves white puberulous; petiole 2mm; peduncles 1-flowered, axillary in several nodes, 3–4mm; bracts in 2 pairs, distal leafy, elliptic, 7 × 3mm; calyx inconspicuous; corolla white, tube 16mm, lobes 5–6 oblong-elliptic, 7 × 3mm; anthers exserted; style bifurcate, exserted. Semi-deciduous forest; 700–710m.

Distr.: SE Nigeria, Cameroon, Gabon, Congo (Kinshasa) and Angola (Cabinda) [Guineo-Congolian].

IUCN: LC

Ndanan 1: Cheek 11534 3/2004; 11954 3/2004; Cheek in MF 233 10/2002.

Aulacocalyx jasminiflora Hook.f. subsp. jasminiflora

Shrub or small tree, to 4(–12)m; stems appressed yellow hairy; leaves often 3 per node at branch junctions, elliptic, c.12 × 5cm, acumen 1.5cm, base cuneate, lateral nerves 3–4 pairs, midrib hairs as for stems, secondary nerves densely scalariform; petiole 5mm; flowers 3–10; calyx golden silky hairy. Semi-deciduous forest; 700m.
Distr.: Sierra Leone to Congo (Kinshasa) [Guineo-Congolian].
IUCN: LC
Ndanan 1: Cheek in MF 289 10/2002.
Note: plot voucher MF289 is sterile, Cheek ix.2010.

Bertiera adamsii N. Hallé

Prostrate herb, to 7cm; ascending stems short, reproducing with stolons, nodes dense; stipules membranous broadly oblong, 9mm, including 4mm triangular apex, brown, with white silky hairs from midrib and margin; leaves obovate-oblanceolate, 5–9 × 2–3.5cm, obtuse, base decurrent, lateral nerves c. 5 pairs, hairy on both surfaces; petiole 1.5–2cm; fruit axillary, 1 or 2, orange-red, succulent, 7mm; calyx remains erect, 6mm. Forest; 700m.
Distr.: Ghana, Cameroon, Gabon, Equatorial Guinea (Rio Muni), CAR and Congo (Kinshasa) [Guineo-Congolian].
IUCN: VU
Ndanan 1: Cheek 11060 10/2002; Harvey 132 10/2002; 194 10/2002.

Bertiera aequatorialis Hallé

Fl. Gabon 17: 50 (1970).
Shrub, 0.6–2.5m; numerous parallel side-branches to 50cm with terminal inflorescences; long white spreading hairy; stipule narrow elliptic, 10mm; leaf-blade elliptic, 7–12 × 3.5–6cm, acumen 0.5cm, base obtuse, lateral nerves 5–6 pairs; inflorescence c. 5cm, with c. 3 curved, spike-like branches; flowers single; calyx lobes 1–2mm; fruit red, globose, 5mm. Semi-deciduous and secondary forest; 680–710m.
Distr.: Cameroon, Gabon, Equatorial Guinea (Rio Muni) [lower Guinea].
IUCN: NT
Mefou Proposed National Park: Nana 51 3/2004; **Ndanan 1:** Cheek 11029 10/2002; 11547 3/2004; Harvey 192 10/2002; **Ndanan 2:** Darbyshire 243 fr., 3/2004.

Bertiera aethiopica Hiern

Fl. Gabon 17: 40 (1970).
Shrub, 2–4m, densely long, spreading to sub-appressed hairy on stems and leaves; stipules long triangular, 6mm; leaf-blade oblong-elliptic, 10–13 × 4.5–5.5cm, acumen weak, acute, base rounded to obtuse, lateral nerves c. 9 pairs; inflorescence terminal, pendant spike 12–25cm; fruits in interrupted clusters, globose and ribbed, hairy, 4mm, ripening purple, juicy. Semi-deciduous forest and *Raphia* swamp; 600–710m.

Distr.: Cameroon, Gabon, Congo (Kinshasa), CAR [Congolian].
IUCN: LC
Mefou Proposed National Park: Cheek in MF 33 3/2004; Etuge 5249 3/2004; Nana 66 3/2004; **Ndanan 1:** Cheek 11062 10/2002; 11501 3/2004; Harvey 184 10/2002.

Bertiera bracteolata Hiern

Liana, 7m; stipule base triangular, 2mm, mucro 2mm; leaves elliptic-oblong or oblanceolate-oblong, c. 10 × 3.5cm, acumen 0.5cm, base acute-decurrent, nerves 7, obscure, puberulent below on midrib; petiole 0.5cm; flowers c. 20, pendulous, in slender branched panicles, terminal on spur shoots, white; corolla tube 7mm, lobes 1mm; fruit blue. Evergreen forest; 700–710m.
Distr.: Sierra Leone to Gabon [upper & lower Guinea].
IUCN: LC
Mefou Proposed National Park: Nana 52 3/2004; **Ndanan 1:** Cheek 11685 3/2004; 11880 3/2004.

Bertiera breviflora Hiern

Shrub, 2m; stipules sheathing at base, limb elliptic, 5mm, acumen 2mm, slender; leaves similar to *B. bracteolata*; flowers c. 50, panicles broader than long, with 4–6 main, spike-like branches to 8cm; fruits red. Evergreen forest; 710m.
Distr.: Sierra Leone to Congo (Kinshasa) [Guineo-Congolian].
IUCN: LC
Ndanan 1: Cheek 11559 3/2004.

Bertiera racemosa (G.Don) K.Schum. var. racemosa

Shrub, 3–6m; stipule foliose, c. 3.5cm; leaves lanceolate or elliptic, to 28 × 11cm, acumen 0.5cm, base obtuse or rounded; petiole c. 1cm; flowers c. 100, pendulous, terminal, c. 25cm, branches 10–15, 1–2cm long, 5-flowered, white; calyx glabrous; corolla tube 7mm, lobes 2mm, bud apex glabrous; fruit glabrous, green. Evergreen forest; 700m.
Distr.: Guinea (Conakry) to Congo (Kinshasa) [Guineo-Congolian].
IUCN: LC
Ndanan 1: Tadjouteu 567 3/2004.

Bertiera retrofracta K.Schum.

Shrub, 2–3m, resembling *B. racemosa* var. *racemosa*, but leaves oblanceolate, base cuneate; calyx, corolla, ovary and fruit entirely and densely pubescent. Evergreen forest; 700–710m.
Distr.: SE Nigeria, Bioko & SW Cameroon [lower Guinea].
IUCN: NT
Mefou Proposed National Park: Cheek in MF 82 3/2004; **Ndanan 1:** Cheek 11034 10/2002; 11601 3/2004; Harvey 179 10/2002; **Ndanan 2:** Cheek 11167 10/2002.

Calycosiphonia spathicalyx (K.Schum.) Robbr.
Bull. Jard. Bot. Nat. Belg. 51: 373 (1981); F.T.E.A.
Rubiaceae: 727 (1988); Hawthorne W., F.G.F.T. Ghana:
35 (1990).
Syn. *Coffea spathicalyx* K.Schum.
Shrub, or tree, to 2–4m, glabrous; stems mid-brown;
stipule transversely oblong-triangular, 2mm, awn stout,
4mm; leaf-blade oblong-elliptic, 8–12 × 3–4cm,
acumen slender, 2cm, often curved, lateral nerves 6–7
pairs, quaternary nerves reticulate, prominent, domatia
absent; petiole c. 1cm; inflorescences axillary, in both
axils of each node; peduncle 5mm, concealed by 3,
sheathing bract pairs; flowers white, petal lobes 7, 1cm;
fruit yellow, globose, 1.5cm, several-seeded. *Raphia*
swamp; 680–720m.
Distr.: Ivory Coast, Ghana, Nigeria, Cameroon, Gabon,
Congo (Brazzaville), Congo (Kinshasa), Angola, Sudan
[tropical Africa].
IUCN: LC
Ndanan 1: Onana 2861 3/2004; **Ndanan 2:** Cheek
11924 3/2004; Darbyshire 231 fr., 3/2004.

Chassalia cristata (Hiern) Bremek.
Liana; stems glossy, glabrous; stipule 2mm, cylindrical
with short simple or bifid awn 0.5mm; leaves elliptic to
13 × 5cm, acumen 2cm, base acute, lateral nerves 6
pairs, weakly looping near margin, domatia absent,
glabrous; petiole 1.5cm; inflorescence 5–8cm, densely
flowered, densely puberulent; peduncle 3cm; flower
buds pink, 10mm, longitudinally ridged. Lowland
evergreen forest; 600m.
Distr.: Nigeria to Tanzania [Guineo-Congolian].
IUCN: LC
Mefou Proposed National Park: Etuge 5213 3/2004.

Chassalia manningii O.Lachenaud ined.
Stout, unbranched shrub, 45–115cm; stems puberulous;
stipules ovate, bifid, c. 12 × 10mm, indumentum as
blade; leaves clustered in funnel at apex, oblanceolate,
to 35 × 13cm, acumen 1cm, base cuneate, abruptly
obtuse, lateral nerves c. 12 pairs forming a strong,
looping submarginal nerve, upper surface glabrous,
lower surface with patent hairs, dense on the nerves;
petiole 1–2.5cm, indumentum as blade; infructescence
erect, fleshy, pink, 7 × 5cm, lateral branches to 7cm,
stout; fruits black, bilobed, to 6 × 8mm; endocarp bony,
brown, PGS elliptic. Lowland evergreen forest; 700m.
Distr.: Cameroon and Nigeria [lower Guinea].
IUCN: EN
Ndanan 1: Harvey 124 10/2002.

Chassalia pteropetala (K.Schum.) Cheek
Kew Bull. 55 (4) 887 (2000); Nord. J. Bot. 28: 13–20
(2010).
Syn. *Psychotria pteropetala* K.Schum.
Erect sparsely branched shrublet, or shrub, 0.2–1.5m;
stems minutely and densely white papillose-shortly hairy
and a ridge at each side; stipules deeply bifid, more so

with age so that older stipules can be u-shaped; leaves
elliptic, to 15 × 7cm, acumen 1cm, base acute-decurrent,
lateral nerves 6–8 pairs, domatia and hairs absent;
inflorescence 2–5cm, subpaniculate; corolla white, in bud
5mm, conspicuously winged; fruit black, on fleshy red
infructescence. Lowland evergreen forest; 700–710m.
Distr.: SE Nigeria, Cameroon, Equatorial Guinea
(Bioko & Rio Muni) [lower Guinea].
IUCN: NT
Mefou Proposed National Park: Nana 31 3/2004;
Onana 2802 3/2004; **Ndanan 1:** Cheek 11249 10/2002;
11500 3/2004; 11611 3/2004; Gosline 430 10/2002;
Harvey 125 10/2002; **Ndanan 2:** Cheek 11097
10/2002.
Note: since described as a *Psychotria* a century ago,
this remained an obscure species until the paper of
Lachenaud & Jongkind (Nordic J. Bot. 28: 13–20
(2010), which delimited the species, showing that it is
scattered throughout the evergreen forest belt of
Cameroon and adjoining SE Nigeria, Cheek vi.2010.

Chassalia subherbacea (Hiern) Hepper
Erect subshrub, 15–30cm tall, gregarious; stems
unbranched or little branched, with a ridge on each
side, glabrous or with very few, short hairs; stipules
ovate-triangular to 7mm, apex shortly bifid; leaves
elliptic, 3–10 × 1.5–10cm, acumen 0.5cm, base acute-
obtuse, lateral nerves 5–10 pairs, domatia absent,
glabrous; petioles 6–12mm; inflorescence dense,
sessile, subcapitate, c. 1cm; corolla buds white, 5mm;
infructescence 4cm, vivid red, fleshy; fruit black, 7mm,
bilobed. Evergreen forest; 600–700m.
Distr.: Sierra Leone, Ivory Coast, Ghana, Nigeria,
Cameroon, Bioko, Gabon, CAR, Congo (Brazzaville),
Congo (Kinshasa) [upper & lower Guinea].
IUCN: NT
Mefou Proposed National Park: Etuge 5258 3/2004;
Ndanan 2: Cheek 11084 10/2002.

Chazaliella coffeosperma (K.Schum.) Verdc. subsp.
coffeosperma
Kew Bull. 31(4): 813 (1977).
Shrub, 1m, glabrous; bark sloughing, white; stem with
ridge on each side; stipules caducous; leaf-blade mid
green, lanceolate to oblong, 9–16 × 3.5–5cm, obtuse-
rounded with abrupt acumen 1.5cm, mucronate, base
acute, lateral nerves 4–8 pairs, looping together
5–10mm from the margin, tertiary nerves conspicuous;
petiole drying black, 6–15mm; panicles terminal, 3–5 in
fascicle, each 3–5-flowered; peduncles 4mm, buds
yellow. Evergreen forest; 700m.
Distr.: Cameroon endemic [lower Guinea].
IUCN: VU
Ndanan 1: Tadjouteu 579 3/2004.
Note: restricted to the Dengdeng-Sangmélina-Eseka-
Bipindi area. Likely to be found in Gabon, Cheek
ix.2010. The subspecies of *C. coffeosperma* are in the
opinion of Lachenaud, xi.2010, not worth maintaining.

Chazaliella macrocarpa Verdc.

Kew Bull. 31(4): 794 (1977).
Shrub, 5–70cm, glabrous; bark sloughing; stems brown, with ridge on each side, branched; stipules caducous; leaf-blade mid-green above, veins white, elliptic, slightly curved, 8–12 × 3–5cm, acumen 1cm, base cuneate, lateral nerves 6–8 pairs, looping 5mm from margin to form a marginal nerve, with a weaker straight intra-marginal nerve and tertiary connecting nerves, veinlets ramifying, prominent, domatia absent; petiole 10–12mm; panicles terminal, subfasciculate, sessile, dense; pedicels 1–2mm; fruit red, ellipsoid, 15 × 8mm. Evergreen forest; 700m.
Distr.: Cameroon, Gabon, Congo (Brazzaville), Congo (Kinshasa) [lower Guinea & Congolian].
IUCN: NT
Ndanan 1: Cheek 11247 10/2002.
Note: Cheek 11247 is a first record for Cameroon. The leaves and habit are smaller than usual so further evaluation is needed. Flowering material is needed. Cheek ix.2010.

Chazaliella obovoidea Verdc. subsp. *villosistipula* Verdc.

Kew Bull. 31(4): 808 (1977).
Shrub, 1.5m; bark sloughing, dark brown, minutely puberulent; stems with a ridge on each side; stipules broadly triangular, 2mm, hairy; leaf-blade elliptic, 4–7.5 × 1.8–3.2cm, acumen long, pointed, 1cm, base decurrent, lateral nerves 8–9 pairs, forming a looping intramarginal nerve in the distal half, midrib orange-brown, prominent below; petiole 6–8mm; inflorescence terminal, single, peduncle patent, 1cm, puberulent, flowers few, subcapitate; pedicel 2mm; fruit ellipsoid, stipitate, 9 × 6mm. Evergreen forest; 700–710m.
Distr.: Cameroon & Congo (Brazzaville) (lower Guinea & Congolian).
IUCN: VU
Mefou Proposed National Park: Cheek in MF 106 3/2004; Nana 56 3/2004; **Ndanan 1:** Cheek 11037 10/2002; 11261 10/2002; 11570 3/2004; 11607 3/2004; 11736 3/2004; 11791 3/2004; Harvey 177 10/2002; 206 10/2002; Onana 2197 10/2002; Tadjouteu 596 3/2004.
Note: Dengdeng to Bipindi area; likely to occur in Congo (Kinshasa), Cheek ix.2010. According to Lachenaud xi.2010 also in Gabon, CAR, Congo (Kinshasa).

Chazaliella oddonii (De Wild.) Petit & Verdc. var. *cameroonensis* Verdc.

Kew Bull. 31: 802 (1977).
Shrub, 0.5–2m; bark sloughing, minutely puberulent to glabrous; stem with a ridge on each side; stipule bases persistent, thickened, pale brown; leaf-blade elliptic, 7–14 × 3–7cm, apex rounded, acumen 0.5cm, abrupt, base decurrent, lateral nerves 6–7 pairs, domatia absent, reticulum conspicuous below; petiole to 2cm; inflorescence terminal; peduncles 3, 1–1.5cm, each 10–15-flowered, glabrous to puberulent; flowers yellow, 3mm wide; fruit ellipsoid, 5mm, red or pale pinkish, translucent, 5-lined. Forest; 600–710m.
Distr.: Liberia to Congo (Kinshasa) [lower Guinea & Congolian].
IUCN: LC
Mefou Proposed National Park: Etuge 5246 3/2004; **Ndanan 1:** Cheek 11828 3/2004; Cheek in MF 370 10/2002; Gosline 424 10/2002; Harvey 153 10/2002; Onana 2195 10/2002; 2920 3/2004; **Ndanan 2:** Cheek 11090 10/2002.
Note: widespread in the forests of Cameroon, known from only c. 20 specimens.

Chazaliella sciadephora (Hiern) Petit & Verdc. var. *sciadephora*

Kew Bull. 31: 790 (1977).
Syn. *Psychotria sciadephora* Hiern
Shrub, 2m; bark sloughing, drying jet black, glabrous; stipule broadly triangular, 7mm; leaves elliptic, c. 20 × 6.5cm, papery, drying dark brown, crumpled, lateral nerves 8 pairs, matt, lacking domatia; flowers white, drying black with white throat hairs; infructescence with several axillary peduncles, to 5cm; fruits ripening yellow or red, later dark brown or black. Forest; 700m.
Distr.: Guinea (Conakry) to Gabon [upper & lower Guinea].
IUCN: LC
Ndanan 1: Cheek 11232 10/2002; 11256 10/2002; Harvey 176 10/2002.
Note: although widespread in W Africa, its habitat is under increasing threat. Cheek 11232 is probably this taxon, Cheek ix.2010.

Corynanthe pachyceras K.Schum.

Tree, to 20m, glabrous; leaves drying brown-green, elliptic to oblanceolate, to 17 × 8cm, acumen 1cm, cuneate, lateral nerves 6–10 pairs, domatia pits elliptic, large; petiole 2cm; panicle terminal, 12cm, dense, each with hundreds of flowers; peduncle 5cm; flowers white, 4mm; corolla lobes with globular terminal appendage; style exserted; stigma spherical; fruit dry, elliptic, 7mm, 2-valved; seeds winged. Forest; 700m.
Distr.: Sierra Leone to Congo (Kinshasa) [Guineo-Congolian].
IUCN: LC
Ndanan 1: Harvey 144 10/2002.
Note: closely related to *Pausinystalia*.

Dictyandra arborescens Welw. ex Hook.f.

Tree, to 7m; bole with conical spines; stipules broadly triangular, 5mm; leaves obovate-elliptic, 12–20 × 5–9cm, acuminate, domatia hairy; cymes lax; calyx lobes elliptic, 1cm, one margin thickened, recurved; corolla 3.5cm, silky tomentose outside; fruits fleshy; calyx foliose. Forest; 700–710m.
Distr.: Guinea (Conakry) to Uganda [Guineo-Congolian].
IUCN: LC
Ndanan 1: Cheek 11010a 10/2002; 11553 3/2004; Cheek in MF 368 10/2002.

Diodia sarmentosa Sw.
F.T.E.A. Rubiaceae: 336 (1976).
Syn. *Diodia scandens* sensu Hepper F.W.T.A. 2: 216 (1963).
Straggling herb, to 5m, scabrid; stipule cupular 1mm, with 5 aristae, 3mm; leaves ovate or ovate-lanceolate, 2.5–5 × 1–2.5cm, pubescent below, scabrid above; petioles to 7mm; axillary fascicles sessile; flowers white, 4-merous; fruit ellipsoid, 3–4mm, didymous, dry. Forest; 700–710m.
Distr.: pantropical.
IUCN: LC
Ndanan 1: Cheek 11600 3/2004; **Ndanan 2:** Darbyshire 187 fl., 3/2004.

Gardenia vogelii Hook.f. ex Planch.
Shrub, 2m; stipule sheathing; leaves papery, elliptic or obovate, c. 20 × 7cm, long-acuminate, acute, margin sometimes deeply toothed; petiole 3mm; corolla tube 12–15cm, lobes 3–8.5cm; fruit cylindrical, 10 × 1cm, sessile; calyx persistent. Forest & farmbush; 550–710m.
Distr.: Liberia to Zimbabwe [Guineo-Congolian].
IUCN: LC
Mefou Proposed National Park: Etuge 5197 3/2004; **Ndanan 1:** Cheek 11625 3/2004.

Geophila afzelii Hiern
Stoloniferous herb, with short vertical shoots, hairy; leaves ovate, 3–6 × 2–5.5cm, subacute, cordate; stipules bifid; petioles 3–15cm long, hairy; inflorescence terminal; peduncle 0.2–5cm; bracts involucral, ovate; bracteoles narrowly lanceolate; flowers white; fruit watery, inflated, red. Forest; 600–710m.
Distr.: Guinea (Conakry) to Congo (Kinshasa) [Guineo-Congolian].
IUCN: LC
Mefou Proposed National Park: Cheek in MF 110 3/2004; Etuge 5239 3/2004; **Ndanan 1:** Cheek 11235 10/2002; 11564 3/2004.

Geophila obvallata Didr. subsp. *obvallata*
Creeping herb; stems below ground or at surface; stipules entire; leaves triangular-ovate to ovate-reniform, 3–5 × 2–4cm, apex acute, base cordate, glabrous above, glabrous or slightly pubescent around midnerve beneath; petioles erect, 5–10cm, with two rows of hairs; inflorescence terminal; peduncle erect, 1.5–7cm, glabrous; bracts green, leafy, 7mm, glabrous; flowers white; fruits watery inflated, bright blue or purple, globose, 1cm (live). Farmland and along stream edges; 693–700m.
Distr.: Guinea (Bissau), Guinea (Conakry), Sierra Leone, Liberia, Ivory Coast, Ghana, Nigeria, Cameroon, Equatorial Guinea, Gabon, Congo (Brazzaville) [Guineo-Congolian].
IUCN: LC
Ndanan 1: Harvey 191 10/2002; Onana 2922 3/2004.

Geophila repens (L.) I.M.Johnston
Prostrate stoloniferous herb; stems and petioles patent-puberulent; stipule semi-circular, membranous, patent, 1mm, glabrous; leaf-blade ovate-orbicular, to 3 × 3cm, rounded, basal sinus 5mm, lower surface dirt-gathering; petiole 3–7.5cm; peduncle erect, 4.5cm; fruit ovoid, 8mm, glossy red. Forest; 700m.
Distr.: pantropical.
IUCN: LC
Ndanan 2: Cheek 11117 10/2002.

Hallea stipulosa (DC.) Leroy
Adansonia (Sér. 2) 15: 65 (1975); F.T.E.A. Rubiaceae: 447 (1988); Hawthorne W., F.G.F.T. Ghana: 29 (1990).
Syn. *Mitragyna stipulosa* (DC.) Kuntze
Tree, 35m; stipule oblong to suborbicular, 4–8 × 2.5–5cm, pubescent; leaves obovate, 12–45 × 8.5–26cm, rounded, acute, nerves to 20; petiole to 5cm; inflorescence of 3–10 terminal capitula, each 5–10mm diam. Swamp forest; 700m.
Distr.: Gambia to Congo (Kinshasa) [Guineo-Congolian].
IUCN: VU
Mefou Proposed National Park: Cheek 13115 2/2006; **Ndanan 1:** Cheek 11226 10/2002.
Note: *Cheek* 11226, midrib pilose below, domatia absent. No juvenile material at Kew so not well matched, Cheek vii.2006.

Heinsia crinita (Afzel.) G.Taylor
Syn. *Heinsia scandens* Mildbr. *nomen.* Wiss. Ergebn. Deutsch. Zentr.-Afr. Exped. 1910–11, 2: 64 (1922).
Shrub, to 7m, appressed grey-puberulous; stipule caducous; leaves elliptic, 9 × 3cm, acuminate, 4-nerved, tertiary nerves normal to midrib; petiole 1cm; flowers terminal, 1 to few; corolla white with yellow hairs in throat, tube c. 2cm; fruit globose, 2cm diam., many-seeded; sepals accrescent, green, 2cm. Evergreen forest; 700m.
Distr.: Guinea (Conakry) to Angola [Guineo-Congolian].
IUCN: LC
Ndanan 1: Cheek 11181 10/2002; Harvey 158 10/2002; **Ndanan 2:** Darbyshire 192 fr., 3/2004.

Hekistocarpa minutiflora Hook.f.
Gregarious herb, 1.5m, puberulent; stipule leafy, triangular, 7mm; leaves membranous, narrowly elliptic, c. 13 × 3cm, acuminate, acute, 8–10-nerved; petiole 1cm; inflorescence axillary, tuning-fork shaped, stalk 5mm, branches c. 3cm; corolla white, tube 3mm. Evergreen forest; 700–710m.
Distr.: SE Nigeria & Cameroon [lower Guinea].
IUCN: LC
Ndanan 1: Cheek 11555 3/2004; **Ndanan 2:** Cheek 11087 10/2002.

Hymenocoleus hirsutus (Benth.) Robbr.
Bull. Jard. Bot. Nat. Belg. 45: 288 (1975); F.T.E.A. Rubiaceae: 115 (1976).
Syn. *Geophila hirsuta* Benth.

Stoloniferous herb with horizontal stems, drying blackish; leaves pubescent above, ovate to ovate-oblong, 2–5 × 1–4cm; stipules shortly divided; inflorescence terminal, without conspicuous leafy bracts, subsessile (rarely pedunculate); flowers white; calyx tube short, lobes much longer than the tube, oblong, hairy; fruit orange, watery inflated; pyrenes with 2 lateral grooves. Forest; 700–710m.
Distr.: Senegal to Congo (Kinshasa) & Tanzania [Guineo-Congolian].
IUCN: LC
Ndanan 1: Cheek 11238 10/2002; 11844 3/2004.

Hymenocoleus libericus (A.Chev. ex Hutch. & Dalziel) Robbr.
F.T.E.A. Rubiaceae: 115 (1976); Bull. Jard. Bot. Nat. Belg. 47: 19 (1977).
Syn. *Geophila liberica* Hutch. & Dalziel
Succulent decumbent herb, stem rooting, apex ascending to 20cm; stems sparsely appressed hairy; stipule caducous; leaf-blade oblong, 9 × 3.5cm, apex acute-acuminate, base subcordate to obliquely obtuse, lateral nerves c. 5 pairs, nerves minutely puberulent; petiole 1.3–2cm; infructescence terminal, sessile; fruits globose, c. 1cm, bright orange, 2-seeded, fleshy; calyx lobes <1mm. Evergreen forest; 700m.
Distr.: Liberia to Uganda [Guineo-Congolian].
IUCN: NT
Ndanan 1: Harvey 119 10/2002; 185 10/2002; **Ndanan 2:** Cheek 11109 10/2002.

Hymenocoleus nervopilosus Robbr. var. *orientalis* Robbr.
Bull. Jard. Bot. Nation. Belg., 47(1–2): 12 (1977).
Gregarious herb, with rooting horizontal stems, ascending to 15–20cm at apex; stems long-hairy; stipules green, triangular, 4mm, with apical bristle 5mm, long hairy; leaf-blade drying green with white nerves above, oblanceolate, 7 × 2.5cm, obtuse-acute, abruptly rounded at base, lateral nerves c. 12 pairs, tertiary nerves scalariform, raised, bristled; inflorescence terminal, sessile, c. 1cm; calyx lobes erect, green, narrowly triangular, 4–5mm, long brown hairy; corolla white, tube 3–4mm, lobes 1.5mm. Evergreen forest; 700–720m.
Distr.: Cameroon & Gabon [lower Guinea].
IUCN: VU
Ndanan 1: Cheek 11506 3/2004; Harvey 81 10/2002; Onana 2848 3/2004.
Note: according to Lachenaud xi.2010 also in Rio Muni and Congo (Brazzaville).

Hymenocoleus neurodictyon (K.Schum.) Robbr. var. *orientalis* (Verdc.) Robbr.
F.T.E.A. Rubiaceae: 116 (1976); Bull. Jard. Bot. Nat. Belg. 47: 9 (1977).
Prostrate, rooting herb; stems fleshy, appressed puberulent; stipule 3–4mm, deeply bifurcate; leaves obovate, 5.5 × 3.5cm, apex rounded or obtuse, base abruptly rounded, lateral nerves puberulent, 7–10 pairs; petiole 0.8–2cm; inflorescence terminal, sessile; fruits bright orange, globose 5mm; calyx lobes 3mm. Evergreen forest; 700m.
Distr.: Liberia, Ivory Coast, Ghana, Cameroon, Congo (Kinshasa), Tanzania [Guineo-Congolian].
IUCN: NT
Mefou Proposed National Park: Cheek in MF 113 3/2004; **Ndanan 2:** Cheek 11109a 10/2002.
Note: the taxonomy of *H. neurodictyon* sensu lato requires further study, more than one species may be involved, Lachenaud xi.2010.

Ixora guineensis Benth.
Syn. *Ixora breviflora* Hiern
Syn. *Ixora talbotii* Wernham
Shrub, 2–3m; leaves coriaceous, oblong, elliptic or obovate-elliptic, c. 14 × 6cm, subacuminate, obtuse, nerves 8, drying yellow-brown below; petiole 1cm; inflorescence sessile, with minute spreading hairs; peduncle c. 2cm; pedicels 3mm; flowers white, corolla tube 25mm. Forest; 550–640m.
Distr.: Liberia, Nigeria to Congo (Brazzaville) [upper & lower Guinea].
IUCN: LC
Mefou Proposed National Park: Etuge 5194 3/2004; **Ndanan 2:** Darbyshire 286 fl., 3/2004.

Keetia acuminata (De Wild.) Bridson
Kew Bull. 41: 985 (1986).
Climber, 3m, pale brown scurfy; stipule ovate-acuminate, 9 × 6mm, white pubescent, midrib prominent; leaves oblong-elliptic, c. 10 × 6cm, acumen short, base rounded, softly pubescent below, nerves 7–9; petiole 1.5cm; fruiting peduncle 3cm, partial-peduncle 3cm; fruit 1 × 1.4cm. *Hypselodelphys* scrub; 640m.
Distr.: Cameroon to Congo (Kinshasa) [lower Guinea & Congolian].
IUCN: LC
Ndanan 2: Darbyshire 306 fr., 3/2004.
Note: *Darbyshire* 306 has atypical fruits wider than long, retuse, Cheek ix.2010.

Keetia hispida (Benth.) Bridson '*setosum*'
Kew Bull. 41: 986 (1986).
Syn. *Canthium hispidum* Benth.
Syn. *Canthium setosum* Hiern F.T.A. 3: 141 (1877).
Climber, 3m; stems glabrous; stipule not seen; leaves drying red below, thickly papery, obovate-elliptic, c. 17 × 8cm, acuminate, unequally truncate-cordate, thinly pilose on both surfaces, domatia inconspicuous, nerves 7–9; petiole 1cm; fruiting peduncle 0.5cm, partial-peduncles 1cm; fruit 1.6 × 1.4cm, ridged. Evergreen forest; 710m.
Distr.: Sierra Leone to Congo (Kinshasa) [Guineo-Congolian].
IUCN: LC
Ndanan 1: Cheek 11776 3/2004.

Note: Cheek 11776 is unique at Kew in the hooked hairs on midrib, secondary nerves and petiole, Cheek ix.2010.

Lasianthus batangensis K.Schum.

Erect shrub, 0.3–2m, pilose; leaves elliptic or slightly obovate, 15–20 × 5–9cm, acutely acuminate, cuneate, midrib pilose above and below; petiole 2–6cm; flowers axillary, several, subsessile, blue or white; corolla 3mm; fruits globose, 8mm, blue. Forest; 700m.
Distr.: Sierra Leone to Congo (Kinshasa) [Guineo-Congolian].
IUCN: LC
Mefou Proposed National Park: Cheek in MF 85 3/2004; **Ndanan 1:** Cheek 11206 10/2002; Harvey 180 10/2002; Tadjouteu 578 3/2004.
Note: *plot voucher MF85 is probably Lasianthus batangensis*, Cheek ix.2010.

Leptactina arnoldiana De Wild.

Fl. Gabon 17: 73 (1970).
Monopodial shrub, 2–6m, glabrous; stipules sheathing, inflated to 1 × 2cm; leaf blades opposite, equal, obovate to oblanceolate, to 42 × 13cm, acumen 0.5cm, base cuneate-decurrent, lateral nerves up to 25 pairs, sparse-scalariform, domatia absent; petiole 1–2cm; inflorescence a fascicle, terminal on a horizontal axillary stem, c. 35cm, naked but for two leaf pairs at the apex; corolla white, tube 9–12cm, lobes 6–9cm; pedicels 3cm; sepals narrowly oblong, 4cm (in fruit); fruit cyclindric 2.5 × 0.8cm, ridged, many-seeded. Forest edge; 710m.
Distr.: Cameroon, Gabon, Cabinda, Congo (Kinshasa) [lower Guinea & Congolian].
IUCN: NT
Ndanan 1: Cheek 11616 3/2004.

Leptactina involucrata Hook.f.

Pl. Syst. Evol. 145: 114 (1984).
Syn. *Dictyandra involucrata* (Hook.f.) Hiern
Shrub, or tree, 10m, glabrous; stipule folicaeous, 1cm; leaves papery, elliptic, 27 × 12cm, acuminate, acute, nerves 10–12; petiole 2cm; flowers white, 10–20 in terminal panicles, 3cm; calyx lobes 8mm; corolla white pubescent, lobes 2mm; fruit ovoid, 2 × 0.7cm; calyx lobes 1.5cm; seeds numerous. Evergreen forest; 650–710m.
Distr.: Sierra Leone to Rio Muni [upper & lower Guinea].
IUCN: LC
Ndanan 1: Cheek 11677 3/2004; Darbyshire 260 fr., 3/2004; Harvey 91 10/2002.

Leptactina laurentiana Dewèvre

Bull. Soc. Roy. Bot. Belgique 34(2): 95. (1895).
Liana, to 4m; stem twisting, terete, woody, glabrous, purple-brown; stipule persistent, to 10 × 7mm, apex gradually awn-like; leaf-blade elliptic, c. 10 × 4cm, acumen 1cm, base cuneate-decurrent, lateral nerves c. 7 pairs, white tuft domatia, quaternary nerves

conspicuous, raphide rods dense on lower surface when dry; petiole 1cm; inflorescence 5cm, terminal on short (20cm) lateral branches, c. 9-flowered; pedicels 3mm, exceeded by bracts; calyx lobes narrowly oblong, 15mm; corolla tube 40mm, lobes 20mm, white, night opening? Forest edge; 710m.
Distr.: Cameroon, Gabon, Congo (Brazzaville) & Congo (Kinshasa) [lower Guinea].
IUCN: NT
Ndanan 1: Cheek 11599 3/2004; 11801 3/2004.

Massularia acuminata (G.Don) Bullock ex Hoyle

Shrub, 4m; stems glabrous; stipule semi-circular, c. 4 × 1.5cm; leaves obovate-elliptic, c. 30 × 10cm, acuminate, subcordate, nerves 15–20, subsessile; flowers 5–10, axillary, panicle branched, 5cm; corolla pink and white, tube 12cm, lobes 8mm; fruit spherical-rostrate, 8 × 6cm, green, fleshy; few-seeded. Evergreen forest; 600–710m.
Distr.: Guinea (Conakry) to Congo (Kinshasa) [Guineo-Congolian].
IUCN: LC
Mefou Proposed National Park: Etuge 5265 3/2004; Nana 57 3/2004; **Ndanan 1:** Cheek 11668 3/2004; Cheek in MF 383 10/2002; Onana 2917 3/2004; 2918 3/2004; **Ndanan 2:** Darbyshire 277 fl., fr., 3/2004.
Note: Onana 2917 has galled fruit, leaves unusually green, axes atypically densely hairy. Very confusing. Cheek ix.2010.
Local Name: Oyebe (Nana 57).
Uses: MATERIALS — Other Materials — stems used to make traps (Nana 57).

Morelia senegalensis A.Rich. ex DC.

Shrub or small tree, to 6m; stems glabrous; stipule oblong, apex rounded, 5mm, including 1mm awn; leaves elliptic-oblong, or slightly obovate, c. 16 × 7, acumen 1cm, base rounded to obtuse, lateral nerves 3–5 pairs, sparsely hairy, domatia absent, lower surface with pale cast, upper drying dark brown; petiole 1.5–2cm; panicles axillary, 5cm, c. 15-flowered; flowers white; petal lobes 8mm, tube short; fruit globose, 8mm crowned with calyx tube 3–4mm, 5-toothed. Forest edge; 600m.
Distr.: Senegal to Sudan [Guineo-Congolian].
IUCN: LC
Mefou Proposed National Park: Etuge 5173 fl., 3/2004.

Morinda mefou Cheek

The Plants of Mefou proposed National Park, Yaoundé: A Conservation Checklist: 76 (2011).
Liana, scrambling on trees to 6m tall; internodes (3–)4–6.5(–8)cm long, 3–4(–5)mm diam., glabrous or sparsely to densely hairy; stipule sheath (1–)2–3 × 4–6(–10)mm the basal half swollen, thickened, the apical half submembranous with conspicuous white rod-like raphides visible at the surface; leaves opposite, equal; blades elliptic or oblong-elliptic, rarely ovate (5.7–)7–10(–11.2) × (2.8–)3.2–5.8(–6)cm, acumen

0.9–1.2cm long, gradually tapering, base rounded or truncate, rarely obtuse, quaternary nerves reticulate, raised on the lower surface, indumentum dense on the lower surface midrib, remainder sparsely hairy, upper surface glabrous or moderately hairy, lateral nerves on each side of the midrib 5–9, parting at 45° from the midrib, forming a looping marginal nerve; petiole 5–8(–9)mm × 0.75–1mm, glabrous or hairy; inflorescence and flowers unknown; infructescence a syncarp comprising 3–4 united fruits, subglobose-lobed, ripening yellow or orange, hard, 2.2–3.1cm diam., subglabrous to long hairy; individual fruits subovoid, the free part volcano shaped, c. 15 × 15mm; calyx throat aperture c. 4mm; disc persisting, annular, flat, 3mm diam.; seed 10 × 8 × 5mm, U-shaped, folded transversely. Lowland evergreen semi-deciduous forest, forest edge or secondary; 600m.
Distr.: Cameroon, Gabon, Congo (Brazzaville) [lower Guinea].
IUCN: VU
Mefou Proposed National Park: Etuge, 5214 3/2004.

Nauclea vanderguchtii (De Wild.) Petit
Tree, c. 30m; stems with ants; stipule ovate, 20–40 × 10–20mm, caducous; leaves oblong-ovate, 21–35 × 8–18cm, glabrous, base cordate, nerves 9–13; petiole 12–30mm; inflorescence spherical, 4–6cm diam. Evergreen forest; 700m.
Distr.: Nigeria, Cameroon to Congo (Kinshasa) [lower Guinea & Congolian].
IUCN: LC
Ndanan 1: Cheek 11244 10/2002; 11989 3/2004; Tadjouteu 586 3/2004.
Note: for *Tadjouteu* 586 the long leaves and large compound fruits are characteristic. No ants (*Crematogaster*) recorded here. *Cheek* 11244 is possibly immature *N. vandeveltii* but may also be larger than usual *N. diderichii* fruits, Cheek vii.2006.
Local Name: Akondok (Tadjouteu 586).
Uses: MATERIALS — Wood — timber.

Oldenlandia lancifolia (Schumach.) DC. var. *lancifolia*
Herb, 30–60cm; glabrous; internodes 2–5cm, cylindric; stipule bifid; leaves sessile, narrowly lanceolate, 2–6 × 0.2–0.7cm; flowers 1–2 per axil; pedicel 8–18mm; corollas white, sometimes pale pink or mauve, 2–3mm; fruit 3–4mm. Wet areas in forest and *Raphia* swamp; 650–710m.
Distr.: tropical Africa.
IUCN: LC
Ndanan 1: Cheek 11749 3/2004; **Ndanan 2:** Darbyshire 209 fl., fr., 3/2004.

Otomeria micrantha K.Schum.
Bot. Jahrb. Syst. xxiii: 423 (1896).
Erect, annual herb, 0.6m; axes puberulent in two lines; stipules reduced to 3–5 filaments, each 3mm; leaf-blade lanceolate, to 9 × 3.75cm, acute, base obtuse, then abruptly decurrent, lateral nerves 6–8 pairs; petiole 1cm; inflorescence terminal, spikes 8cm, interupted; flowers sessile, paired; corolla pink, tubular 4mm, 4-lobed; fruit with sepals accrescent, very unequal, foliose, to 7mm. Forest edge, river edge in forest; 710m.
Distr.: Cameroon, Gabon, Congo (Kinshasa) [Congolian].
IUCN: LC
Ndanan 1: Cheek 11542 3/2004.
Note: *Otomeria micrantha* is partially cited in F.W.T.A 2: 215 (1963) as follows: *Otomeria* sp. A, *Otomeria micrantha* (?) of Verdcourt, partly, not of K.Schum.

Oxyanthus formosus Hook.f. ex Planch.
Shrub, or tree, 3–10m; stipule lanceolate, 2–3 × 1–1.5cm; leaves coriaceous, oblong to 25 × 12cm, short-acuminate, unequally obtuse; petiole 1cm; corolla tube 12 × 0.1cm; lobes 2.5 × 0.1cm; fruit ellipsoid, 5 × 2.5cm; pedicel 2.5cm. Evergreen forest; 700–710m.
Distr.: Mali to Uganda [Guineo-Congolian].
IUCN: LC
Ndanan 1: Cheek 11568 3/2004; Gosline 403 10/2002; Tadjouteu 601 3/2004.

Oxyanthus pallidus Hiern
Shrub, 3–6m; stipule ovate-orbicular, 15 × 9mm, long acuminate; leaves coriaceous, elliptic, 20–25 × 10–15cm, acuminate, obtuse, decurrent; petiole 1cm; corolla tube 17–18cm, lobes 1.7–1.8cm; fruit spherical, 2.5cm; pedicel 1.5cm. Evergreen forest; 679–710m.
Distr.: Ivory Coast to Congo (Kinshasa) [Guineo-Congolian].
IUCN: LC
Mefou Proposed National Park: Nana 28 3/2004;
Ndanan 1: Cheek 11619 3/2004; 11837 3/2004; 11909 3/2004; Harvey 168 10/2002; 171 10/2002; Onana 2900 3/2004.

Oxyanthus speciosus DC. subsp. *speciosus*
Syn. *Oxyanthus speciosus* DC. subsp. *globosus* Bridson Kew Bull. 34: 115 (1979); F.T.E.A. Rubiaceae: 529 (1988).
Shrub, or tree, 5–12m; stipule oblong-acuminate; leaves elliptic, c. 20 × 9cm, acuminate, unequally obtuse; petiole 1.5cm; corolla tube 3.5cm, lobes 1.5cm; fruit ellipsoid, 2.5 × 1.5cm. Forest; 700–710m.
Distr.: Senegal to Zambia [Guineo-Congolian].
IUCN: LC
Ndanan 1: Cheek 11185 10/2002; 11526 3/2004; 11681 3/2004; Harvey 135 10/2002; Onana 2201 10/2002; 2817 3/2004; **Ndanan 2:** Cheek 11081 10/2002.

Parapentas setigera (Hiern) Verdc.
Straggling herb, pubescent; stipules with 3–6 filiform lobes to 2.5mm; leaves ovate to elliptic, 2.7–3.7 × 1–2cm, glabrescent; petiole 2–11mm; fascicles axillary; corolla white, tube 5mm, lobes deltoid, 2mm; fruit 3mm, dehiscing; seeds numerous. Forest; 700m.
Distr.: Guinea (Conakry) to Congo (Kinshasa)

[Guineo-Congolian].
IUCN: LC
Ndanan 1: Harvey 211 10/2002.

Pauridiantha liebrechtsiana (De Wild. & T.Durand) Ntore & Dessein
Bot. J. Linn. Soc. 141(1): 113 (2003).
Shrub, 2.5m; stems glabrous; leaves narrowly elliptic, blade 8.5–12 × 2.3–3.2cm, acumen 1.5cm, lateral nerves c. 8 pairs, domatia hairy; petiole c. 1cm; cymes 1–2 per axil, each 3(–5-flowered), 2cm; peduncle 5–13mm; buds purple, 1–2mm; inflorescence with bicoloured hairs. *Raphia* swamp and river edge; 600–650m.
Distr.: Cameroon, CAR, Congo (Brazzaville), Congo (Kinshasa) [Congolian].
IUCN: LC
Mefou Proposed National Park: Etuge 5168 3/2004; **Ndanan 2:** Darbyshire 298 fl., 3/2004.
Note: newly recorded for Cameroon here. Characteristic of inundated forest, Cheek, iv.2010.

Pauridiantha pyramidata (K. Krause) Bremek.
Bot. Jahrb. Syst. 71: 211 (1940).
Shrub, c. 1m; stems appressed grey puberulent; leaves lanceolate, 6–7 × 2–2.5cm, acute-acuminate, lateral nerves c. 8 pairs, domatia hairy; inflorescence axillary, contracted, hairy, 5mm; peduncle 1–2mm; 6–8-flowered; pedicels 1mm; calyx lobes ± as long as corolla; corolla yellow, tube 3mm, hairy; fruit globose, 4mm, purple, many-seeded. Evergreen, periodically inundated forest; 710–720m.
Distr.: Cameroon, CAR, Congo (Brazzaville), Congo (Kinshasa), Gabon [Congolian].
IUCN: LC
Ndanan 1: Cheek 11649 3/2004; Onana 2846 3/2004.

Pavetta bidentata Hiern var. *bidentata*
Shrub, 2(–4)m, glabrous; floriferous twigs to 30cm; leaves thinly leathery, mostly narrowly oblong-elliptic, to 25 × 6cm, acute, cuneate, midrib drying orange, lateral nerves 15 pairs, domatia pits elongate, hairy, nodules elongate along midrib, rare in blade, tertiary venation just visible; petiole to 3.5cm; inflorescence 4cm across, subglabrous; corolla white, tube 4–10 × 1–2mm, lobes 5–10mm; fruit globose, pink or white with green stripes, then black; seeds 1–2, concave. Forest; 680–720m.
Distr.: SE Nigeria to Congo (Kinshasa) [Guineo-Congolian].
IUCN: LC
Ndanan 1: Cheek 11201 10/2002; Harvey 175 10/2002; Onana 2860 3/2004; Tadjouteu 618 3/2004; **Ndanan 2:** Darbyshire 242 fl., 3/2004.
Note: *Harvey* 175 is probably this taxon, Cheek ix.2010.

Pavetta calothyrsa Bremek.
Ann. Missouri Bot. Gard. 83: 107 (1996).
Shrub, to 3m, glabrous; leaves opposite, glossy above,

elliptic, 15–25 × 7–10cm, acumen indistinct, base acute-decurrent, lateral nerves 7 pairs, domatia absent; petioles 3–4(–6)cm; stipule sheath 4mm, awn 1.5mm; inflorescence terminal, c. 6 × 7cm, white puberulent, dense-flowered; flower bud 7mm long; fruit globose, 8mm, glossy; calyx lobes square, 4, erect, as long as tube. Edge of gallery forest; 600m.
Distr.: Cameroon, Gabon & Congo (Kinshasa) [lower Guinea & Congolian].
IUCN: LC
Mefou Proposed National Park: Etuge 5287 3/2004.
Note: we follow Manning (Ann. Miss. Bot. Gard. 83: 107 (1996)) in restoring *Pavetta calothyrsa* from synonymy under *P. nitidula*. This species has been reported from Gabon and Congo (Kinshasa) as well as Cameroon, Cheek vi.2009.

Pavetta camerounensis S.Manning subsp. *camerounensis*
Ann. Missouri Bot. Gard. 83: 108 (1996).
Shrublet, 30cm; branches 10–14cm, horizontal from each node, naked apart from a single terminal pair of leaves and the inflorescence, hairs absent, but epidermis rough, flaking, gathering soil; stipule broadly triangular, 5mm; leaves equal in pair, narrow-elliptic, 20 × 7cm, acumen c. 1cm, base unequally acute, lateral nerves 10–12 on each side of the midrib, forming a looping marginal nerve, domatia absent, quaternary nerves reticulate, visible with naked eye; petiole (1–)1.5–2cm; inflorescence subcapitate, sessile, terminal on horizontal side branches, c.1cm diam., glabrous; pedicels 3–4mm; sepal lobes <0.5mm. Evergreen forest; 700–710m.
Distr.: Cameroon & Equatorial Guinea [lower Guinea].
IUCN: NT
Ndanan 1: Cheek 11678 3/2004; Cheek in MF 349 10/2002.

Pavetta gabonica Bremek.
Ann. Missouri Bot. Gard. 83: 113 (1996).
Shrub, 1.5m; main axis with bark grubby white, flaking; leafy stems dark brown, minutely papillate, mat; stipule 2 × 3mm, awn long; leaves opposite, equal, elliptic, 15 × 6cm, acumen 0.5cm, base acute; petiole c. 3.5mm; inflorescence 1–5.5cm across; calyx lobes 0.75mm; flowers yellow; corolla tube 2–4mm, lobes 3–5mm; fruit orange, globose, 6mm. Stream banks in forest; 700m.
Distr.: Cameroon & Gabon [lower Guinea].
IUCN: NT
Ndanan 1: Gosline 402 10/2002.

Pavetta owariensis P.Beauv. var. *glaucescens* (Hiern) S.Manning
Ann. Missouri Bot. Gard. 83: 136 (1996).
Syn. *Pavetta glaucescens* Hiern
Shrub, to 15m, minutely papillate-puberulent; floriferous twigs 30–40cm; leaves elliptic-oblong, 12–20 × 5–8cm, acumen 1cm, cuneate, lateral nerves 8 pairs, domatia present (pits), bacterial nodules

inconspicuous; petiole 3cm; panicle 10cm across; calyx lobes to 1.5mm; corolla white; fruit 8mm diam.; calyx lobes not persisting. Forest; 710m.

Distr.: SE Nigeria to Angola [lower Guinea].

IUCN: LC

Ndanan 1: Cheek 11510 3/2004.

Note: *Cheek* 11510 is probably this (note very wide inflorescence) but flowers not yet open so not sure, Cheek ix.2010.

Pavetta subg. *Baconia* sp. 1 Mefou

Shrub or small tree, 4–6m; stems densely minutely puberulous; leaf-blades equal in pair, elliptic, (5.5–)11–15 × (2.8–)3.5–6cm, acumen 1cm, base unequally acute, margin often crenate, lateral nerves (5–)8–10 pairs on each side of the midrib, not uniting, domatia white hairy at nodes along the length of both midrib and secondary nerves, secondary nerves with indumentum as stem, quaternary nerves raised, forming conspicuous reticulum on lower surface; bacterial nodules scattered, as linear "vein knots"; petioles variable on each specimen, 0.7–3.5cm; inflorescences terminal on apparent lateral branches, 10–15cm, bearing only two leaf-pairs near apex, indumentum as stems, flower mass moderately dense, 8–11cm across; flowers white, jasmine-scented, 4-merous; sepal lobes quadrate 1.5 × 1mm; corolla tube 4mm, lobes 6–7mm, throat white hairy; anthers exserted; style exserted 7–11mm, clavate; fruit globose, 6mm, green (immature). Forest; 600–710m.

Distr.: Cameroon [Cameroon endemic].

Mefou Proposed National Park: Etuge 5256 3/2004; Nana 61 3/2004; **Ndanan 1:** Cheek 11683 3/2004.

Note: unusual in subg. *Baconia* for the broad inflorescences and domatia along the secondary nerves. Probably a new species to science. Cheek ix.2010.

Pentodon pentandrus (Schum. & Thonn.) Vatke

Glabrous herb; stems 50cm long or more, succulent, creeping, apex erect; leaves opposite, equal, narrowly oblong to lanceolate, c. 6–7 × 1.5cm, acute, base cuneate, nerves inconspicuous, leaf bases sessile, forming a sheath around the stem 2mm long with the stipule, from which an awn 1–2mm projects; panicles axillary, c. 10cm; peduncle 4cm; rachis with 4 nodes equally separated, each with 2–3 flowers; pedicels 1cm; corolla white, 3mm, caducous; fruit green fleshy, 4mm. Aquatic habitats; 710m.

Distr.: tropical Africa.

IUCN: LC

Mefou Proposed National Park: Cheek in RP 84 3/2004.

Poecilocalyx setiflorus (Good) Bremek.

Fl. Gabon 12: 233 (1966).

Shrub, 1–4m; hairs dense, spreading, 1mm, dull white on stem, midrib, petiole, stipule and fruit; stipule lanceolate, 10 × 4mm, with a tooth 1–3mm on each side, or entire, subsessile, axillary, 1–2 per axil; leaf blade bullate, obovate or oblong-elliptic, 8–13 × 3–4.5cm, acumen 1cm, with mucro 1–2mm, base abruptly rounded, lateral nerves 10–12 pairs, domatia absent, tertiary nerves scalariform; petiole 5mm; fruit globose, 4mm, with 5–7 erect green calyx lobes 4 × 1mm. Evergreen forest; 600–710m.

Distr.: Cameroon, Gabon & Cabinda [lower Guinea & Congolian].

IUCN: VU

Mefou Proposed National Park: Cheek in MF 75 3/2004; Etuge 5241 3/2004; **Ndanan 1:** Cheek 11205 10/2002; 11233 10/2002; 11608 3/2004.

Note: of conservation status since at K, 1 specimen in Cameroon (Mbalmayo) and 2 in Gabon (Massif du Chaillu & Riv. Eteke) & 1 in Cabinda, Cheek x.2010. According to Lachenaud xi.2010 other specimens occur.

Pseudosabicea sthenula N. Hallé

Fl. Gabon 12: 208 (1966).

Prostrate herb, covering several m^2; internodes 8–9cm, stems with scattered 3mm long brown hairs; stipules 1–1.5cm, dentate; leaves single, not paired, elliptic 6–9 × 3–5cm, subacuminate, base cordate, unequal, lateral nerves 6–7 pairs, lower surface white; petiole 1–5.5cm, hairs as stem; flowers in sessile axillary clusters, white; petal lobes 5; fruit ellipsoid, 1–2cm; calyx lobes spreading, brown, 4mm. Evergreen forest; 710m.

Distr.: Gabon & Cameroon [lower Guinea].

IUCN: VU

Ndanan 1: Cheek 11557 3/2004.

Note: *Cheek* 11557 is the first record in Cameroon at Kew. Previously endemic to Gabon where only four locations are recorded (Fl. Gabon 12: 208 (1966). *Pseudosabicea* has recently been included in *Sabicea* by some authors, Cheek ix.2010.

Psilanthus mannii Hook.f.

Shrub, 4m; stems whitish-grey, glabrous; stipule triangular, 1.5 × 1.5mm; leaves papery, elliptic-oblong, c. 16 × 6cm, acuminate, acute, nerves 6; petiole 1cm; flowers 1–few, axillary, subsessile, white; corolla tube 6cm, lobes 3cm; fruit top-shaped, 2 × 2cm, glossy, black; with green, strap-shaped calyx lobes 4 × 0.5cm; seeds 2. Evergreen forest; 710m.

Distr.: Guinea (Conakry) to Congo (Kinshasa) [Guineo-Congolian].

IUCN: LC

Ndanan 1: Cheek 11615 3/2004.

Psychotria anetoclada Hiern

F.T.A. 3: 206 (1877).

Shrub, 1–2m, glabrous; stipule triangular 2mm; leaf blades elliptic-oblong or obovate, 5.5–4 × 2.5–6cm, acumen to 0.7cm, base cuneate, lateral nerves 6–9, domatia absent; petiole 5–20mm; inflorescence terminal, umbellate, sessile to pedunculate; fruiting pedicel 4-7 mm, green to whitish, fleshy; fruits globose to ellipsoid, 1cm, red, usually enlarged to galls by 4x this size; calyx persistent, c.1.5 mm, truncate. Forest;

600–710m.
Distr.: Cameroon, Equatorial Guinea (Rio Muni), Gabon, Congo (Kinshasa) [lower Guinea & Congolian].
IUCN: LC
Mefou Proposed National Park: Etuge, 5279 3/2004; Nana 60 3/2004; **Ndanan 1:** Cheek, 11045 10/2002; 11498 3/2004; 11838 3/2004; Cheek in MF 255 10/2002.
Note (Mefou): this species has been wrongly synonymised with *P. subobliqua* by Petit (1964). It differs inter alia in having a truncate, not lobed, calyx. It is a very common species in Cameroon, commoner than *P. subobliqua*. Lachenaud xi.2010.

Psychotria bifaria Hiern var. *bifaria*
Erect shrublet, 0.5–1m; stem 4-angled, glabrous apart from two densely puberulent purple lines; stipule 5mm; leaves papery, drying pale grey below, bacterial nodules punctate, inconspicuous, blade elliptic, to 9 × 4cm, usually much smaller, acumen 1cm, obtuse, lateral nerves 8 pairs; petiole 0.5cm; panicle sessile, diffuse, 1.5cm, 5-flowered; fruit pendulous, fleshy, fusiform, 8mm, red. Forest; 660–700m.
Distr.: Bioko, Cameroon & Gabon [lower Guinea].
IUCN: LC
Ndanan 1: Tadjouteu 595 3/2004; **Ndanan 2:** Darbyshire 193 fl., 3/2004; 271 fl., 3/2004.

Psychotria camptopus Verdc.
Kew Bull. 30: 259 (1975).
Syn. *Cephaelis mannii* (Hook.f.) Hiern
Tree, 3–5m, glabrous; stipules 2 × 1.5cm; leaves leathery, obovate, 26 × 13cm, acumen obtuse, 1cm, acute, lateral nerves 14 pairs; petiole 6cm; peduncle 1–3m, pendulous, red; flowers white outside, yellow within, 1cm, 10–20, enveloped in red fleshy bracts; involucre 3 × 5cm, glabrous. Forest; 600–720m.
Distr.: SE Nigeria, Bioko, SW Cameroon [W Cameroon highlands].
IUCN: NT
Mefou Proposed National Park: Etuge 5262 3/2004; **Ndanan 1:** Cheek 11850 3/2004; Harvey 116 10/2002; Onana 2819 3/2004; Tadjouteu 603 3/2004.

Psychotria ebensis K.Schum.
Bull. Jard. Bot. Brux. 34: 157 (1964).
Syn. *Psychotria coeruleo-violacea* K.Schum.
Shrub, c. 1m, most surfaces with spreading red-brown 1mm hairs, these fringing the leaf, bract and sepal edges; stipules oblong c. 8 × 7mm, connate at base, with two abrupt awns 5–6mm; leaf blades opposite, equal, obovate to elliptic, c. 14 × 7cm, acumen 1cm, base acute, lateral nerves c. 12 pairs; petiole 1.5cm; panicle terminal c. 8 × 10cm, moderately diffuse; bracts elliptic, to 6mm; corolla blue, 6mm in bud; petals appendaged inside; sepal lobes 4mm; fruit 1cm. Evergreen forest; 679m.
Distr.: Nigeria, Cameroon, Equatorial Guinea (Rio Muni) and Gabon [lower Guinea].
IUCN: LC

Ndanan 1: Onana 2909 3/2004.

Psychotria fimbriatifolia R.D.Good
Bull. Jard. Bot. Brux. 34: 162 (1964).
Syn. *Psychotria brenanii* sensu Hepper & Keay, pro parte
Creeping, prostrate shrublet; flowering stems ascending, stem hairs dense, dark brown, crinkled, patent; stipule sheath 4–5mm, lobe triangular with abrupt 10mm bristle, hairs as stem; leaves oblong-elliptic or oblanceolate, 4–7.5 × 2–3.5cm, apex obtuse-apiculate, 2mm, base abruptly subtruncate or subcordate, lateral nerves c. 10 pairs, hairs as stem densest on midrib; petiole 4–6mm, hairy; inflorescence terminal, subsessile, capitate, c. 1 × 2cm; involucral bracts dense, triangular, 5 × 5mm, hairy; flowers 10–20; calyx lobes filiform, 4mm, brown hairy; corolla white. Secondary forest; 655–660m.
Distr.: Liberia to Congo (Kinshasa) [lower Guinea & Congolian].
IUCN: NT
Mefou Proposed National Park: Etuge 5307 3/2004; **Ndanan 2:** Darbyshire 280 3/2004.

Psychotria globosa Hiern var. *globosa*
Monopodial herb, to 0.6m, sparsely puberulent, glabrescent; stipule oblong-elliptic, 12mm; leaves drying black, elliptic or obovate, 13 × 6cm, subacuminate, acute, lateral nerves 8–10 pairs; petiole 1cm; panicle capitate, dense, flat-topped, 3cm; flowers white, 9mm; fruit ovoid, 1cm, red, smooth. Forest; 700m.
Distr.: Nigeria, Cameroon, Equatorial Guinea (Bioko & Rio Muni) [lower Guinea].
IUCN: NT
Ndanan 1: Gosline 431 10/2002.

Psychotria latistipula Benth.
Shrub, 0.3–2m, glabrous; stipule ovate, 1.2cm, bifurcate to half its length, glabrous; leaves thinly papery, drying dark brown below, elliptic, 16 × 7cm, subacuminate, decurrent, 20-nerved; petiole 2cm; inflorescence diffuse, 10–20cm; bracts lanceolate, 1cm, patent; flowers white, 4mm; infructescence pendulous; fruit globose, 4mm, red, ridged; pedicel not fleshy. Forest; 600–700m.
Distr.: Nigeria, Bioko, Cameroon & Gabon [lower Guinea].
IUCN: LC
Mefou Proposed National Park: Etuge 5144 3/2004; Nana 59 3/2004; **Ndanan 1:** Harvey 195 10/2002.

Psychotria leptophylla Hiern
Shrub, 0.5–1m, glabrous; stipules 7mm, bifurcate, densely brown hairy in lower half; leaves drying bright green, blade elliptic, 25 × 10cm, subacuminate, decurrent, lateral nerves 12 pairs, bacterial nodules mostly punctate; petiole to 6cm; panicle to 6cm; peduncle to 10cm, 2-winged; flowers white, 3mm;

corolla 4-lobed; fruit globose, red, 5mm. Forest; 600–710m.
Distr.: SE Nigeria, Bioko, Cameroon, Gabon & Congo (Kinshasa) [lower Guinea & Congolian].
IUCN: LC
Mefou Proposed National Park: Nana 73 3/2004; **Ndanan 1:** Cheek 11509 3/2004; Tadjouteu 597 3/2004.

Psychotria longicalyx O.Lachenaud sp. nov.
Shrub 2m, glabrous; glabrous, stem drying black-brown, numerous longitudinal grooves; stipule caducous, longitudinal ridges absent; leaf blades equal, in pairs, leathery, drying black-brown, obovate, c. 16 × 9.5cm, acumen broad and blunt, 3mm, base cuneate, lateral nerves c. 7 pairs, forming a weak, looping marginal nerve, quaternary nerves reticulate, not very conspicuous, domatia absent; petiole 2–2.5cm; infructescence terminal, sessile, subglobose, 3cm; fruits c. 12, ovoid, 1cm; persistent calyx 10mm, lobes slender, for half that length. Evergreen forest; 700m.
Distr.: Cameroon, Gabon, Congo (Brazzaville), Congo (Kinshasa) [lower Guinea].
IUCN: NT
Ndanan 2: Cheek 11173 10/2002.

Psychotria peduncularis (Salisb.) Steyerm. var. *hypsophila* (K.Schum. & K.Krause) Verdc.
Kew Bull. 30: 257 (1975).
Syn. *Cephaelis peduncularis* Salisb. var. *hypsophila* (K.Schum. & K.Krause) Hepper
Shrub, 1–5m, glabrous; leaves elliptic, to 15 × 8cm, acumen 0.5cm, acute, lateral nerves 12–15 pairs; petiole 2–3cm; stipule translucent, bifurcate, 1 × 0.8cm; inflorescence capitate; peduncle 2–4cm, nodding in accrescence, glabrous; involucral bracts fleshy; flowers 10–15, white, 5mm; infructescence umbellate; bracts fallen; pedicels white, 1.5cm; berries blue, 7mm. Forest; 640–710m.
Distr.: Bioko & Cameroon [W Cameroon Uplands].
IUCN: LC
Ndanan 1: Cheek 11654 3/2004; Harvey 204 10/2002; **Ndanan 2:** Cheek 11151 10/2002; Darbyshire 285 fl., 3/2004.
Note: the status of this and other varieties of *P. peduncularis* needs further study, Lachenaud, ii.2010.

Psychotria rubripilis K. Schum.
Bot. Jahrb. Syst. xxxiii. 369 (1903); Bull. Jard. Bot. Nat. Belg. 36: 162 (1966).
Shrub, 0.5m, erect; most surfaces with spreading white (drying red) crinkled hairs 1–2mm long; stipule 8mm, deeply bifid, the distal 2/3 awned; leaf-blades opposite, equal, oblanceolate or elliptic, 10–4.5cm, thinly papery, acumen 0.5–1cm, base obliquely truncate to subcordate, lateral nerves c. 10 pairs, bacterial nodules black spots on lower surface; petiole 4–5mm; infructescence terminal, subsessile; fruit globose, 4mm; calyx lobes 5, strap-like 4mm. Secondary forest and *Raphia* swamp; 600m.

Distr.: Cameroon, Gabon, W Congo (Kinshasa) [Guineo-Congolian].
IUCN: VU
Mefou Proposed National Park: Etuge 5270 3/2004.
Note: although distinctive and widespread, very rare. Etuge 5270 is only the ninth record known to the first author, Cheek x.2010. According to Lachenaud xi. 2010 also in Rio Muni, CAR and Congo (Brazzaville).

Psychotria sycophylla (K.Schum.) Petit
Bull. Jard. Bot. Brux. 34: 213 (1964).
Shrub, 1–3m, glabrous; stem stout, matt grey, hollow; stipule caducous, triangular 7 × 7mm, mucro 1mm; leaf-blades opposite, equal, narrowly oblong to elliptic, thickly leathery, drying dark green to dark brown above, yellow-brown below, to 21 × 9cm, apex acute to rounded, acumen abrupt, submucro 0.5cm, base acute, shortly decurrent, lateral nerves obscure, c. 10 pairs; petiole thick, 4–6cm; inflorescence terminal, shortly branched, 2cm diam. with peduncle erect, 4–11cm, 2mm wide; branches one pair, each to 1.5cm or not visible, when apparently capitate, 2cm diam.; flowers 30–50, calyx subtruncate, corolla buds 3mm, green at anthesis; infructescence to 9cm wide; fruit ellipsoid or ovoid, red, 8–10mm, ridged. Forest; 600–710m.
Distr.: Cameroon [Cameroon endemic].
IUCN: NT
Mefou Proposed National Park: Etuge 5243 3/2004; **Ndanan 1:** Cheek 11797 3/2004; **Ndanan 2:** Cheek 11125 10/2002.
Note: the type is from Yaoundé. Only about 10 locations are known between Edéa and DengDeng.

Psychotria thonneri (De Wild. & T.Durand) O.Lachenaud ined.
Monopodial shrub, 0.7–1.5m, glabrous; stems drying black; stipules oblong, 1.5 × 1.1cm, bifid, with 5mm awns; leaf-blades opposite, equal, clustered at stem apex, drying black-green, elliptic, c. 24 × 13cm, acumen 0.6cm, base obtuse-acute, lateral nerves c. 18 pairs, forming a marginal nerve, domatia absent; petiole c. 3cm; inflorescence terminal, sessile, enclosed in an involucre of c. 4 green bracts with reflexed tips, 3 × 5cm; infructescence becoming axillary; peduncle 1.5cm; involucral bracts falling; young fruits with calyx lobes narrowly triangular, 7mm, ripe fruit dark metallic blue, ellipsoid, longitudinally ridged, 7mm; pedicels blue or white. Evergreen forest; 680–700m.
Distr.: Cameroon, Gabon, Equatorial Guinea, CAR, Congo (Kinshasa) [lower Guinea & Congolian].
IUCN: NT
Ndanan 1: Cheek 11258 10/2002; Gosline 433 10/2002; **Ndanan 2:** Darbyshire 245 fr., 3/2004.

Psychotria venosa (Hiern) Petit
Tree, (2.5–)8–12m, glabrous; stipule broadly ovate, 1cm, sheathing; leaves drying dark brown, elliptic, to

18 × 10cm, acumen 0.5cm, obtuse, lateral nereves 12 pairs, finely puberulent below, domatia pits; petiole to 2.5cm, slightly winged; panicle flat-topped, 15cm wide; peduncle 8cm; fruit globose, 5mm, red, faintly ridged. Forest & farmbush; 600–710m.
Distr.: Nigeria to Congo (Kinshasa) [Guineo-Congolian].
IUCN: LC
Mefou Proposed National Park: <u>Etuge 5158</u> fl., 3/2004; **Ndanan 1:** <u>Cheek 11648</u> 3/2004.

Psychotria sp. nov. aff. *fernandopoensis* Petit

Shrub, or tree, 2.5–3m, glabrous; stem drying grey-green, with fine longitudinal grooves; stipule suboblong, 8 × 6mm, shortly bifid or entire, margin black ciliate; leaf blades opposite, grey-green above, nerves orange-green below, narrowly elliptic-oblong, c. 15 × 5cm, acumen weak, 0.5cm, base acute, shortly decurrent, lateral nerves c. 12 pairs, quaternary nerves broadly reticulate, domatia present as small tufts; petiole 1–2cm; inflorescence terminal, sessile capitate to pedunculate with two short branches; peduncle to 1.5cm long, 5–10-flowered; calyx 1.5mm wide, cupular; fruit ellipsoid, red 1cm, not ridged; calyx remains minute. Riverbed and *Raphia* swamp; 550–650m.
Distr.: Cameroon [Cameroon endemic].
IUCN: EN
Mefou Proposed National Park: <u>Etuge 5195</u> 3/2004; **Ndanan 2:** <u>Darbyshire 215</u> fl., fr., 3/2004.
Notes: = Ngameni Ka[m]ga 59 (BR, P) & J and A Raynal 10756 (P), Lachenaud viii.2007.

Rothmannia hispida (K.Schum.) Fagerlind

Tree, 5–20m; stems, nerves and calyx hispid; leaves papery, elliptic, c. 15 × 6cm, acuminate, drying black; flowers white; calyx tube 1.5cm, limb 3cm, teeth 2cm; corolla basal tube 12 × 0.6cm, upper tube 3 × 2.5cm, lobes 1.5 × 1cm, outer surface grey silky hairy. Evergreen forest; 600–700m.
Distr.: Guinea (Conakry) to Congo (Kinshasa) [Guineo-Congolian].
IUCN: LC
Mefou Proposed National Park: <u>Cheek in MF 65</u> 3/2004; <u>Etuge 5242</u> 3/2004; <u>Onana 2800</u> fr., 3/2004; **Ndanan 1:** <u>Cheek 11059</u> 10/2002; <u>Cheek in MF 284</u> 10/2002; <u>Onana 2203</u> 10/2002; **Ndanan 2:** <u>Cheek 11119</u> 10/2002.
Local Name: Endone (Cheek 11119).

Rothmannia longiflora Salisb.

Shrub, or tree, to 5m, glabrous; leaves papery, elliptic, c. 12 × 5cm, acuminate, drying green, nerves 4–5 pairs; flowers green, blotched purple and white; calyx tube 6mm, limb 9mm, teeth 1.5mm; corolla basal tube 14.5 × 0.9cm, upper tube c. 3 × 3cm, lobes ovate, 2 × 2cm, outer surface densely puberulent. Evergreen forest; 600–710m.
Distr.: Guinea (Bissau) to Uganda [Guineo-Congolian].
IUCN: LC

Mefou Proposed National Park: <u>Etuge 5230</u> 3/2004; **Ndanan 1:** <u>Cheek 11680</u> 3/2004.

Rothmannia lujae (De Wild.) Keay

Tree, 15m; internodes drying black, glabrous, 4-angled, with long swelling above the node (ant-house); stipule semi-orbicular, 5mm, mucronate; leaf-blade elliptic, 17–20 × 10–11cm, apex rounded or subacuminate, base obtuse, lateral nerves 6–9 pairs, looping near margin, subscalariform, domatia absent; petiole 5mm; fruit ellipsoid, 15 × 10cm with 5 longitudinal ridges and minute scales, hard, brown. Evergreen forest; 710m.
Distr.: Nigeria, Cameroon, Gabon and Congo (Brazzaville) [lower Guinea].
IUCN: LC
Ndanan 1: <u>Cheek 11692</u> 3/2004.

Rothmannia octomera (Hook.f.) Fagerlind

Shrub, 2.5m; stems and leaves densely patent pubescent; stipule acutely triangular, 3mm; leaves 3 per node, elliptic or obovate, papery, c. 15 × 8–9cm, acumen 1cm, base unequally acute to subcordate, lateral nerves c. 7 pairs, domatia absent; petioles 1cm; fruit axillary, cylindric-ellipsoid, green, spotted white, 7 × 2.5cm, excluding the 8 filiform sepal lobes 4.5cm. Forest; 700m.
Distr.: Nigeria, Cameroon, Bioko, Gabon, Congo (Brazzaville), Congo (Kinshasa) and CAR [Guineo-Congolian].
IUCN: LC
Ndanan 1: <u>Cheek 11053</u> 10/2002; <u>Tadjouteu 623</u> 3/2004.

Rothmannia talbotii (Wernham) Keay

Tree, 5m, resembling *R. hispida* and *R. urcelliformis* but leaves whitish below, obovate-oblanceolate, 12–24 × 4–9cm, acuminate, cuneate, lateral nerves 7–9 pairs, venation obscure below, rusty pubescent when young; calyx densely brown tomentellous, lobes linear, 5–6mm; corolla tube 20–25cm, densely shortly velutinous outside, 2.5cm wide at mouth, lobes 4.5cm; fruit ellipsoid, 5-ridged, 6 × 3.5cm. Forest; 600–700m.
Distr.: Nigeria to Angola (Cabinda) [lower Guinea].
IUCN: LC
Mefou Proposed National Park: <u>Etuge 5273</u> 3/2004; **Ndanan 1:** <u>Gosline 445</u> 10/2002.

Rothmannia urcelliformis (Hiern) Bullock ex Robyns

Shrub, or tree, to 20m; stems glabrescent; leaves elliptic, c. 13 × 5cm, acuminate, white-tuft domatia, puberulent below, 7–8 pairs nerves; flowers white, blotched purple; calyx tube 0.3cm, limb 2cm, teeth 1.5cm; corolla basal tube 2.5 × 0.4cm, upper tube 5 × 4cm, lobes 3 × 1.5cm. Evergreen forest; 700m.
Distr.: Guinea (Conakry) to Mozambique [tropical Africa].
IUCN: LC
Ndanan 2: <u>Gosline 413</u> fr., 10/2002.

Rothmannia whitfieldii (Lindl.) Dandy

Tree, 12m; leaves leathery, glabrous, elliptic, drying brown below, c. 24 × 8cm, acumen 1.5cm, acute-obtuse, lateral nerves 8 pairs, venation conspicuous below; petiole 1.5cm; calyx lobes 15–66mm; corolla tube long velutinous, 3–17cm; fruit subglobose, 7cm, 10-ridged. Forest; 600m.
Distr.: tropical Africa.
IUCN: LC
Mefou Proposed National Park: Etuge 5228 3/2004.

Rutidea olenotricha Hiern

Climber, densely and shortly brown pubescent; leaves elliptic, oblong or oblanceolate, to 15 × 8cm, shortly acuminate, obtuse to rounded, nerves 7–8, domatia large, bright brown hairy, extending along secondary nerves; petiole 10mm; inflorescence with numerous branches; corolla yellow, tube 5mm; fruit yellow, 6mm. Evergreen forest; 700m.
Distr.: Sierra Leone to Congo (Kinshasa) [Guineo-Congolian].
IUCN: LC
Ndanan 1: Cheek 11011 10/2002; Tadjouteu 621 3/2004.

Rutidea smithii Hiern subsp. *smithii*

Climber, grey puberulent or glabrescent; stipule awn 8mm; leaves papery, drying matt black-brown above, grey-brown below, elliptic or elliptic-obovate, 10–17 × 4–8cm, shortly acuminate, acute, nerves 7–9, with bright white hairy domatia extending to tertiary nerve junctions; petiole 1–2.5cm; inflorescence with numerous branches; corolla white, tube 3mm; fruit green, 6mm. Evergreen forest; 700m.
Distr.: Sierra Leone to Kenya [tropical Africa].
IUCN: LC
Ndanan 2: Cheek 11123 10/2002.

Rytigynia membranacea (Hiern) Robyns

Shrub, 2m, glabrous; stems purple, lenticels white; leaves elliptic, c. 11 × 5cm, acumen 0.5cm, lateral nerves 7–10, domatia hairy; petioles 10mm; inflorescence axillary; peduncle 5mm; pedicels 6, extending in fruit to 10mm; fruit 5-lobed, 1cm. Evergreen forest; 693m.
Distr.: SE Nigeria, Cameroon, Gabon & Congo (Brazzaville) [lower Guinea & Congolian].
IUCN: NT
Ndanan 1: Onana 2921 3/2004.
Note: Onana 2921 is this taxon, or close, Cheek iv.2010.

Rytigynia sp. aff. membranacea of Mefou (Hiern) Robyns

Shrub, 1–4m, possibly deciduous; stems with epidermis flaking, glabrous; leaves elliptic, c. 6 × 2.5cm, acumen 1cm, lateral nerves 3–4 pairs, axils webbed, hairs sparse; petiole 3mm; inflorescence axillary; peduncle 1mm; bract cupular, splitting; flowers 3–15; pedicels 4mm, glabrous; corolla yellow, tube 3 × 2.5mm, lobes 2mm; style head winged, truncate, broadest at apex.

Semi-deciduous forest; 600–710m.
Distr.: Cameroon & CAR [lower Guinea & Congolian].
Mefou Proposed National Park: Etuge 5117 3/2003; **Ndanan 1:** Cheek 11707 3/2004; 11860 3/2004.
Note: differs from *R. membranacea* and related species in the flaking bark, smaller few-nerved leaves and shorter pedicels with flowers half the size. Possibly sp. nov. Leaves all unfolding when collected, possibly deciduous, Cheek, iv.2010.

Sabicea calycina Benth.

Climber, puberulous; leaves membranous, oblong, c. 9 × 4cm, acuminate, cordate, whitish-green below, nerves 9; petiole to 3.5cm; flowers 10–15; peduncle c. 6cm; bracts ovate, 1.2cm; calyx lobes elliptic, purple, 1.2cm; corolla white, tube 2cm. Forest; 600–710m.
Distr.: Sierra Leone to Congo (Kinshasa) [Guineo-Congolian].
IUCN: LC
Mefou Proposed National Park: Nana 71 3/2004; **Ndanan 1:** Cheek 11528 3/2004; Onana 2894 3/2004.
Note: Cheek 11528 is tentatively placed in this taxon, Cheek iii.2010.

Sabicea venosa Benth.

Climber, appressed white pubescent; leaves thinly papery, elliptic, to 9 × 3.5cm, acuminate, rounded to obtuse, nerves 16; petiole 1cm; flowers 5–10; peduncle 0.5cm; branches 0.5cm; bracts inconspicuous; calyx lobes 3mm; corolla white, tube 5mm; fruit 1cm, globose, white. Evergreen forest; 600–710m.
Distr.: Senegal to Congo (Kinshasa) [Guineo-Congolian].
IUCN: LC
Mefou Proposed National Park: Etuge 5121 3/2003; **Ndanan 1:** Cheek 11216 10/2002; 11757 3/2004; **Ndanan 2:** Darbyshire 274 fl., fr., 3/2004.
Note: Cheek 11216 has far fewer secondary nerves, hairs on nerves below sparser, strigose. Possibly a variant of *S. venosa*, Cheek iii.2010.

Sacosperma paniculatum (Benth.) G.Taylor

Climber, to 3m or more, glabrous; leaves papery, elliptic or oblong, c. 9 × 4cm, subacuminate, acute, nerves 8–10, inconspicuous; petiole 1cm; inflorescence c. 30 × 15cm, diffuse with 4 pairs of branches; fruits dry, elliptic, 5mm long, seeds numerous, winged. Lake edge; 710m.
Distr.: Gambia to Congo (Kinshasa) [Guineo-Congolian].
IUCN: LC
Ndanan 1: Cheek 11642 3/2004.

Schumanniophyton magnificum (K.Schum.) Harms

Monopodial treelet, to 5m; leaf-like branches opposite, 1.5m long, bearing 3 leaflet-like leaves each c. 1.5 × 0.6m; flowers sessile in erect clusters from branch ends; corolla white, tube 9cm; fruit globose, 4cm. Forest; 700m.
Distr.: S Nigeria to Congo (Kinshasa) [lower Guinea &

Congolian].
IUCN: LC
Ndanan 2: <u>Cheek 11124</u> 10/2002; <u>Darbyshire 197</u> fr., 3/2004.

Sherbournia bignoniiflora (Welw.) Hua

Woody climber, to 10m; stems glabrous; stipule elliptic, 7mm; leaves elliptic, c. 11 × 6cm, subacuminate, obtuse-truncate; petiole 1cm, hairy; flowers axillary, 1–3, violet and white; calyx tube 5 × 7mm, lobes elliptic, 15mm; corolla tube 25mm, lobes orbicular, 10mm. Evergreen forest; 600–700m.
Distr.: Sierra Leone to Zambia [Guineo-Congolian].
IUCN: LC
Mefou Proposed National Park: <u>Etuge 5221</u> 3/2004; **Ndanan 2:** <u>Cheek 11110</u> 10/2002.

Sherbournia hapalophylla (Wernham) Hepper

Climber, stem twisting; stems glabrescent, matt grey; bark flaking transversely; stipule elliptic, 10 × 3–4mm; leaf-blades elliptic, 12–15.5 × 5–5.5cm, subacuminate, base cuneate, lateral nerves 8 pairs, densely appressed white hairy on lower surface; petiole 1.7–2.2cm; inflorescence axillary, 3cm, few-flowered; fruits globose, 3cm, orange, barely ridged; calyx 7mm, lobes 1mm. Evergreen forest; 700m.
Distr.: Nigeria & Cameroon [lower Guinea].
IUCN: NT
Ndanan 2: <u>Cheek 11163</u> 10/2002.

Tricalysia amplexicaulis Robbr.

Bull. Jard. Bot. Nat. Belg. 57: 158 (1987).
Erect shrub, 1–2m, glabrous; stems unbranched, internodes c. 8cm; stipules triangular, 4mm, awn 1–2mm; leaf-blade oblanceolate-oblong, 13–19 × 3.5–5.5cm, acumen 0.5cm, base abruptly rounded, lateral nerves c. 8 pairs, inconspicuous; petiole 0.5–1.5mm; inflorescences axillary, c. 0.5cm, concealed in trapped debris; corolla white, 3mm; fruit orange, glossy, ellipsoid, 1cm. Evergreen forest; 700–710m.
Distr.: Cameroon [Cameroon endemic].
IUCN: VU
Mefou Proposed National Park: <u>Cheek in MF 46</u> 3/2004; **Ndanan 1:** <u>Cheek 11208</u> 10/2002; <u>11492</u> 3/2004; <u>11569</u> 3/2004; <u>Onana 2830</u> 3/2004.
Note: according to Lachenaud xi.2010 also in Gabon.

Tricalysia crepiniana De Wild. & T.Durand

Bull. Jard. Bot. Nat. Belg. 49: 329–335 (1979).
Shrub, or small tree, 4m, densely white patent puberulent; stipule transversely oblong, apex triangular 1.5mm, bristle 8mm; leaf-blade oblong-elliptic, 12 × 4.5cm, acumen 1.5cm, base obtuse, lateral nerves 5 pairs, weakly scalariform, domatia white tufted; petiole 3–5mm; inflorescences axillary at several nodes; peduncle 3–4mm with 3 pairs of hairy, narrowly triangular bracts; immature fruits green, globose, 4mm; calyx divided to base, lobes 5, 3mm. Evergreen forest; 600m.
Distr.: Cameroon, Congo (Brazzaville), Congo

(Kinshasa), Angola (Cabinda) [lower Guinea & Congolian].
IUCN: NT
Mefou Proposed National Park: <u>Etuge 5289</u> 3/2004.

Tricalysia gossweileri S.Moore

Bull. Jard. Bot. Nat. Belg. 49: 300 (1979).
Shrub, 1.5–3m, glabrous; leaves papery, drying pale grey-green, elliptic or obovate, c. 17 × 6.5cm, acuminate, acute, nerves 4–5; petiole 7mm; inflorescences 3–5-flowered, sessile; bracts and bracteoles cup-shaped; pedicel concealed; calyx 2mm, shortly lobed; corolla white, tube 3mm; fruit ellipsoid to 1.5 × 1cm, violet flushed white. Forest; 700–720m.
Distr.: Cameroon to Angola [lower Guinea].
IUCN: LC
Ndanan 1: <u>Cheek 11567</u> 3/2004; <u>Harvey 161</u> 10/2002; <u>Onana 2852</u> 3/2004; **Ndanan 2:** <u>Cheek 11105</u> 10/2002.
Note: 22 sites known (Robbrecht op. cit. 1979: 301).

Tricalysia oligoneura K.Schum.

Shrub, 1–2m, branched; stems sparsely appressed, white hairy; stipule narrowly triangular-aristate, 3mm; leaves oblong, 10 × 3.5cm, acumen abrupt, 1–1.5cm, curved, base acute, lateral nerves 3–4 pairs, domatia absent, tertiary nerves sparse, inconspicuous; petiole 5mm; inflorescence axillary, 5–10mm; peduncles 3-flowered, axis partly concealed by 3 pairs of cupular bracts; pseudopedicels present; young fruit with persistent cylindric calyx tube 1mm, densely puberulent, awns 5, each 1mm; ripe fruit globose, 5mm, orange. Forest; 700–710m.
Distr.: Nigeria, Cameroon, CAR and Congo (Kinshasa) [Guineo-Congolian].
IUCN: LC
Ndanan 1: <u>Cheek 11888</u> 3/2004; <u>Tadjouteu 616</u> 3/2004.

Tricalysia pallens Hiern

Syn. *Tricalysia pallens* Hiern var. *gabonica* (Hiern) N.Hallé Fl. Gabon 17: 310 (1970).
Shrub, or tree, (2–)5–6(–14)m, appressed white-puberulent; leaves obovate or elliptic, c. 10 × 3.5cm, acuminate, base acute-decurrent, nerves 5, inconspicuous, domatia pits rimmed with white hairs; petiole 3mm; inflorescences contracted, sessile, 10–20 flowers per node; fruit globose, 5mm, red; calyx, bracts and bracteoles cup-like, densely white puberulent. Forest; 600–700m.
Distr.: Liberia to Mozambique [tropical Africa].
IUCN: LC
Mefou Proposed National Park: <u>Etuge 5146</u> fr., 3/2004; **Ndanan 1:** <u>Cheek in MF 281</u> 10/2002; <u>Tadjouteu 614</u> 3/2004.

Trichostachys aurea Hiern

Syn. *Trichostachys zenkeri* De Wild. Pl. Bequeart. 6: 83 (1932).
Monopodial erect shrub, to 0.5m, ciliate-hairy; new

shoots arising from base of stem; stipule narrowly triangular, 7–9mm; leaves obovate, to 16 × 6cm, acute-acuminate, cuneate; sessile or petiole 2mm; inflorescence terminal; peduncle 2cm; flowers 30–50 in a subglobose head 1–2cm. Forest; 700m.
Distr.: Sierra Leone to Cabinda [upper & lower Guinea].
IUCN: LC
Ndanan 1: Cheek 11207 10/2002.

Uncaria africana G. Don
Climber, with pairs of hooks 2cm long at nodes, puberulent; stems square; leaves elliptic-oblong, 11 × 5cm, acuminate, obtuse, nerves 5, drying pink below; petiole 1cm; inflorescence globular. Evergreen forest; 700–710m.
Distr.: Guinea (Bissau) to Uganda [Guineo-Congolian].
IUCN: LC
Ndanan 1: Cheek 11217 10/2002; 11697 3/2004.

Virectaria procumbens (Sm.) Bremek.
Herb, 20cm, straggling; stems with 2 lines of pubescence; stipules triangular, 3mm, entire or cleft; leaves ovate-oblong to subspathulate, 1–6 × 0.5–3.5cm; inflorescence terminal, few flowered; flowers white; calyx lobes spathulate; corolla 1cm; fruit a capsule, with a single raised disc 1mm long. Forest; 710m.
Distr.: Guinea (Bissau) to Congo (Kinshasa) [Guineo-Congolian].
IUCN: LC
Ndanan 1: Cheek 11750 3/2004.

RUTACEAE

T. Heller (K), C. Couch (K), E. Fenton (K) & M. Cheek (K)

Fl. Cameroun 1 (1963).

Citropsis articulata (Spreng.) Swingle & M.Kellerm.
Monopodial shrub, 1.5–3m, glabrous; axillary spines 2cm; leaves 15–30cm, 1–2-jugate, terminal leaflet largest, elliptic, to 18 × 7cm, obscurely crenate, petiole and rachis broadly-winged, to 2cm wide; panicles axillary, to 2cm; fruit globose, 1.5cm. Forest; 600m.
Distr.: Sierra Leone to Gabon [upper & lower Guinea].
IUCN: LC
Mefou Proposed National Park: Etuge 5181 3/2004.

Zanthoxylum buesgenii (Engl.) P.G.Waterman
Taxon 24: 363 (1975).
Syn. *Fagara buesgenii* Engl.
Shrub 0.5–2.5m; stems thorny, densely covered in bronze-coloured hairs; leaves to 35cm, compound, imparipinnate; leaflets 9, suboblanceolate-elliptic, 8.5–11.5 × 4.5–5cm, acumen 1cm, base attenuate, 5–7 nerves on each side of the midrib, abaxial surface pubescent particularly on the veins, small thorns also present on the main vein, adaxial surface pubescent,

particularly along the veins; inflorescence a panicle; fruits ± globose, approx. 0.8cm, pale green; seed ± globose, black, shiny. Semi-deciduous forest; 679–710m.
Distr.: S Nigeria, Cameroon [lower Guinea].
IUCN: NT
Ndanan 1: Cheek 11814 3/2004; Onana 2915 3/2004;
Ndanan 2: Gosline 415 10/2002.
Note: 6 specimens known from Cameroon alone (Fl. Cameroun).

Zanthoxylum heitzii (Aubrév. & Pellegr.) P.G.Waterman
Taxon 24: 363 (1975).
Syn. *Fagara heitzii* Aubrév. & Pellegr. Fl. Cameroun 1: 60 (1963).
Tree, glabrous; dbh 1.2–1.5m; stems with thorns; leaves imparipinnate; leaflets >20, oblong, 13 × 3.5cm, acumen 1.5cm, base sub-oblique, 12–16 nerves on each side of the midrib; inflorescence a panicle to 20cm; flowers 1(–2)mm. Forest; 710m.
Distr.: Cameroon, Gabon [Congolian].
IUCN: NT
Ndanan 1: Cheek 11699 3/2004.
Note: 10 Specimens know from Cameroon alone (Fl. Cameroun). Cheek 11699 is tentatively placed in this taxon, Couch xi.2005.

Zanthoxylum thomense A.Chev. ex P.G.Waterman
Taxon 24: 365 (1975).
Syn. *Fagara welwitschii* Engl. Fl. Cameroun 1: 74 (1963).
Tree, with spines; leafy stems with white lenticels and a few spines, glabrous; leaves alternate, estipulate, pinnately compound, 2-jugate (sapling); petiole spiny, base blask, constricted; leaflets elliptic or oblanceolate-elliptic, acumen 1.5cm, base unequally obtuse-acute, margin inconspicuously serrate, lateral nerves c. 12 pairs, yellow, unifying to form a very faint looping marginal nerve, midrib with a few spines, lower surface white-green, translucent glands sparse and small; petiolule black, 1mm. *Raphia* swamp; 710m.
Distr.: tropical Africa.
IUCN: LC
Mefou Proposed National Park: Cheek in RP 59 3/2004.
Note: treated as a synonym of *Z. rubescens* in F.W.T.A. but maintained by Letouzey in Fl. Cameroun 1: 84 (1963) with some hesitation, Cheek x.2010.

SAPINDACEAE

H. Fortune-Hopkins (K), J.-M. Onana (YA) & M. Cheek (K)

Fl. Cameroun 16 (1973).

Allophylus lastoursvillensis Pellegr.
Bull. Soc. Bot. France 100: 189 (1953); Fl. Gabon 23: 36 (1973).

Shrub or climber; leaves trifoliolate, blades chartaceous, median leaflet elliptic or broadly elliptic, to 16 × 7.5cm with petiolule to 1.5cm, lateral leaflets narrowly ovate or ovate, markedly asymmetric at base, to 12.5 × 6cm, almost sessile (petiolule to 3mm), margins dentate, upper surface glabrous except midrib, lower surface with tufts of hairs, including domatia; petiole to 8.5cm, with hooked hairs; axis of inflorescence unbranched, puberulent; flowers white, in small cymes; filaments hairy; fruit 0.7cm diam., glabrous. Track-side, semi-deciduous and evergreen forest; 700m.
Distr.: Cameroon, Gabon, Congo (Kinshasa) [lower Guinea & Congolian].
IUCN: LC
Ndanan 1: Cheek 11036 10/2002; 11222 10/2002; Harvey 150 10/2002; **Ndanan 2:** Cheek 11095 10/2002.
Note: Cheek 11036 and Harvey 150 are tentatively placed in this taxon. They are a reasonable match with Cheek 11095, Hopkins x.2010.

Allophylus spp. indet. of Mefou

A variety of sterile collections all with trifoliolate leaves and either glabrous or puberulent stems; leaflet blades glabrous above, sometimes with hairs on midrib, glabrous or subglabrous below, domatia small or absent, margins dentate, lateral leaflets asymmetric at base, ± sessile. *Raphia* swamp and evergreen forest; 700–710m.
Mefou Proposed National Park: Cheek in RP 52 3/2004; 58 3/2004; **Ndanan 1:** Cheek 11672 3/2004; Cheek in MF 335 10/2002.
Note: some or all may be juvenile forms of *A. lastoursvillensis*, Hopkins x.2010.

Allophylus sp. 1 of Mefou

Sterile liane?; woody stems 3mm diam.; leaves trifoliolate; leaf-blades chartaceous, median leaflet broadly elliptic, to 14.5 × 7.5cm with petiolule 1cm, lateral leaflets oblong-elliptic, ± symmetric at base, to 9.5 × 4.5cm with petiolule 0.5cm, margins entire, upper surface glabrous except midrib, lower surface subglabrous, domatia absent; petiole to 8.5cm, puberulent. Forest; 700m.
Ndanan 1: Cheek in MF 249 10/2002.
Note: plot voucher MF249 might be *Allophylus talbotii* Baker f. which has trifoliolate leaves with entire margins, but fertile material would be needed to confirm this determination, Hopkins x.2010.

Blighia welwitschii (Hiern) Radlk.

Tree, to 40m; young stems minutely velutinous; leaves paripinnate, 3–4-jugate; leaflets opposite; leaflet-blades oblong-elliptic or obovate-elliptic, 12–25 × 4–8cm with petiolules c. 0.5cm, glabrous above except for sunken midrib, sparsely hairy below on veins, secondary veins 15–20 pairs, straight to within 3mm of margin, tertiary veins irregularly scalariform, venation prominent beneath, intervenium often with small vein-knots on either side of midrib; petiole and rachis minutely velutinous; inflorescence an axillary raceme; capsules 3-angled, orange-red, ± fleshy, interior surface of locules velutinous; seeds blackish. Forest, *Raphia* swamp; 600–700m.
Distr.: Sierra Leone to Cameroon, Gabon, Congo (Kinshasa), Angola, CAR and Angola [Guineo-Congolian].
IUCN: LC
Mefou Proposed National Park: Cheek in MF 121 3/2004; Etuge 5260 3/2004; **Ndanan 1:** Cheek in MF 216 10/2002.

Cardiospermum halicacabum L.

Slender herbaceous climber with pubescent stems; leaves bi-ternate with blade 5–8(–12) × 7–14cm; leaflets chartaceous, toothed/lobed, pubescent on both surfaces, domatia lacking; petiole 1–2cm, densely pubescent; inflorescence an axillary umbellate cyme; peduncle c. 12cm long, pilose, with a pair of opposite tendrils shortly below flowers; flowers white, c. 3mm diam.; fruit a papery inflated sac, 4cm diam., densely pubescent when young. Farmbush, roadside; 710m.
Distr.: widespread in tropics and subtropics [pantropical].
IUCN: LC
Ndanan 1: Cheek 11708 3/2004.

Chytranthus angustifolius Exell

Fl. Gabon 23: 100 (1973); Fl. Cameroun 16: 100 (1973).
Syn. *Chytranthus bracteosus* Radlk.
Seedling; leaves paripinnate, 6-jugate; leaflets opposite, oblong-elliptic, to 18.5 × 4cm, with acuminate apex, glabrous above except for impressed midrib, minutely hairy below (x 20); rachis and petiole slightly striated, minutely pubescent. Forest near swamp-edged water course; 700m.
Distr.: Ivory Coast and Cameroon to Congo (Kinshasa) and Cabinda [lower Guinea].
IUCN: LC
Mefou Proposed National Park: Cheek in MF 55 3/2004.
Note: *plot voucher* MF55 is tentatively placed in this taxon, Hopkins x.2010.

Chytranthus atroviolaceus Baker f. ex Hutch. & Dalziel

Syn. *Chytranthus brunneo-tomentosus* Gilg ex Radlk.
Monopodial treelet, c. 3m; leaves 3-jugate, 60cm, paripinnate; leaflets leathery, elliptic to 30 × 10cm, acumen 1.5cm, obtuse, lateral nerves c. 15; petiole 20cm; inflorescences numerous from ground to 1m up trunk, spicate, c. 15cm; flowers 4mm, densely purple pubescent. Forest; 710m.
Distr.: Sierra Leone to Congo (Brazzaville) [upper & lower Guinea].
IUCN: NT
Ndanan 1: Cheek 11714 3/2004.
Note: known from about 15 sites.

Chytranthus gilletii De Wild.

Fl. Cameroun 16: 89 (1973).

Syn. *Chytranthus* sp. A sensu Keay

Monopodial treelet, 3m; with leaves mainly towards apex; young stems brown tomentose; leaves paripinnate, 6–7-jugate; leaflets mostly opposite; leaflet-blades oblong, 17.5–33 × 4–7.5cm, glabrous above and below, midrib indented above, rachis and petiole slightly striated, minutely pubescent; inflorescence cauliflorous; fruits in clusters, 4 × 4cm when fresh, square-turbinate in outline, sharply 3-lobed in cross-section, exocarp bright orange-red, minutely pubescent, mesocarp pithy white, interior of locules glabrous; seeds brown. *Raphia* swamp; 650m.

Distr.: S Nigeria, Cameroon, Gabon, CAR, Congo (Brazzaville), Congo (Kinshasa) [lower Guinea & Congolian].

IUCN: LC

Ndanan 2: Darbyshire 259 fr., 3/2004.

Note: Darbyshire 259 matches the description in Fl. Cameroun but little material at K for comparison. Differs from *C. edulis* Pierre only by its glabrous locules, Hopkins x.2010.

Chytranthus talbotii (Baker f.) Keay

Plant, 3m high; leaves paripinnate, 6-jugate; leaflets opposite, oblong-elliptic, to 22.5 × 6.5cm, with acuminate apex, glabrous above and below, midrib prominent above, rachis and petiole striated, glabrous. Forest near swamp-edged water course; 700m.

Distr.: SE Nigeria, Cameroon, Equatorial Guinea, Gabon, Congo (Brazzaville), Congo (Kinshasa) [lower Guinea].

IUCN: NT

Mefou Proposed National Park: Cheek in MF 103 3/2004.

Note: plot voucher MF103 is probably this taxon, Hopkins iv.2010.

Eriocoelum macrocarpum Gilg ex Engl.

Tree, 20m, glabrescent; leaves punctate, 2–3-jugate, paripinnate, 30cm; uppermost leaflets largest, obovate to 30 × 14.5cm, basal leaflets 1cm, resembling stipules, shortly acuminate, soon glabrous below; inflorescence spike-like, 10–20cm; flowers c. 3mm; fruit hard, woody, orange, smooth, depressed globose, to 3 × 4cm, valves 3, hairy inside. Forest; 700–710m.

Distr.: SE Nigeria to Congo (Kinshasa) [lower Guinea & Congolian].

IUCN: LC

Ndanan 1: Cheek 11820 3/2004; Cheek in MF 328 10/2002.

Note: plot voucher series MF328 has tentatively been placed in this taxon, it is a sapling to 1.5m, Hopkins iii.2010.

Eriocoelum sp. 1 of Mefou

Sapling, 4m; leaves paripinnate, 6-jugate; leaflets mostly subopposite, largest in mid-part of leaf, oblong-elliptic or elliptic, to 30 × 10cm, shortly acuminate, basal leaflets markedly above base, not stipule-like;

both surfaces glabrous except on venation and shiny, midrib impressed above, shortly hairy, rachis ridged; petiole flattened above, both dark brown pubescent. Forest; 700m.

Ndanan 1: Cheek in MF 319 10/2002.

Eriocoelum sp. 2 of Mefou

Seedling, 1m; with apical bud brown velutinous; leaves paripinnate, 4-jugate; leaflets mostly subopposite, uppermost leaflets largest, obovate-elliptic, 19 × 8.5cm, shortly acuminate, basal leaflets 6cm long, resembling stipules, blades glabrous above and below except on midrib, midrib prominent above; rachis and petiole striate, puberulent. Forest; 700m.

Ndanan 1: Cheek in MF 231 10/2002.

Laccodiscus klaineanus Pierre ex Engl.

Fl. Cameroun 16: 160 (1973); Fl. Gabon 23: 160 (1973). Small tree, 2–8m; leaves paripinnate, 4-jugate; leaflets mostly subopposite, uppermost largest, elliptic, to 38 × 15cm with petiolules <5mm, basal leaflets reduced, close to base of rachis, c. 8cm long, stipule-like, blades ± glabrous above and below, midrib prominent above, pilose, margin toothed especially distally, teeth spine-like; rachis and petiole pilose, glabrescent; inflorescence paniculate; capsules 3-lobed. Evergreen semi-deciduous forest; 710m.

Distr.: Cameroon & Gabon [lower Guinea].

IUCN: NT

Ndanan 1: Cheek 11849 3/2004.

Note: Cheek 11849 has comparatively broader, less hairy leaflets than is typical for this species, Hopkins x.2010.

Lecaniodiscus cupanioides Planch. ex Benth.

Small tree, 6+m; young stems fawn-puberulent; leaves paripinnate, 3–6-jugate; leaflets subopposite or opposite, elliptic or oblong elliptic, largest per leaf 9–17 × 4–7.5cm, acute at apex, glabrous above and below except for minute hairs on venation, midrib impressed above, secondary veins spreading, weakly developed ones alternating with well developed ones; petiolules striate, swollen-triangular at base; rachis and petiole fawn-puberulent; inflorescences axillary, racemose; fruits ovoid, 1 × 0.7cm, pointed at apex, brown-velutinous. Forest and second growth; 600–700m.

Distr.: Senegal to Kenya and S to Angola [tropical Africa].

IUCN: LC

Mefou Proposed National Park: Etuge 5149 3/2004; Onana 2809 3/2004; **Ndanan 1:** Cheek in MF 296 10/2002.

Paullinia pinnata L.

Woody liana, to 25m; leaves c. 12cm, 2-jugate, rachis winged; leaflets elliptic, to 11 × 6cm, obscurely toothed, apex rounded; petiole c. 10cm; inflorescence tendriliform, as long as leaves, spicate; flowers white, 3mm; fruit red, 3-lobed, obovoid, 4 × 1cm, stipitate;

seed white and red, 0.5cm. Forest & farmbush; 700m.
Distr.: tropical Africa & America [pantropical].
IUCN: LC
Mefou Proposed National Park: Cheek in RP 73
3/2004; Nana 34 3/2004; **Ndanan 1:** Cheek 11050
10/2002; Cheek in MF 315 10/2002; Darbyshire 182 fl.,
fr., 3/2004; Harvey 152 10/2002; 96 10/2002.

SAPOTACEAE

Y.B. Harvey (K)

Fl. Cameroun 2 (1964).

Baillonella toxisperma Pierre
Tree, to 12m tall; c. 80cm dbh; bole nearly black,
ridged, hard; slash yellow, white exudate; leaves
clustered at ends of branches; stipules lanceolate; blades
20–30 × 6–10cm, rounded with acuminate apex, cuneate
at base, young leaves with chestnut pubescence,
subglabrous when mature although hairs persistent on
midrib; inflorescence of dense flowering fascicles at the
branch tips; pedicels 2–3cm, pubescent; calyx c. 1cm
long, with 8 lobes, 4 inner and 4 outer, pubescent on
exterior surface; corolla with 8 lobes, each with 2 dorsal
appendages longer than the lobes (5.5mm); tube 2.5mm
long; lobes c. 4mm long; fruits large, spherical, c. 6.5cm
diam., grey-green; 1–2-seeded in a yellowish-white
pulp; seeds ellipsoid, c. 4.2 × 2.5 × 2cm, ventral scar
nearly the entire length of the convex ventral face.
Evergreen forest & secondary forest growth; 679–710m.
Distr.: Nigeria to W Congo (Kinshasa) [lower Guinea].
IUCN: VU
Ndanan 1: Cheek 11537 3/2004; Onana 2919 3/2004.
Uses: MATERIALS — Wood — timber (Cheek 11537).

Englerophytum magalismontanum (Sond.) T.D.Penn.
Pennington T., Gen. of Sapot.: 252 (1991).
Syn. *Bequaertiodendron magalismontana* (Sond.)
Heine & J.H.Hemsley
Tree or shrub, to 35m; immature growth with dense
ferrugineous indumentum; stipules to 1cm; leaves
elliptic-oblong, 4–20 × 1.5–6cm, coriaceous, apex
obtuse, base cuneate to rounded, upper surface
glabrous, under surface with silky indumentum,
becoming grey, lateral nerves numerous, closely
parallel; petiole 0.5–3cm; flowers densely fascicled
at nodes on older branches; pedicels to 1cm, pubescent;
sepals ovate, 2–3 × 2–2.5mm; corolla whitish, tube to
2.5mm; lobes to 3.5 × 2.5mm; fruit red, obovoid, to 2.5
× 1.8cm; seed ellipsoid, to 1.7 × 1.4cm. Forest; 700m.
Distr.: Guinea (Conakry) to South Africa [tropical
Africa].
IUCN: LC
Mefou Proposed National Park: Cheek in MF 79
3/2004.
Note: *plot voucher* MF79 is tentatively placed in this
taxon, sterile specimen too immature, Harvey x.2010.

Omphalocarpum procerum P.Beauv.
Tree, to 20 m; cauliflorous; leaves oblanceolate,
10–18.5 × 3–6.5cm; petioles 10–30mm; flowers
subsessile; many imbricate bracts below the calyx,
velutinous, fawn-coloured; corollas to 30mm long,
lobes to 20mm; fruits depressed, globose to 160 ×
100cm; seeds to 40 per fruit, c. 4cm long, with
longitudinal scar, chestnut brown. Forest; 700m.
Distr.: Benin, Ghana, Nigeria, Cameroon, Gabon,
Congo (Kinshasa) [Guineo-Congolian].
IUCN: LC
Ndanan 1: Harvey 215 10/2002.

Synsepalum brevipes (Baker) T.D.Penn.
Pennington T., Gen. of Sapot.: 249 (1991).
Syn. *Pachystela brevipes* (Baker) Baill. ex Engl.
Medium tree with dense crown, to 35m, bole often
fluted; young growth with dense, short-appressed hairs;
leaf blades oblanceolate to obovate, 9–26 × 3.5–10cm,
acuminate, obtuse or emarginate, cuneate, lower surface
with greyish pubescence; petioles to 1cm long; flowers
in dense clusters in older leaf axils; pedicels to 2mm
long; sepals to 4 × 3mm; corolla yellowish green or
cream, tube to 2mm, lobes to 4.5 × 2.5mm; fruits
(orange–)yellow, ellipsoid, beaked, to 2.5cm long; seeds
ellipsoid, to 2cm, shiny brown. Forest; 710m.
Distr.: Guinea (Conakry) to Sudan, E Africa and south
to Mozambique & Angola [tropical Africa].
IUCN: LC
Ndanan 1: Cheek 11635 3/2004; 11653 3/2004.
Note: *Cheek* 11635 and 11653 are sterile and are
tentatively placed in this taxon, Harvey i.2006.

SCROPHULARIACEAE

E. Fischer (University of Koblenz, Germany),
S. Ghazanfar (K) & I. Darbyshire (K)

Note: according to APG III (2009) *Artanema* and
Torenia are considered as Linderniaceae and *Bacopa*
and *Scoparia* as Plantaginaceae, Fischer xi.2010.

Artanema longifolium (L.) Vatke
Erect herb, 1m, glabrous; stem 4-angular; leaves opposite,
narrow elliptic, c. 15 × 4cm, gradually long acute, base
cuneate-decurrent, serrate; petiole 1cm, winged; raceme
terminal, 15cm; pedicel 0.5cm; corolla 3cm, purple; fruit
globose, 1cm. Swampy roadside in forest; 700m.
Distr.: Liberia to Uganda & tropical Asia [palaeotropics].
IUCN: LC
Ndanan 1: Harvey 78 10/2002.

Bacopa crenata (P.Beauv.) Hepper
Prostrate rooting, branching herb; stems spongy,
swollen, terete, glabrous; leaves opposite, estipulate,
simple, sessile, narrowly obovate to 12 × 5mm, apex
rounded, base cuneate, margin crenate, nerves not
discernable; flowers axillary, single; pedicel 5mm;

calyx 2-lobed, ovate, 5mm, enclosing the white, 5-lobed corolla. *Raphia* swamp; 710m.
Distr.: tropical Africa.
IUCN: LC
Ndanan 1: <u>Cheek 11745</u> 3/2004.

Scoparia dulcis L.
Erect herb, to 0.5m, puberulent; leaves opposite, elliptic, c. 1.5cm, decurrent, deeply serrate; petiole to 1cm; flowers 4mm, axillary, white, symmetrical; pedicels 3mm; fruit globose, 2mm. Roadside weed; 600m.
Distr.: pantropical.
IUCN: LC
Mefou Proposed National Park: <u>Etuge 5247</u> 3/2004; **Ndanan 1:** <u>Onana 2863</u> 3/2004.

Torenia thouarsii (Cham. & Schltdl.) Kuntze
Erect annual herb, glabrous, 15cm; leaves opposite, simple, estipulate; blade ovate, 5–25 × 7–14mm, acute, base truncate, margin serrate, lateral nerves 3 pairs; petiole 3–10mm; flowers axillary, single; pedicel 2mm; calyx 5mm, winged; corolla white, 2-lipped, exserted; fruit ellipsoid, 9 × 3mm, concealed in calyx. Roadside; 600–710m.
Distr.: pantropical.
IUCN: LC
Mefou Proposed National Park: <u>Cheek in RP 11</u> 3/2004; <u>Etuge 5125</u> fl., 3/2004.

SIMAROUBACEAE

C. Couch (K) & M. Cheek (K)

Quassia africana (Baill.) Baill.
Tree, 3–4m; lacking scented or coloured exudate or translucent gland dots, stems corky white; leaves alternate, pinnately compound, 45cm; petiole 15cm; leaflets opposite, 1–2(–3) pairs and terminal, elliptic 12–6 × 6–7cm, acumen 1–1.5cm, base obtuse then abruptly decurrent, lateral nerves 8 pairs; flower spike terminal, erect, 15 × 3cm, dense; pedicel 5mm; flowers white; petals 5, free, 7mm; anthers 10, sweetly scented; fruits apocarpus, ripening red, 2cm, stipitate. Forest; 679–700m.
Distr.: Cameroon to Congo (Kinshasa) [Congolian].
IUCN: LC
Ndanan 1: <u>Onana 2898</u> 3/2004; **Ndanan 2:** <u>Cheek 11136</u> 10/2002.
Local Name: Ovuin (Ewondo) (Cheek 11136).
Uses: MEDICINES (Cheek 11136).

SOLANACEAE

J.-M. Onana (YA)

Brugmansia suaveolens (Humb. & Bonpl. ex Willd.) Bercht. & Presl
Fl. Rwanda 3: 359 (1985).
Syn. *Datura suaveolens* Humb. & Bonpl. ex Willd.
Shrub or small tree, 2–5m; leaves large, up to 40 ×

15cm; flowers 25–30cm, pendulous, white, fragrant; calyx 5-toothed at apex, very inflated, not appressed to the thin pipe-like lower part of the corolla-tube; fruit fusiform. Cultivated, often as hedging; 680m.
Distr.: native of Brazil, introduced throughout the tropics [pantropical].
IUCN: LC
Ndanan 2: <u>Darbyshire 233</u> 3/2004.

Capsicum annuum L.
Annual or biennial herb; leaves broadly lanceolate to ovate, apex acutely acuminate, 5–8 × 2–5cm; inflorescences axillary, 3–8 flowered; pedicels to c. 1.5cm, nodding at maturity, dilated distally; calyx obscurely 5-toothed, 10-ribbed; corolla rotate-campanulate, deeply 5-lobed, usually straight, white or greenish; fruits solitary, a ± elongated berry, 1cm to more. Cultivated in gardens and farms, sometimes escaping into forest understorey; 600–700m.
Distr.: widely dispersed throughout the tropics, cultivated and sometimes naturalised [pantropical].
IUCN: LC
Mefou Proposed National Park: <u>Etuge 5143</u> fr., 3/2004; **Ndanan 1:** <u>Cheek in MF 346</u> 10/2002.
Local Name: Ododobola (wild pepper) (Etuge 5143).

Nicotiana tabacum L.
Robust annual, to 2m; upper leaves oblong-lanceolate to elliptic, 8–15 × 1.5–6cm; inflorescence a terminal cyme; flowers viscid-glandular outside; corolla tubular, c. 4cm long, lobes acute, white, cream or pinkish. Cultivated; 600m.
Distr.: native of S America, widely cultivated in warmer parts of the world [pantropical].
IUCN: LC
Mefou Proposed National Park: <u>Etuge 5284</u> 3/2004.

Solanum terminale Forssk.
Syn. *Solanum terminale* Forssk. subsp. *inconstans* (C.H.Wright) Heine Bothalia 25: 49 (1995).
Syn. *Solanum terminale* Forssk. subsp. *sanaganum* (Bitter) Heine Fl. Rwanda 3: 382 (1985).
Woody climber, to c. 10–15m; leaves elliptic, c. 12 × 5cm, acuminate, glabrous; petiole 1–2cm; inflorescence terminal or lateral, paniculate or cymose, very rarely spicate, 10–20cm; flowers c. 5 × 8mm; petals purple; staminal tube yellow. Forest; 679–710m.
Distr.: tropical Africa.
IUCN: LC
Ndanan 1: <u>Cheek 11597</u> 3/2004; <u>Onana 2897</u> 3/2004; <u>Tadjouteu 575</u> 3/2004.

Solanum torvum Sw.
Shrub, to 3m; stems occasionally armed, densely stellate hairy; leaves large, ± elliptic, to 10–16 × 4–12cm, subscabrid, lobate-sinuate to subentire; inflorescence of corymbose cymes, 2–5(–14)-flowered; corolla white (rarely purple), to 2.5cm; fruit c. 1cm,

globose, dirty brown, occasionally drying black. A common weed in farmbush or forest; 650–700m.
Distr.: pantropical.
IUCN: LC
Ndanan 1: Harvey 74 10/2002; Tadjouteu 629 4/2004.

STERCULIACEAE

M. Cheek (K) & J. Arcate (K)

Cola flaviflora Engl. & K.Krause
Small tree, 2–8m; leaves orbicular in outline, to 35cm, digitately 3-lobed, green below; petiole to 20cm; inflorescence axillary, sessile; flowers urceolate, yellow-green, <1cm; fruit follicles elongate, each 6 × 2.5 × 2cm, bright red. Forest; 700m.
Distr.: SE Nigeria & W Cameroon [lower Guinea].
IUCN: NT
Ndanan 1: Tadjouteu 593 3/2004.

Cola flavo-velutina K.Schum.
Shrub or small tree, branching; stems soon corky, white, cauliflorous; leaves alternate, simple, both long and short on some stems; leaf blade elliptic, 5–12 × 1.5–4.75cm, acumen 1cm, base cuneate, lateral nerves 5–7 pairs, white, very sparsely puberulent, not uniting in a marginal nerve; petioles patent brown puberulent, 1–25mm, swollen at apex; stipules distinctly white, hardened, divided into 3–4 awns, 5–7mm long; flowers fasciculate, yellow, to 1cm; fruit with carpels 5, flattened ellipsoid, densely hairy, shortly rostrate, equal, held like a wheel, to 7cm diam. Forest; 700m.
Distr.: Ghana, Nigeria, Cameroon and Gabon [upper & lower Guinea].
IUCN: LC
Mefou Proposed National Park: Cheek in MF 100 3/2004.
Note: plot voucher MF100 is probably *Cola flavo-velutina*, white stems, indurated white, ridged deeply divided stipules. Cheek x.2010.

Cola lateritia K.Schum. var. *lateritia*
Tree, 15–30m, minutely appressed brown stellate hairy; leaves leathery, ovate or orbicular in outline, to 24 × 20cm, shallowly 3–5-lobed, or subentire, apex rounded or subacuminate, base cordate, sinus obtuse, nervation palmate, minutely and sparingly stellate hairy; petiole c. 14cm; inflorescences on leafy stems; fruit carpels globose, red, glabrous, c. 6 × 6.5cm, verrucate. Forest; 700–710m.
Distr.: SE Nigeria to Congo (Kinshasa) [lower Guinea and Congolian].
IUCN: LC
Ndanan 1: Cheek in MF 234 10/2002; 256 10/2002; 355 10/2002; **Ndanan 2:** Cheek 11932 3/2004.
Note: Cheek 11932 is probably this taxon, fertile material is needed to confirm. *Pterygota macrocarpa* is similar! Plot vouchers MF234, MF256 and MF355 are

probaby also a juvenile form of this taxon, the apical cluster of grey slender stipules does not fit *Pterygota* but neither is it certain that it is this *Cola,* Cheek x.2010.

Cola letouzeyana Nkongm.
Adansonia (Sér. 2) 7: 337 (1986).
Shrub or tree, 2–3m; stems black, corky-ridged; leaves alternate, simple; leaf-blades oblanceolate-oblong, to 17 × 6cm, acumen 1.25cm, base shortly cordate, lateral nerves 9–11 pairs, white, uniting, forming a looping marginal nerve, puberulent, older leaves densely galled with dense brown tomentose domes 2mm wide arising from both surfaces of the gall; petioles 2–4mm, hairs as stem; stipules persistent, filiform, 7mm; flowers single on leafy stems, pale yellow, axillary; pedicel filiform, 2cm; perianth 4mm diam. Forest; 680–710m.
Distr.: Cameroon [Cameroon Endemic].
IUCN: VU
Mefou Proposed National Park: Cheek in MF 35 3/2004; **Ndanan 1:** Cheek 11192 10/2002; 11630 3/2004; 11951 3/2004; **Ndanan 2:** Darbyshire 248 3/2004.

Cola nitida (Vent.) Schott & Endl.
Tree, 12–25m; leaves simple, elliptic-oblong, c. 16 × 6cm, gradually acuminate, c. 1.5cm, base obtuse to rounded, entire, glabrous; petiole 0.5–5cm, variable on one stem; inflorescence a panicle to 4cm, below the leaves, densely hairy; flower 1.5cm, white, centre purple. Farmbush and around villages; 600–700m.
Distr.: Sierra Leone to Cameroon, cultivated widely in tropics [upper & lower Guinea].
IUCN: LC
Mefou Proposed National Park: Etuge 5296 3/2004; **Ndanan 1:** Harvey 131 10/2002.
Note: Harvey 131 is possibly this species, but flowers not seen (perhaps the YA specimen is fertile?). Etuge 5296 is possibly this taxon, fertile material needed, fruit not seen. Cheek x.2010.
Uses: FOOD — Seeds — seeds edible (planted for cola nuts).

Cola pachycarpa K.Schum.
Tree, to 6m; leaves alternate, digitate, leaflets 7, elliptic-oblong, to 23 × 11cm, acumen 1cm, obtuse-decurrent, lateral nerves 12–15 pairs; petiolules to 8cm; stipules 1.5cm; cauliflorous, flowers in fascicles; pedicel 2cm; corolla cylindrical, 1.5cm, deep pink. Forest; 700–710m.
Distr.: SE Nigeria to Congo (Kinshasa) [lower Guinea & Congolian].
IUCN: LC
Ndanan 1: Cheek 11604 3/2004; 11830 3/2004; **Ndanan 2:** Cheek 11085 fr., 10/2002.
Local Name: Ekom (Cheek 11085).
Uses: FOOD — Seeds — seeds edible (Cheek 11085).

Cola verticillata (Thonn.) Stapf ex A.Chev.

Tree, 7m; leafy stems stout, 8mm diam., mid brown, longitudinally fissured, with sparse white stellate scales; internodes 8–27cm; leaf-blade elliptic, to 24 × 9.5cm, acumen 0.7cm, base acute, lateral nerves 6–9 pairs not uniting at margin, indumentum as stem, lower surface glossy, quaternary nerves reticulate, conspicuous; petioles (2.5–)5–10cm, indumentum as stem, apex swollen; stipules not seen. Forest; 710m.

Distr.: Ghana to Congo (Kinshasa) and Angola (Cabinda) [upper & lower Guinea].

IUCN: NT

Ndanan 1: Cheek 11905 3/2004.

Note: Cheek 11905 is tentatively placed in this taxon, flowering and fruiting material needed to be certain. This will be a range extension to the SE if confirmed, Cheek x.2010.

Cola sp. 1 of Mefou

Treelet, 4m; branches sparse, long, dense, short brown hairy; stipules linear-strap-like 12mm, fairly persistent; leaves simple, alternate, both long and short petioled on each stem; blade oblanceolate, to 20 × 6.5cm including an abrupt, 2cm acumen, base cuneate, margin undulate, lateral nerves to 9 pairs; petioles 0.5–7cm. Evergreen forest; 700–710m.

Ndanan 1: Cheek 11193 10/2002; 11735 3/2004.

Note: resembles *C. griseiflora* De Wild. but that has early caducous stipules and lateral nerves 5–7 pairs. Fertile material needed. Cheek x.2010.

Leptonychia lasiogyne K.Schum.

Nat. Pflanzenfam. Nachtr. I: 241 (1897).

Shrub, 2–3m; leafy stems densely stellate hairy with pale brown hairs 0.25mm diam., soon falling; leaves alternate; leaf-blade oblong or elliptic, 11–15 × 3–5.5cm, acumen 0.5–1.5cm, base rounded to obtuse, weakly tri-nerved, the laterals extending 1/3 the blade length, other laterals 3–5 pairs, uniting into a looping marginal nerve, tertiary nerves reticulate, domatia elliptic, glabrous, lower surface with apparently glabrous but with minute thinly scattered stellate hairs and surface glands; petioles 1–1.5cm, distal half swollen; stipules caducous; inflorescences axillary, 2–3-flowered, 1cm long, buds golden hairy; flowers 1cm diam., greenish white. Evergreen forest, *Raphia* swamp; 640–720m.

Distr.: Cameroon endemic.

IUCN: DD

Ndanan 1: Cheek 11541 3/2004; 11661 3/2004; Onana 2837 3/2004; Tadjouteu 587 3/2004; 613 3/2004; **Ndanan 2:** Darbyshire 305 fl., fr., 3/2004.

Note: species of *Leptonychia* in Cameroon are poorly defined and need revision. This species is based on material collected in Yaoundé and maybe restricted to the Yaoundé area although it has been applied to many other specimens from more distinct locations. Cheek x.2010.

Leptonychia subtomentosa K.Schum.

Nat. Pflanzenfam. Nachtr. I: 241 (1897).

Shrub, 3m; stems persistently softly tomentose with brown 1mm hairs extending to petiole and nerves; leaves alternate, needle-like, 6–7mm; leaf-blade elliptic-oblong, 12–18 × 6–9cm, excluding the 2–4cm slender acumen, base shallowly cordate to obtuse, weakly 3-nerved at base, lateral nerves extending 1/3 the length of the blade, lateral nerves 6–7 pairs, weakly uniting near the margin, tertiary nerves scalariform, domatia absent; petiole 4–8mm; flowers resembling *L. lasiogyne*; fruits 1.5cm, densely long brown tomentose. Forest; 550–720m.

Distr.: Cameroon, Central Region [Cameroon Endemic].

IUCN: VU

Mefou Proposed National Park: Etuge 5190 3/2004; **Ndanan 1:** Cheek 11727 3/2004; 11911 3/2004; Harvey 121 10/2002; Onana 2871 3/2004; **Ndanan 2:** Cheek 11933 3/2004.

Note: restricted to the Yaoundé area. First recorded there by Zenker and Staudt. Cheek x.2010.

Leptonychia sp. 1 of Mefou

Shrub, 2–4m, soon glabrous; leaves alternate, simple, fairly even in size, elliptic-oblong, to 18 × 7.5cm, acumen 1.5cm, base acute, 3-nerved at base, other lateral nerves 3–4 pairs, domatia elliptic, surrounded by a line of hairs, quaternary veins prominent; petiole 1–1.5cm, the distal half swollen, black. Forest; 700m.

Ndanan 1: Cheek in MF 251 10/2002; 290 10/2002; 303 10/2002; 363 10/2002; 369 10/2002.

Note: flowering material is needed to identify this species. The genus *Leptonychia* in Cameroon needs revision. Cheek x.2010.

Scaphopetalum zenkeri K.Schum.

Gregarious shrub, or small tree, 2–8(–14)m; boles dark brown-black; leafy stems densely shortly, grey-brown patent puberulent, extending to petiole; leaves alternate, simple; leaf-blade narrowly or very narrowly elliptic, very variable in size, 12–26 × 3.2–6cm, acumen 2cm, base obtuse, weakly 3-nerved at base, laterals extending 1/4 blade length, remaining, lateral nerves 8–10 pairs, tertiary nerves imperfectly scalariform; petiole 5–8mm; stipule narrowly triangular, 5 × 1–2mm, appressed hairy; flowers 2–3 on peduncles 2mm, on leafy stems or leafless stems; flower buds globose to 5-angular, red, 5mm diam. Forest; 550–720m.

Distr.: SE Nigeria and Cameroon [lower Guinea].

IUCN: LC

Mefou Proposed National Park: Cheek in MF 4 3/2004; Etuge 5187 fl., 3/2004; **Ndanan 1:** Cheek 11209 10/2002; 11603 3/2004; Onana 2847 3/2004; **Ndanan 2:** Darbyshire 230 fl., 3/2004.

Note: plot voucher MF4 is probably this taxon, Cheek x.2010.

Sterculia oblonga Mast.

Syn. *Eribroma oblonga* (Mast.) Pierre ex A.Chev. Fl. Gabon 2: 19 (1961); Keay R.W.J., Trees of Nigeria: 129 (1989).

Deciduous tree, 20–40m; buttresses c. 60 × 60cm; bole 60cm dbh, white, with rectangular plates; slash pink, inner pale yellow; leaves (flushing) oblong or obovate-oblong, c. 6 × 2cm, subacuminate, acute, lateral nerves c. 14 pairs, midrib and petiole densely hairy, glabrescent. Semi-deciduous forest; 700–710m.
Distr.: Ivory Coast to Gabon [upper & lower Guinea].
IUCN: LC
Mefou Proposed National Park: Cheek in RP 81 3/2004; **Ndanan 2:** Cheek 11142 10/2002.
Note: plot voucher RP81 is a sapling and probably this taxon, Cheek x.2010.
Local Name: Eyon (Ewondo) (Cheek 11142).
Uses: MATERIALS — Wood — used as beams in the construction of houses (Cheek 11142).

Sterculia rhinopetala K.Schum.

Tree, to 40m, buttressed, glabrous; leaves drying dark brown above, pale brown below, elliptic-oblong, to 20 × 8cm, subacuminate, rounded, lateral nerves 12 pairs, petiole 6cm; panicles to 18 × 4cm, puberulent; perianth 3–7mm, puberulent; follicles oblong, 6 × 3cm. Forest; 700–710m.
Distr.: Ivory Coast, Ghana, Nigeria & Cameroon [upper & lower Guinea].
IUCN: LC
Ndanan 1: Cheek 11068 10/2002; 11794 3/2004; Cheek in MF 203 10/2002; 298 10/2002.

Sterculia tragacantha Lindl.

Tree, 10–25m; leaves simple oblong, oblong-obovate or elliptic, c. 19 × 11cm, rounded to subacuminate, base obtuse, stellate-velvety below; petiole c. 5cm; stipules caducous; panicles axillary, slender, 15cm; flowers pink, 0.7cm; perianth lobes adhering at apex; follicles five c. 8.5 × 2.5cm, golden-brown hairy. Secondary forest; 700–710m.
Distr.: Mali to Mozambique [tropical Africa].
IUCN: LC
Mefou Proposed National Park: Cheek in MF 19 3/2004; Cheek in RP 25 3/2004; 35 3/2004; **Ndanan 1:** Cheek 11627 3/2004; Cheek in MF 209 10/2002; 377 10/2002.

Triplochiton scleroxylon K.Schum.

Tree, to 60m; bole whitish brown with buttresses concave, projecting 2m; leaves orbicular in outline, c. 12 × 15cm, 5-lobed by c. $^1/_3$ the radius, lobes subtriangular, blade base truncate; petiole c. 5cm. Semi-deciduous forest; 700–710m.
Distr.: Guinee (Conakry) to Congo (Kinshasa) [Guineo-Congolian].
IUCN: LC
Ndanan 1: Cheek 11027 10/2002; 11706 3/2004.
Uses: MATERIALS — Wood — timber (Cheek pers. comm.).

THYMELAEACEAE

M. Cheek (K)

Fl. Cameroun 5 (1966).

Dicranolepis disticha Planch.

Shrub, 1–3.5m; resembling *D. vestita*, but flowers ± glabrous (not seen at Mefou); fruits globular, 1cm, glabrous, without rostrum. Forest; 700m.
Distr.: Guinea (Conakry) to Congo (Kinshasa) [Guineo-Congolian].
IUCN: LC
Ndanan 1: Cheek 11033 10/2002.

Dicranolepis vestita Engl.

Shrub or tree, (2–)3–8m; leaves obliquely oblong-elliptic, c. 8 × 2.5cm, acumen 1.5cm; flowers erect, sessile, 1–several in old leaf axils; flowers 2.5cm wide, white, fragrant; calyx tube c. 2cm, densely white appressed-hairy, dilated at base; petals 10, entire; anthers sessile; stigma capitate; fruit pendulous, 2.5 × 1.5cm including a 1cm robust rostrum, densely white hairy. Forest; 700–720m.
Distr.: SE Nigeria, Bioko & Cameroon [lower Guinea].
IUCN: NT
Ndanan 1: Cheek 11605 3/2004; Onana 2192 10/2002; 2829 3/2004; Tadjouteu 602 3/2004.

Octolepis casearia Oliv.

Shrub or tree, 1.5–3m; bark white, sloughing, glabrous; leaves alternate to 25 × 8cm, acuminate, acute; petiole 1cm; inflorescence axillary, 1–4 flowered fascicles; pedicels 0.5cm; flowers white, tepals 4, 5mm; fruit ovoid, 1.5cm, 4-valved. Forest; 700m.
Distr.: SE Nigeria to Gabon [lower Guinea].
IUCN: NT
Ndanan 1: Tadjouteu 617 3/2004.

TILIACEAE

M. Cheek (K) & J. Arcate (K)

Ancistrocarpus densispinosus Oliv.

Climber, puberulent; leaves elliptic, c. 15 × 7cm, acuminate, rounded, finely serrate, nerves thinly puberulent below; stipule entire, 4mm; inflorescence leaf-opposed, c. 10-flowered; flowers yellow, 2.5cm diam.; stamens in bundles; fruit globose, 4.5cm including dense, glabrous hooked spines c. 1cm. Forest; 700m.
Distr.: Nigeria to Congo (Kinshasa) [lower Guinea & Congolian].
IUCN: LC
Ndanan 1: Cheek 11002 10/2002.

Desplatsia dewevrei (De Wild. & T.Durand) Burret

Tree, 10m; trunk 30cm dbh, stems minutely puberulent; leaves oblong-elliptic, to c. 24 × 10cm, long acuminate, base unequally truncate, deeply serrate, drying brown

below, glabrous; stipules triangular, 2mm puberulent; fruit oblong-ellipsoid to 12 × 12cm. Forest; 700–710m.
Distr.: Ivory Coast to Congo (Kinshasa) [Guineo-Congolian].
IUCN: LC
Ndanan 1: Cheek 11702 3/2004; Tadjouteu 591 3/2004; **Ndanan 2:** Cheek 11113 10/2002.
Local Name: Ngeg (Cheek 11113).
Uses: SOCIAL USES — Unspecified Plant Parts — medicine for those who steal (Cheek 11113).

Desplatsia subericarpa Bocq.

Shrub, 3–6m; slender, dark brown pubescent; leaves c. 17 × 6cm, obscurely serrate, glabrous below; stipule 5mm, digitately divided; flowers pink, 1cm; fruit c. 5 × 5cm. Forest; 700–710m.
Distr.: Sierra Leone to Congo (Kinshasa) [Guineo-Congolian].
IUCN: LC
Ndanan 1: Cheek 11737 3/2004; Tadjouteu 569 3/2004; 590 3/2004; **Ndanan 2:** Cheek 11089 10/2002.
Local Name: Mell feneing (Cheek 11089).
Uses: MEDICINES — Unspecified Medicinal Disorders — unspecified plant parts (Cheek 11089).

Duboscia macrocarpa Bocq.

Syn. *Duboscia viridiflora* (K.Schum.) Mildbr.
Tree, 30m; 40–65cm dbh, bole crooked, slight buttresses, slash fibrous, oxidizing in 4 minutes; stems softly stellate brown pubescent; leaves obovate-oblong, to 18 × 7cm, long acuminate, rounded or cordate, serrate-dentate, greyish-white below, densely, softly tomentose; inflorescence leaf-opposed, subumbellate, many-flowered; flowers greenish-white 1cm diam.; fruit obovoid, 3.5 × 2.5cm, woody, 6–8 ribbed. Forest; 680m.
Distr.: Ivory Coast to Congo (Kinshasa) [Guineo-Congolian].
IUCN: LC
Ndanan 2: Darbyshire 234 fr., 3/2004.

Glyphaea brevis (Spreng.) Monach.

Shrub, 7m; stems brown with scattered stellate hairs; leaves obovate or oblong, c. 24 × 9cm, long acuminate, obtuse to rounded, serrate finely hairy below; petiole 3.5cm, bipulvinate; inflorescence leaf-opposed, few-flowered; flowers 3cm, yellow; fruit subcylindrical, c. 7 × 1cm, ridged, rostrate, indehiscent, green. Farmbush; 600–700m.
Distr.: tropical Africa.
IUCN: LC
Mefou Proposed National Park: Etuge 5131 3/2004; **Ndanan 1:** Cheek 11026 10/2002; Cheek in MF 238 10/2002; 243 10/2002; **Ndanan 2:** Darbyshire 225 fl., 3/2004.
Note: plot voucher MF238 is probably this taxon, Cheek x.2010.
Local Name: Ivei (Etuge 5131).

Grewia pubescens P.Beauv.

Syn. *Grewia mollis* Juss.
Shrub, 2m or climber, 3m; stem densely shortly grey pubescent; leaves alternate; stipules caducous, linear, flat, 5mm; blade lanceolate to oblong, 10 × 4cm, subacuminate, base obtuse, rounded, lower surface velvety grey-white, 3-nerved from base, scalariform, margin finely serrate; petiole 1cm; inflorescence 3-flowered, sub-axillary, 5–7cm; peduncle 3cm; pedicel 2cm; flowers pale pink, 4cm wide; calyx white tomentose, lobes 2.5cm at anthesis; stamens free, numerous, 2.5cm, exserted; fruit 1.5cm diam., 4-lobed. Semi-deciduous forest; 700m.
Distr.: Guinea (Conakry), Sierra Leone, Ivory Coast, Liberia, Ghana, Benin, Togo, Nigeria, Cameroon; CAR, Gabon, Congo (Brazzaville), Congo (Kinshasa), Sudan, Uganda, Angola [tropical Africa].
IUCN: LC
Ndanan 1: Cheek 11078 10/2002; 11944 3/2004; Harvey 111 10/2002.

Microcos barombiensis (K.Schum.) Cheek

The Plants of Kupe, Mwanenguba and the Bakossi Mountains: 414 (2004).
Syn. *Grewia barombiensis* K.Schum.
Climber; stems brown puberulent; leaves ovate-elliptic, c. 17 × 9cm, acuminate, rounded or cordate, entire, subglossy above, glabrous below apart from nerves; stipules c. 6mm, bifurcate; inflorescence terminal, 10 × 7cm; main bracts digitately divided to base; flowers in umbels of 2–3, subtended by 5–6 epicalycular bracteoles; flowers white, 13mm diam.; fruit spindle-shaped, 3 × 1.5cm glossy red, with white thinly scattered hairs. Forest; 640–710m.
Distr.: Ivory Coast to Angola [upper & lower Guinea].
IUCN: NT
Ndanan 1: Cheek 11546 3/2004; Tadjouteu 636 4/2004; **Ndanan 2:** Darbyshire 309 fr., 3/2004.
Note: *Microcos* is not always maintained as distinct from *Grewia* in Africa despite Burret's excellent revisionary work of 1926 (Notizbl. Bot. Gart. Berlin 9), Cheek 2004.

Triumfetta cordifolia A.Rich. var. *tomentosa* Sprague

F.T.E.A. Tiliaceae: 87 (2001).
Shrubby herb, to 3m, puberulent; leaves slightly 3-lobed, basal leaves drying black, ovate, c. 13 × 10cm, long-acuminate, base cordate to obtuse, coarsely dentate, lower surface completely obscured by dense but short, brownish-white hairs; inflorescence raceme-like, sparingly branched; flowers 8mm; fruit drying black, globose, 10mm including hooked bristles to 4mm, very sparsely to densely white hairy. Roadside; 600–700m.
Distr.: Nigeria to Zimbabwe [lower Guinea, Congolian & S Africa].
IUCN: LC

Mefou Proposed National Park: Etuge 5269 3/2004; **Ndanan 1:** Cheek in MF 404 10/2002; **Ndanan 2:** Cheek 11080 10/2002.

Triumfetta sp. 1 of Mefou

Herb, 1m; stem and petiole glabrous but for a line of hairs; leaves alternate, ovate-lanceolate, 10 × 6cm, acuminate, cordate, serrate-dentate, glabrous on both surfaces apart from very sparse minute bright white bushy hairs, basal tooth glands absent; petiole 3–4cm; inflorescence terminal, 20cm, 2–3 branches; flowers clustered, yellow, 8mm; fruits reported as glabrous, hooked. Evergreen forest; 700m.
Ndanan 2: Cheek 11138 10/2002.
Local Name: Okon (Ewondo) (Cheek 11138).

ULMACEAE

J.-M. Onana (YA), X. van der Burgt (K), M. Cheek (K) & C. Couch (K)

Fl. Cameroun 8 (1968).

Celtis adolfi-friderici Engl.

Tree, to 30m; mature leaves asymmetric oblong-elliptic, 10.5–17 × 6.5–7cm, acumen obtuse, 1.2cm, base acute, margin entire, paired basal lateral nerves extend almost to the apex, subsequent 2 lateral nerve pairs in upper half of leaf, alternate, tertiary nerves reticulate, not prominent, lower surface slightly scabridulous, upper surface glossy, glabrous; petiole stout, 1.5cm; fruit globose, 1.7cm diam., smooth to foveolate; style bases persistent. Semi-deciduous forest; 700–710m.
Distr.: Ivory Coast to Sudan & Congo (Kinshasa) [Guineo-Congolian].
IUCN: LC
Ndanan 1: Cheek 11039 10/2002; 11667 3/2004; 11798 3/2004; 11913 3/2004.

Celtis gomphophylla Baker

F.T.E.A. Ulmaceae: 6 (1966); Fl. Zamb. 9(6): 9 (1991).
Syn. *Celtis durandii* Engl.
Small tree, to 7m; stems glabrous; leaves ovate-oblong, 9.5–19 × 3.5–8cm, acumen 1.8cm, acute, base obtuse, ± asymmetrical, margin serrate in upper third (immature) to entire (mature), lateral nerves 5(–6) pairs, alternate after basal (or rarely second) pair, basal pair not extending into upper half of leaf, veins yellow and prominent below, surfaces scabridulous when immature, soon glabrous; fruit in axillary fascicles of 2–3; pedicels 0.4–1cm, globose or ovoid, 0.6–0.7cm diam.; calyx persistent, lobes triangular, 1mm; paired styles persistent, c. 1mm, simple. Wet (& semi-deciduous) forest; 680m.
Distr.: Ivory Coast to Congo (Kinshasa), Zimbabwe & Madagascar [tropical & subtropical Africa].
IUCN: LC
Ndanan 2: Darbyshire 235 fr., 3/2004.

Celtis mildbraedii Engl.

Tree; leafy stems minutely and densely brown papillate-puberulent; leaves alternate, estipulate; blade thickly papery, elliptic, 9–16 × 4.5–8cm, acumen 0.5–1cm, apex rounded, with 1mm mucro, base asymmetric, obtuse to acute, margin serrate, teeth 3–4mm, (0–)2–10 per side, 3-nerved from base, basal laterals not reaching upper half, remaining laterals 3–5 pairs on each side of the midrib, very weakly scalariform, quaternary nerves finely reticulate, raised, glossy; petiole 5–10mm, indumentum as stem. Forest; 700–710m.
Distr.: Ivory Coast to Tanzania [tropical Africa].
IUCN: LC
Ndanan 1: Cheek 11734 3/2004; Cheek in MF 221 10/2002; 337 10/2002.

Celtis tessmannii Rendle

J. Bot. 53: 297 (1915); F.T.A. 6(2): 7 (1917).
Tree, 30–40m; dbh 80–100cm; bark grey; stems pubescent; leaves elliptic, 10 × 5cm, acumen 1cm, base oblique, tri-nerved, entire, abaxial surface with scabrid hairs; petiole 4mm; inflorescences axillary or terminal to sub-terminal (5–)8cm; petals pale green; drupe ovoid, 1.5–2 × 1.2–1.5cm, scabrid; Semi-deciduous forest; 700m.
Distr.: Cameroon, Equatorial Guinea, Gabon, Congo (Brazzaville), CAR, Congo (Kinshasa) [lower Guinea & Congolian].
IUCN: LC
Ndanan 1: Cheek 11040 10/2002; Cheek in MF 246 10/2002; 264 10/2002; 271 10/2002; Onana 2813 3/2004; 2816 3/2004.
Note: 12 Specimens known from Cameroon alone (Fl. Cameroon).

Celtis zenkeri Engl.

Tree, to 15m, crown conical, compact; immature stems puberulent; mature leaves subsymmetrical, elliptic, 10.5–12 × 4.2–5cm, acumen to 1.7cm, acute, base acute, margin entire, lateral nerves 4–5 pairs, alternate beyond basal pair, yellowish below, tertiary nerves markedly parallel, lower surface softly puberulent, upper surface glabrous; petiole c. 6mm, puberulent; fruits axillary, paired or single; peduncle c. 1cm; pedicels 3mm; fruit globose, 6mm, rugulose; paired styles more or less persistent, divaricating. Semi-deciduous forest; 700m.
Distr.: Guinea (Conakry) to Angola & Tanzania [tropical Africa].
IUCN: LC
Ndanan 1: Cheek in MF 323 10/2002.

Holoptelea grandis (Hutch.) Mildbr.

Tree, to 30–40m, glabrous; dbh 80–100cm; leaves elliptic, 8 × 5cm, acumen 0.5cm, base attenuate, 5–6 nerves on each side of the midrib, entire; petiole 1.5cm; stipules 2mm; fruit flat, dry, winged, obovate, c. 3 × 4cm, apex notched; styles 2, filiform, persistant. Semi-deciduous forest; 710m.

Distr.: Ivory Coast, Togo, Equatorial Guinea, CAR, Congo (Kinshasa), Sudan, Uganda [tropical Africa].
IUCN: LC
Ndanan 1: <u>Cheek 11864</u> 3/2004.
Note: fallen fruits only (*Cheek* 11864).

Trema orientalis (L.) Blume
Syn. *Trema guineensis* (Schum. & Thonn.) Ficalho
Tree, to 8m; young stems densely pubescent; leaves variable, distichous, ovate-lanceolate, 6.5–13.5 × 2.8–5.3cm, apex acuminate, base truncate, margin serrulate, lateral nerves 4(–6) pairs, alternate above basal pair, upper surface scabrid, lower surface scabrid or sparsely pubescent to densely pubescent; cymes axillary, c. 10–20-flowered; peduncle 0–0.5cm; flowers white, c. 2mm; sepals broadly elliptic, obtuse, puberulent; fruit globose, 2–3mm diam., green; styles and sepals persistent. Forest & farmbush; 700m.
Distr.: widespread in tropical Africa & Asia [palaeotropics].
IUCN: LC
Mefou Proposed National Park: <u>Nana 33</u> 3/2004; **Ndanan 1:** <u>Cheek in MF 408</u> 10/2002; <u>Darbyshire 180</u> fr., 3/2004; <u>Harvey 200</u> 10/2002.

URTICACEAE

J.-M. Onana (YA), X. van der Burgt (K), C. Couch (K) & M. Cheek (K)

Fl. Cameroun 8 (1968).

Boehmeria macrophylla Hornem.
F.T.E.A. Urticaceae: 44 (1989); Fl. Zamb. 9(6): 108 (1991).
Syn. *Boehmeria platyphylla* D.Don
Shrub, to 2(–3)m; branches glabrous, except when young; leaves opposite, anisophyllous, ovate, 10–13.5 × 5.5–9cm, acuminate, base acute to rounded, margin serrate, basal lateral nerves prominent, upper surface sparsely pubescent, cystoliths punctiform, lower surface glabrescent; petiole to 6cm; spikes axillary, 7–50cm, whip-like, with glomerules of flowers spaced 1–10mm apart; male glomerules 1–2mm; female 2–3mm. Forest & forest edge; 600–710m.
Distr.: tropical Africa & Madagascar, tropical Asia to SW China [palaeotropics].
IUCN: LC
Mefou Proposed National Park: <u>Cheek in RP 24</u> 3/2004; **Etuge 5129** 3/2004; **Ndanan 1:** <u>Cheek 11643</u> 3/2004; **Ndanan 2:** <u>Darbyshire 264</u> 3/2004.

Laportea aestuans (L.) Chew
Fl. Cameroun 8: 121 (1968); F.T.E.A. Urticaceae: 23 (1989); Fl. Zamb. 9(6): 92 (1991).
Syn. *Fleurya aestuans* (L.) Gaudich.
Herb, to 60cm; stems glabrescent; stipules lanceolate; leaves ovate, 4–8 × 3.5–6cm, acuminate, base truncate, margin serrate, surfaces stinging-setulose, lateral nerves 5–8 pairs; petioles 5.5(–8.5)cm, setulose with some

glandular hairs; panicles many-branched on peduncles 4–10cm, glandular-hairy; flowers in clusters, 5mm; female flower pedicels 0.5mm, subwinged; achene flattened ovoid, centre warted. Farmbush & forest; 700m.
Distr.: pantropical.
IUCN: LC
Ndanan 2: <u>Cheek 11096</u> 10/2002.

Laportea ovalifolia (Schumach.) Chew
Fl. Cameroun 8: 131 (1968); F.T.E.A. Urticaceae: 18 (1989); Fl. Zamb. 9(6): 89 (1991).
Syn. *Fleurya ovalifolia* (Schumach.) Dandy
Stoloniferous herb, to 60cm, erect to prostrate, except for male inflorescence; stems stinging-hairy; stipules 7mm, lanceolate; leaves alternate, ovate, 3–9 × 2–5cm, acuminate, base obtuse, margin serrate, cystoliths punctiform; male inflorescence axillary or from stolons, erect, paniculate, c. 13cm on peduncle to 22cm, branches short, flower clusters 0.5–1cm; female inflorescence arising directly from ground or rarely in axils of upper leaves, densely racemose, 1.5(–5)cm, peduncles to 14cm, densely stinging-hairy; achene flattened ovoid with a membranous margin and warted centre surrounded by a ridge. Roadside & forest; 660m.
Distr.: Sierra Leone to Zimbabwe [tropical Africa].
IUCN: LC
Ndanan 2: <u>Darbyshire 334</u> fl., 4/2004.

Pouzolzia guineensis Benth.
Herb, to 1m, leaves to 6.5 × 2cm, base acute; petiole to 2.5cm, densely pubescent; inflorescence c. 0.5cm, few-flowered; bracts and male flowers with prominent hooked hairs; achene 1.5mm, pale brown, often enclosed in persistent perianth. Open forest & roadsides; 700m.
Distr.: Senegal to Ethiopia and Angola [tropical Africa].
IUCN: LC
Ndanan 1: <u>Cheek in MF 352</u> 10/2002.

Procris crenata C.B.Rob.
Unbranched, epiphytic herb, to 40cm; stems succulent, 0.5cm diam. (dried); leaves only in upper section of stems, alternate, narrowly elliptic, 11–16 × 2.5–3cm, acuminate, base cuneate, margin shallowly serrate-crenate, glabrous; petiole 5mm; inflorescences on leafless lower stems, 4–5 together (female) or single (male); peduncle 5mm, succulent, clusters capitate, 1–2mm diam. Forest; 700–710m.
Distr.: tropical Africa, Madagascar, India to Philippines [palaeotropics].
IUCN: LC
Mefou Proposed National Park: <u>Cheek in RP 4</u> 3/2004; **Ndanan 2:** <u>Cheek 11171</u> 10/2002.

Urera gravenreuthii Engl.
Liana, to 8m; stems densely covered in gold/bronze-coloured hairs; leaves elliptic, 12–16 × 7–9cm, 5–7 nerves on each side of the midrib, dentate, adaxial

surface dark green with scabrid hairs, abaxial surface paler green, pubescent (particularly along the veins); petiole 1.5–8cm; panicle 5–7cm; flower buds 1mm diam., red-brown; fruit clusters on leafless parts of the stem. Forest; 700m.

Distr.: Cameroon [lower Guinea (montane)].
IUCN: NT
Ndanan 1: Cheek 11183 10/2002.
Note: 6 spcimens known in Cameroon (Fl. Cameroon).

Urera thonneri De Wild. & T.Durand
F.T.A. 6(2): 256 (1917).
Creeper and climber, often flowering on leafless stems, sometimes on the ground (Darbyshire 269); stems grey, peeling, 4mm diam., longitudinally grooved; leaves alternate; blades lanceolate, papery, to 13 × 6cm, acumen 1.5cm, base rounded, 3-nerved from base, remaining laterals 2–3 pairs, lower surface with white rod raphides resembling appressed hairs; petiole 1.5–2.5cm; infructescence dense, 4–8 × 2–10cm, subsessile, heavily protected by stinging hairs 1–2mm, otherwise glabrous; fruits red or yellow, narrow ovoid, 2mm. Riverine and inundated forest; 660–720m.

Distr.: Cameroon, Congo (Brazzaville) and Congo (Kinshasa) [Congolian].
IUCN: LC
Ndanan 1: Onana 2844 3/2004; **Ndanan 2:** Darbyshire 269 fr., 3/2004.
Note: Onana 2844 and Darbyshire 269 are the first records for Cameroon of this Congolian species? Cheek x.2010.

Urera trinervis (Hochst.) Friis & Immelman
F.T.E.A. Urticaceae: 6 (1989); Fl. Zamb. 9(6): 81 (1991); Fl. Zamb. 9(6): 83 (1991).
Syn. *Urera cameroonensis* Wedd.
Robust climber, glabrous; stems cylindrical; leaves ovate-elliptic, 10.5–14 × 5.5–7.5cm, acumen to 1.7cm, base acute to rounded, margin entire, basal lateral nerve pair prominent, cystoliths inconspicuous; petiole 2–4.5cm; male panicles to 6cm; peduncles 1cm; pedicels 1.5mm; female panicles denser, c. 4cm, with clusters of stinging hairs; flowers sessile. Forest; 710m.

Distr.: Ghana to E & S Africa, Madagascar [tropical Africa & Madagascar].
IUCN: LC
Ndanan 1: Cheek 11877 3/2004.

VERBENACEAE

M. Cheek (K)

Stachytarpheta cayennensis (Rich.) Vahl
Enum. Pl. [Vahl] 1: 208 (1804).
Shrubby herb, to 2m; stems glabrous; leaves ovate or elliptic, 1.8–8 × 0.5–4cm, attenuate into petiole, 1–1.5cm; inflorescence a slender spike, up to 20–25(–34)cm, with some pubescence; bracts linear to triangular-subulate, 4–5mm, acuminate; calyx 4–5mm, with 4 equal teeth; corolla white or mostly pale blue; tube 4–5mm, scarcely exceeding the calyx. Roadsides, fallow; 710m.

Distr.: Sierra Leone to Cameroon, Uganda, Mozambique, Zimbabwe, widespread in tropical America, naturalised throughout the tropics [pantropical].
IUCN: LC
Ndanan 1: Cheek 11484 3/2004.

VIOLACEAE

G. Achoundong (YA) & M. Cheek (K)

Rinorea batesii Chipp
Bull. Misc. Inform. Kew 1923: 297 (1923).
Shrub, to 3m high, glabrous; leaves pale green, thinly leathery, elliptic, 6.5–11 × 3–6cm, acumen 0.5–1cm, base attenuate, 5–6 nerves on each side of the midrib, densely scalariform, margin entire becoming denticulate; petiole 0.8–3cm; inflorescence a panicle, terminal, 2–5cm; flowers yellow, buds 2mm wide. *Raphia* swamp; 550m.

Distr.: Cameroon, CAR, Congo (Kinshasa) [Congolian].
IUCN: LC
Mefou Proposed National Park: Etuge 5193 3/2004.

Rinorea cerasifolia M.Brandt
Bot. Jahrb. Syst. 51(1): 118 (1913).
Shrub, to 2m, glabrous; leaves drying dull brown green, elliptic, 8–13 × 3–5cm, apex acuminate, acumen 1–2cm, entire; petiole 1–1.5cm; inflorescence a panicle terminal, 4–10cm; flower buds bright yellow, c. 5mm; immature fruits pale green. *Raphia* swamp; 650m.

Distr.: Cameroon, Gabon, Congo, CAR [lower Guinea & Congolian].
IUCN: DD
Ndanan 2: Darbyshire 261 fl., 3/2004.

Rinorea claessensi De Wild.
Bull. Jard. Bot. État Bruxelles 6: 156 (1920).
Shrub, to 5m tall, glabrous; leaves elliptic, 10–20 × 5.5–8.5cm, apex cuspidate to acute, 5–8 nerves on each side of the midrib, margin entire to slightly serrate; panicle terminal, 5–7cm, densely pubescent; flowers green, buds 2mm across. Riverbank; 640m.

Distr.: Cameroon, Congo (Kinshasa) [Congolian].
IUCN: NT
Ndanan 2: Darbyshire 289 fr., 3/2004.
Note: Darbyshire 289 is tentatively placed in this taxon by Achoundong viii.2005 (Kew only has specimens of this taxon from Congo (Kinshasa).

Rinorea dentata (P.Beauv.) Kuntze
Shrub or small tree, to 5m; stems puberulent; leaves papery, elliptic, 15.5–19.5 × 6–8cm, acumen 1.7cm, base acute, margin denticulate, lateral nerves 11–13 pairs, midrib puberulous below, laminae glabrous except when young, eglandular; petiole 0.6–1.2cm; panicles terminal, c. 3cm, puberulent, c. 8–10-flowered; peduncle 3cm; sepals triangular, 1.5mm; petals

lanceolate, 4mm, yellow. Forest; 700m.
Distr.: Liberia to Uganda [Guineo-Congolian].
IUCN: LC
Ndanan 1: Harvey 112 10/2002.

Rinorea lepidobotrys Mildbr.

Repert. Spec. Nov. Regni Veg. 18: 96 (1922).
Shrub, to 2.5m, spreading; stem patent-puberulent; leaves bright green, obovate-elliptic, 10–12 × 4–4.5cm, acumen 1.5cm, base cordate, margin subentire, lateral nerves 5–6 pairs; petiole 1–2mm; infructescence spikes erect, 2cm, covered in concave bracts 2–3mm; fruits 3-lobed, smooth, red-green; seeds shiny black. Dense forest; 640m.
Distr.: Cameroon endemic.
IUCN: NT
Ndanan 2: Darbyshire 288 fr., 3/2004.

Rinorea oblongifolia (C.H.Wright) Marquand ex Chipp

Tree, 4–15m; stems glabrous; leaves chartaceous, (oblong–)elliptic, 22–34 × 9–12.5cm, acumen 1.7cm, base acute-cuneate, margin serrulate, lateral nerves 9–12 pairs, glabrous, eglandular; petiole 1.5–5.5cm; panicles terminal, 6–10cm, puberulent, many-flowered; sepals broadly ovate; 2mm; petals triangular, 3–4mm, yellow, densely puberulent outside. Forest; 600–710m.
Distr.: Sierra Leone to Uganda [Guineo-Congolian].
IUCN: LC
Ndanan 1: Cheek 11590 3/2004; Tadjouteu 574 3/2004; **Ndanan 2:** Etuge 5118 3/2003.

Rinorea subintegrifolia (P.Beauv.) Kuntze

Shrub, to 2m, glabrous; leaves pale green, elliptic, c. 13 × 5.5cm, acumen 1cm, base acute, lateral nerves 4–5 pairs, white, conspicuous, margin toothed; petiole 1–2cm; flowers whitish to purplish green, subcylindric, no scent at noon. Forest; 710m.
Distr.: Guinea (Conakry) to Cameroon, Gabon, Congo (Kinshasa) and Tanzania [Guineo-Congolian].
IUCN: LC
Ndanan 1: Cheek 11666 3/2004; 11788 3/2004; 11800 3/2004; 11867 3/2004.

Rinorea subsessilis Brandt

Fl. Afr. Cent. Violaceae: 63 (1969).
Shrub, 1–3m, glabrescent; leaves to 9.5–23.5 × 4.5–8cm, acumen to 2.3cm, base acute, margin subentire, lateral nerves 8–12 pairs, lower surface glandular; inflorescence terminal, (sub)corymbiform, 7 × 10cm; fruit smooth, beak 3mm Forest; 700m.
Distr.: Cameroon & Congo (Kinshasa) [lower Guinea & Congolian].
IUCN: LC
Ndanan 1: Tadjouteu 606 3/2004.

Rinorea welwitschii (Oliv.) Kuntze

Fl. Zamb. 1: 251 (1960); Kew Bull. 16: 418 (1963); Fl. Afr. Cent. Violaceae: 59 (1969); Kew Bull. 36: 121 (1981); Keay R.W.J., Trees of Nigeria: 46 (1989);

Hawthorne W., F.G.F.T. Ghana: 132 (1990).
Syn. *Rinorea elliotii* Engl.
Syn. *Rinorea longicuspis* Engl.
Shrub or small tree, 1.5–9m; leaves (6–)8.5–15(–19) × 3–7cm., obovate to oblanceolate, abruptly acuminate at the apex, cuneate at the base, ± bluntly serrate; petiole 5–20(–30)mm, pubescent; stipules c. 6–8mm, subulate, pubescent, soon caducous; inflorescence compound, terminal, narrowly paniculate to triangular, with few-flowered cymose clusters at the ends of short lateral branches; peduncle shortly brown-pilose; bracts shortly brown-pilose, persistent; pedicels shortly pilose; sepals c. 2mm, ovate-elliptic; petals 3–4mm, yellow, oblong, obtuse, recurved above, slightly unequal, sparsely pubescent or almost glabrous; capsule 10–13mm, dark red-brown, ± erect, 3-lobed, pubescent, ± smooth, coriaceous, several-seeded; seeds c. 4–5mm. Forest, by streams; 658–710m.
Distr.: tropical Africa.
IUCN: LC
Mefou Proposed National Park: Etuge 5301 3/2004;
Ndanan 1: Cheek 11596 3/2004; Onana 2190 10/2002;
Ndanan 2: Cheek 11093 10/2002.

Rinorea zenkeri Engl.

Bot. Jahrb. Syst. 33: 146 (1904).
Shrub, to 3m, brown-black pilose; leaves oblanceolate, 7–15 × 3.5–7.5cm, apex acuminate 0.5–1cm, base cordate, serrulate; petiole 3–5mm; scale leaves green persistent, 4mm; inflorescence terminal, pubescent; calyx purple-green; corolla pale yellow; fruits brownish-red, 1.5cm, densely bristled. Semi-deciduous forest at roadside; 600–710m.
Distr.: Cameroon endemic.
IUCN: NT
Mefou Proposed National Park: Etuge 5140 3/2004;
Ndanan 1: Cheek 11795 3/2004.

VISCACEAE

R.M. Polhill (K)

Fl. Cameroun 23 (1982).

Viscum congolense De Wild.

Polhill R. & Wiens, D. Mistletoes of Africa: 283 (1998).
Globose, dioecious parasitic shrub, < 0.5m; nodes often dilated; leaves highly variable, elliptic-ovate to oblong, 4–6(–10) × 2–4cm, conspicuously triplinerved above and below, base cuneate, margin sometimes crisped; male and female flowers ocurring in triads (occasionally up to 6), <3mm; berries subsessile, 6–9mm, greenish-white, translucent. Forest; 680–700m.
Distr.: Ivory Coast to Congo (Kinshasa), Rwanda, Burundi, Ethiopia, Angola [Guineo-Congolian.].
IUCN: LC
Ndanan 1: Darbyshire 238 fl., fr., 3/2004; **Ndanan 2:** Cheek 11092 10/2002; 11098 10/2002.

VITACEAE

L. Pearce (K), I. Darbyshire (K), M. Cheek (K) & E. Fenton (K)

Fl. Cameroun 13 (1972).

Ampelocissus abyssinica (A.Rich.) Planch.
Fl. Cameroun 13: 14–15 (1972); F.T.E.A. Vitaceae: 3 (1993).
Syn. *Ampelocissus cavicaulis* (Baker) Planch.
Woody climber, to 5m; stems ridged, lightly pubescent; tendrils bifid; leaves deeply 5(–7) digitately lobed, round to ovate-cordate in outline, 25 × 25cm, lobe tips acuminate, margin sparsely dentate, ± glabrous, purplish beneath; petiole 11cm, ridged, minutely pubescent; inflorescence a panicle 10–20cm, with densely pubescent axes; peduncle 7–12cm; flowers pentamerous; calyx cupular, glabrous; corolla green, 2–2.5mm; style broadly conical; fruit oblong-ellipsoid, 1.1.5cm, glabrous. Forest; 710m.
Distr.: Cameroon, Gabon, CAR, Equatorial Guinea, Congo (Kinshasa), Burundi, Sudan, Ethiopia, Uganda, Tanzania, Angola [tropical Africa].
IUCN: LC
Ndanan 1: Cheek 11726 3/2004.
Note: Cheek 11726 is sterile, inflorescence and fruit characters from F.T.E.A (Vitaceae): 1–3 (1993), Pearce x.2010.

Cayratia debilis (Baker) Suess.
Fl. Cameroun 13: 22–24 (1972); F.T.E.A. Vitaceae: 139 (1993).
Syn. *Cissus debilis* (Baker) Planch.
Slender, herbaceous or subwoody climber, to 2m; stems terete, glabrous to sparsely puberulous; tendrils simple or bifid; leaves papery, pedately 5-foliolate; leaflets 5–9.5(–12) × 2.5–6cm, central leaflet ovate-elliptic, c. 9 × 4.5cm, apex acuminate, base obtuse to oblique, lateral leaflets ovate to obovate, asymmetric, margins lightly toothed, principal veins of upper surface bearing thick whitish hairs, lower surface glabrous; petiole 6–11cm; inflorescence a loose cyme, 15–24 × 4.5–14cm; peduncle puberulent; flowers to 2mm; calyx cupular; corolla white, globular in bud; fruit globose, flattened at apex, 3–5mm; seeds 2.5mm, deltoid-cordiform. Primary forest understorey and *Raphia* swamp; 700–710m.
Distr.: Liberia, Ivory Coast, Ghana, Nigeria, Cameroon, Bioko, São Tomé, Gabon, Congo (Kinshasa), Sudan, Uganda, Tanzania, Angola [Guineo-Congolian].
IUCN: LC
Ndanan 1: Cheek 11744 3/2004; Tadjouteu 581 3/2004.

Cissus aralioides (Welw. ex Baker) Planch.
Fl. Cameroun 13: 88–91 (1972).
Climber, to 15m; stems terete, succulent, glabrous, drying yellow-green; tendrils simple or bifid; leaves palmately (3–)5-foliolate, 24–30 × 13–18cm, central

leaflet obovate, 13–18 × 4.5–5.5cm, apex acuminate, base acute, margin finely toothed, lateral leaflets asymmetric, ovate-elliptic, glabrous; petiole 6.5–14cm; inflorescence a compound, many-flowered cyme to 24 × 9.5cm; peduncle glabrescent; flowers 0.35cm; calyx cupular, puberulent; corolla buds rounded at apex; pedicels 0.6cm, minutely puberulent; fruit ellipsoid, 2.5 × 1.5cm, glabrous, green-red; seeds flattened, ellipsoid, 1.8 × 0.9cm, smooth. Undisturbed forest and *Raphia* swamp; 700–710m.
Distr.: Senegal, Guinea (Bissau), Guinea (Conakry), Sierra Leone, Liberia, Ivory Coast, Ghana, Nigeria, Cameroon, Equatorial Guinea, Gabon, CAR, São Tomé, Congo (Kinshasa), Sudan, Uganda, Kenya, Tanzania, Mozambique & Angola [tropical Africa].
IUCN: LC
Mefou Proposed National Park: Cheek in RP 42 3/2004; **Ndanan 2:** Cheek 11131 10/2002.
Note: flowering specimen only.

Cissus barbeyana De Wild. & T.Durand
Closely resembling *C. leonardii*, of which it is sometimes treated as a synonym, but with cylindrical mature stems; leaves less undulate and with finer teeth; flowering pedicels glabrous; fruits larger, 8–9 × 5–7mm. Secondary forest; 700m.
Distr.: Nigeria, Cameroon, Gabon, São Tomé, Congo (Kinshasa), Angola [lower Guinea & Congolian].
IUCN: LC
Ndanan 1: Tadjouteu 615 3/2004.

Cissus barteri (Baker) Planch.
Herbaceous climber; stems angular, glabrous; tendrils simple; leaves ovate, to 15 × 8cm, apex acuminate, base subcordate, margin finely dentate, drying green; petiole to 5.5cm. Forest; 700m.
Distr.: Nigeria to Congo (Kinshasa) [lower Guinea & Congolian].
IUCN: LC
Ndanan 1: Cheek in MF 311 10/2002.
Note: plot voucher MF311 has angular stems. According to Fl. Cameroun 13: 126 (1972) the stems should be cylindrical but otherwise a good match for *C.barteri*, Pearce x.2010.

Cissus diffusiflora (Baker) Planch.
Herbaceous climber, to 7m; stems cylindrical, pubescent; tendrils few, bifid; leaves papery, ovate-lanceolate, lamina 8–12.5 × 2.5–4.5cm, apex attenuate-acuminate, base truncate to shallowly cordate, margin finely dentate, principal veins of lower surface pilose; petiole 0.5–3cm, pubescent; inflorescence to 1cm long; flowers in umbellate clusters; peduncles 1–5mm, pubescent; flowers to 1.5mm long; calyx cupular, buds rounded at apex, yellow, flushed red; pedicel to 4mm, glabrous; fruit pyriform or globose, 0.6 × 0.5cm, glabrous, red-black; seeds c. 0.45cm long, flattened ovoid with a prominent dorsal ridge. Disturbed forest and forest edge; 700m.

Distr.: Guinea (Bissau) & (Conakry), Sierra Leone, Liberia, Ivory Coast, Ghana, Nigeria, Cameroon, Bioko, Gabon, Congo (Kinshasa), Sudan [Guineo-Congolian].
IUCN: LC
Ndanan 1: Cheek 11041 10/2002; Harvey 126 10/2002; 210 10/2002; 76 10/2002.

Cissus dinklagei Gilg & Brandt

Fl. Cameroun 13: 114–116 (1972).
Woody climber, to 30m; stem terete, sparsely pubescent, denser on the nodes; tendrils simple; leaves subcoriaceous, deltoid to elliptic, lamina 7.5–18.5 × 3–9cm, apex acuminate, base truncate to subcordate, margin entire, lower surface glabrous to weakly pubescent, lower surface with domatia in the axes of the primary and secondary nerves; petiole 1.5–8cm, sparsely pubescent; inflorescence in cymes, 4–10cm; peduncles 3–5cm; flowers 4–5mm; calyx cupular; corolla ovoid, densely puberulent; fruits orange, ovoid, 3–4 × 2–3cm, glabrous, warty; seeds ovoid-oblong, 15–17 × 10–12 × 7–7mm, rostrum long, thick, median ridge prominent. Cleared forest patch and *Raphia* swamp; 700m.
Distr.: Cameroon, Gabon, CAR, Congo (Kinshasa), Angola [Guineo-Congolian].
IUCN: LC
Ndanan 1: Cheek 11076 10/2002; **Ndanan 2:** Darbyshire 194 3/2004.
Note: material sterile or with fruit only. Inflorescence characters from Fl. Cameroun 13: 114–116 (1972).
Local Name: Aka'ndig (Cheek 11076).
Uses: ANIMAL FOOD — Unspecified Parts — eaten by monkeys (Cheek 11076).

Cissus glaucophylla Hook.f.

Woody climber; stems, peduncle and underside of leaves orange-mealy; stems terete, finely striate; tendrils forked; leaves ovate, lamina 5–9 × 3–8cm, apex with acumen to 1cm, base cordate, margins undulate with recurved short, deltoid teeth, glabrous, glaucous above; petiole 3–8cm, glabrous; inflorescence cymose, many-flowered, to 15cm; peduncle to 7cm; flowers 1.5–1.7mm; calyx cupular, glabrous to puberulent; corolla ovoid, glabrous to finely puberulent; fruit obovoid to pyriform, to 6 × 4mm, apiculate. Semi-deciduous forest; 700m.
Distr.: Sierra Leone, Liberia, Ivory Coast, Ghana, Nigeria, Equatorial Guinea, Cameroon, Gabon, CAR, Congo (Kinshasa) and Angola [Guineo-Congolian].
IUCN: LC
Mefou Proposed National Park: Nana 41 3/2004.
Note: includes *C. smithiana* (Baker) Planchon of Fl. Cameroun 13: 94–96 (1972). Specimens fruiting only. Inflorescence characteristics taken from Fl. Cameroun 13: 94–96 (1972). Pearce x.2010.

Cissus leonardii Dewit

Fl. Gabon 14: 95 (1968); Fl. Cameroun 13: 116–117 (1972).
Herbaceous climber, to 6m; mature stems subquadrangular, glabrous; tendrils simple; leaves ovate-oblong, lamina 6–12 × 2–6.5cm, acumen to 1.2cm, base truncate to subcordate, margin shallowly undulate-dentate, glabrous, lower surface with domatia in the axes of the primary nerves, drying green-brown below, brown above; petiole 2–11.5cm; inflorescence to 7cm long with several umbellate cymes; peduncle 1–2cm, puberulent; flowers 2–2.5mm; calyx cupular; corolla buds rounded; pedicel 3–5mm, puberulent; fruit obovoid, c. 0.6 × 0.5cm, greenish-pink maturing to maroon-black; seeds c. 0.5cm long, subreniform, several-ridged dorsally. Secondary forest and *Raphia* swamp; 600–710m.
Distr.: Cameroon, Gabon, CAR, Congo (Brazzaville), Congo (Kinshasa) [lower Guinea & Congolian].
IUCN: LC
Mefou Proposed National Park: Cheek in RP 7 3/2004; Etuge 5104 3/2003; **Ndanan 1:** Harvey 98 10/2002.
Note: fruiting specimens only. Inflorescence characters from Fl. Cameroun 13: 116–117 (1972). Pearce x.2010.

Cissus petiolata Hook.f.

Climber, to 15m; stems square, ridged when mature, glabrous; leaves ovate to pentagonal, lamina 7–13 × 7–9cm, apex acuminate, base deeply cordate, margin finely dentate, surfaces glabrescent; petiole to 10cm; inflorescence 3–8cm, of c. 4 many-flowered umbellate cymes; peduncle 1–2cm; flowers 2–3mm; calyx cupuliform, glabrous; corolla oval-oblong, conical at the summit, glabrous; pedicels densely pubescent; fruit subglobose, 0.5–0.9cm diam., tip apiculate, glabrous, green; seeds 0.7cm long, flattened ovoid, subreniform laterally, smooth. Semi-deciduous and riverine forest; 700–710m.
Distr.: Guinea (Conakry), Sierra Leone, Liberia, Ghana, Togo, Benin, Nigeria, Cameroon, Gabon, CAR, Congo (Kinshasa), Burundi, Rwanda, Sudan, Ethiopia, Eritrea, Uganda, Kenya, Tanzania, Malawi, Zambia, Zimbabwe, Mozambique and Angola [tropical Africa].
IUCN: LC
Ndanan 1: Cheek 11052 10/2002; 11900 3/2004; Cheek in MF 406 10/2002.

Cissus planchoniana Gilg

Fl. Cameroun 13: 120–122 (1972).
Herbaceous, glabrescent climber, to 0.3m; stems slender, terete to angular with narrow wings in younger growth; tendrils slim, forked; leaves lanceolate to oblong-ovate, lamina 3–5.5 × 1.5–3cm, apex attenuate to acuminate, base truncate to rounded or shallowly cordate, margins finely toothed; petiole, 5–18mm, channelled above and winged; inflorescence a cyme composed of few-flowered umbelliform cymules, 5–15mm; peduncle 1–5mm; flowers 2–2.5mm; calyx 1mm; corolla ovoid-globular; petals deltoid, 1.2–1.5mm; pedicels 1–2mm; fruit ovoid to globose, 5–7 × 4–5mm, red, ripening black; seeds ovoid-reniform, 5–6 × 3–3.5 × 3mm, rostrum long, median and lateral lines very neat and prominent. *Gilbertiodendron* forest; 710m.
Distr.: Cameroon, Gabon, Congo (Kinshasa), Uganda,

Tanzania [lower Guinea & Congolian].
IUCN: LC
Ndanan 1: <u>Cheek 11558</u> 3/2004.
Note: specimen fruiting, flower characters from Fl. Cameroun 13: 120–122 (1972), Pearce x.2010.

Cissus sp. 1 of Mefou

Liana seedling, to 12cm; stem striate, with densely lanate 0.6cm, cream jointed hairs and sparse 0.2mm red medifixed hairs; leaves cordate, to 8.2 × 6.2cm, margins with numerous fine teeth to 1cm and long cream hairs, both surfaces of leaves with long cream hairs on veins, lower surface with evenly spaced red medifixed hairs; petiole to 4cm, indumentum as for stems. *Raphia* swamp; 710m.
Mefou Proposed National Park: <u>Cheek in RP 45</u> 3/2004.
Note: plot voucher RP45 is a sterile, juvenile specimen. Indumentum does not match that of any of the mature specimens examined at K. Closest to *C. polyantha* Gilg & Brandt and *C. oreophila* Gilg & Brandt, Pearce x.2010.

Cissus sp. 2 of Mefou

Climber, to 3m; stem terete to angular, striate, ± glabrous except for a few long fine orange hairs nearer nodes; tendrils long, forked, glabrous; leaves ovate to pentagonal, 8–11.5 × 6–8cm, apex acuminate, base deeply cordate, margins with small, quite fine teeth; petiole 3.5–5.5cm; leaf veins and petiole with a dense covering of long fine orange hairs and short, coarse red hairs. *Raphia* swamp; 700m.
Ndanan 2: <u>Darbyshire 195</u> 3/2004.
Note: Darbyshire 195 is closest to *C. polyantha* and *C. ruginosicarpus*. *C. polyantha* has a covering of fine reddish hairs on the stems and leaves, whereas Darbyshire 195 has reddish hairs but only on the veins of the leaf undersurface and petioles. *C. ruginosicarpus* has red hairs between the veins of the leaf undersurfaces. Pearce x.2010.

Cyphostemma camerounense Desc.

Fl. Cameroun 13: 69–70 (1972).
Climber; stems cylindric with dense pubescence of erect hairs and long, fine, glandular hairs; tendrils slender, slightly pubescent, forked; leaves digitately 5-foliolate; leaflets obovate-elliptic, 6–15 × 3–6cm, apex obtuse with a short acumen, base cuneate-rounded, margins shallowly crenate, deltoid, terminated by a thick tooth; petioles 4–10cm, terete, striate with numerous glandular hairs; inflorescence 8–12cm, composed of loose, few-flowered umbelliform cymes; peduncle 4–6cm; flowers 2.5–3mm; calyx cupuliform, 0.8mm; corolla cylindric, mamillate and with a tuft of hairs at the apex; pedicels 0.5–1.5mm, densely pubescent; fruits obovoid, 7–10 × 5–8mm, weakly pubescent but with numerous long, fine glandular hairs; seed ovoid-elliptic, 6 × 3.5 × 3.5mm. Semi-deciduous forest; 700m.
Distr.: Cameroon [Cameroon endemic].
IUCN: EN
Ndanan 1: <u>Cheek 11042</u> 10/2002.
Note: 5 specimens only are cited in Fl. Cameroun. Cheek 11042 matches Breteler 642 which was cited in Fl. Cameroun for this taxon. Specimen fruiting, description of inflorescence from Fl. Cameroun 13: 69–70 (1972). Pearce x.2010.

VOCHYSIACEAE

M. Cheek (K)

Erismadelphus exsul Mildbr. var. *platyphyllus* Keay & Stafleu

Tree; coloured exudate absent; leafy stems densely tomentellous; leaves drying midbrown, opposite, simple, estipulate, elliptic, 16 × 8cm, acumen 1.5cm, base acute-obtuse, entire, lateral nerves 8–9 pairs, uniting in a bold looping nerve 4mm from the margin, quaternary nerves reticulate, raised; panicles terminal, brown tomentellous; flowers zygomorphic, 1.25 × 1cm; petals 5; stamens 1; fruit inferior, with unequally enlarged, wing-like papery sepals, 7mm. Forest; 700m.
Distr.: SE Nigeria, Cameroon and Gabon [lower Guinea].
IUCN: NT
Mefou Proposed National Park: <u>Cheek in MF 112</u> 3/2004.

MONOCOTYLEDONAE

ALISMATACEAE

M. Cheek (K), E. Fenton (K) & Y.B. Harvey (K)

Caldesia reniformis (D.Don) Makino

Aquatic herb; bottom-rooting; leaves held above the water; leaf-blades orbicular or elliptic, 5–9 × 5–7.5cm, apex rounded, base very shallowly to deeply cordate, lateral nerves. c. 6 on each side of the midrib; petioles robust, 10–40cm long; inflorescence as tall or taller than leaves, c. 10cm wide, branches with flowers paired; pedicels 1.5cm; flowers 5mm diam., white. River edges and swamps.
Distr.: tropica Africa, Asia and Australia.
IUCN: LC
Mefou Proposed National Park: Sight Record 2 /2001.
Note: known only from photograph taken by Benedict Pollard, 2001.

Ranalisma humile (Kunth) Hutch.

Small herb, to 4(–15)cm; stolons short; leaves ovate to ovate-lanceolate, 0.5–1.5 × 1–3cm (emersed), or linear and up to 0.5 × 10cm (submerged), apex acute, base rounded to cuneate; petiole 1–10cm; inflorescence 1-flowered; peduncles equal petioles in length; bracts 3–5mm; sepals ovate, 4 × 3mm; petals 6 × 4cm, white; stamens, 8–12mm; filaments 2.5mm; carpels numerous, in a globose head, obliquely obovoid in fruit, compressed. Swamps and marshy places; 600m.
Distr.: Senegal to Sudan, Congo (Kinshasa) to Zambia [tropical Africa].
IUCN: LC
Mefou Proposed National Park: Etuge 5167 fl., 3/2004.

AMARYLLIDACEAE

M. Cheek (K), E. Fenton (K) & Y.B. Harvey (K)

Fl. Cameroun 30 (1987).

Crinum jagus (Thomps.) Dandy

Bulbous herb, 0.6–1m; leaves erect, often petiolate, c. 50 × 7cm, c. 30-nerved, margin often undulate; inflorescence above leaves, 3-flowered, with 2 involucral bracts; flowers irregular, infundibuliform, sweetly scented, 7 × 3.5cm, white with green markings; stamens black. Forest, savanna, plantations, usually riverine; 710m.
Distr.: Guinea (Conakry) to Angola, Sudan, Uganda [Guineo-Congolian].
IUCN: LC
Ndanan 1: Cheek 11694 3/2004.
Note: Cheek 11694 has not been confirmed at Kew, unicate voucher collection, Harvey iii.2010.

Crinum purpurascens Herb.

Bulbous herb, to 1m; bulb subglobose, 3–5cm diam.; leaves 20–70(–100) × 1–4cm; scape reddish, 20–50cm; bracts subtending 2–10 subsessile flowers; flowers with perianth segments spreading, c. 7 × 7cm, white; filaments purple apically; stamens orange. Farmbush; 710m.
Distr.: Gambia to Angola [upper & lower Guinea].
IUCN: LC
Ndanan 1: Cheek 11759 3/2004.
Uses: ENVIRONMENTAL USES — Ornamentals — cultivated (Cheek 11759).

Scadoxus multiflorus (Martyn) Raf.

Fl. Cameroun 30: 8 (1987).
Syn. *Haemanthus multiflorus* Martyn
Bulbous herb, 25–80cm; bulb cylindrical, c. 2 × 1.5cm; leaves expanding after flowering, ovate-lanceolate, to 25 × 8cm, base attenuate; inflorescence lateral, 7–25cm, globose, many-flowered; flowers scarlet; pedicels 1.5–3.5cm. Forest edge & semi-deciduous forest; 640m.
Distr.: Senegal to Somalia & to S Africa, also Yemen [tropical Africa].
IUCN: LC
Ndanan 2: Darbyshire 312 fl., 3/2004.

ANTHERICACEAE

I. Nordal (O) & E. Fenton (K)

Chlorophytum alismifolium Baker

Herb, to 15cm, drying green; lamina lanceolate, 7–13 × 1.4–2.3cm, apex acute, base attenuate, lamina at an angle with pseudopetiole; pseudopetiole distinct, 5–10cm, canaliculate. Forest, *Raphia* swamp; 550m.
Distr.: Sierra Leone, Liberia, Ivory Coast, Ghana, Nigeria, Cameroon and Congo (Kinshasa) [Guineo-Congolian].
IUCN: LC
Mefou Proposed National Park: Etuge 5198 3/2004.
Note: Etuge 5198 is a sterile voucher specimen.

Chlorophytum orchidastrum Lindl.

Robust herb, 40–100cm; drying black; lamina elliptic, 20–30 × 6–9cm, apex acuminate, base rounded to truncate; petiole distinct, to 40cm; inflorescence to 70cm, erect, with up to 3 lateral branches; flowers white; tepals 3-veined, c. 5 × 1mm; pedicels 5–8mm, with joint in upper half; bracts c. 6 × 5mm, glabrous; capsule c. 7 × 8mm. Forest; 700–710m.
Distr.: Guinea (Conakry), Sierra Leone, Liberia, Ivory Coast, Ghana, Nigeria, Cameroon, Equatorial Guinea, Gabon, Congo (Kinshasa) [Guineo-Congolian].
IUCN: LC
Ndanan 1: Cheek 11218 10/2002; 11514 3/2004.

ARACEAE

A. Haigh (K), P. Boyce, M. Cheek (K), Seren Thomas (K) & P. Mezili (YA)

Fl. Cameroun 31 (1988).

Amorphophallus staudtii (Engl.) N.E.Br.
Herb; leaf like a tattered umbrella, leaflets acute, not fishtail shaped; petiole to 1.2m, smooth; inflorescence pale dirty cream-white; peduncle very short; spadix base not swollen. Forest; 700m.
Distr.: Cameroon, Equatorial Guinea [lower Guinea].
IUCN: NT
Ndanan 2: Cheek 11102a fl., 10/2002.
Note: known from only 10 sites in Cameroon, where it is likely threatened in the Bamenda Highlands by agricultural encroachment into existing forest patches. Also recorded from Equatorial Guinea, but poorly documented here.

Anchomanes hookeri Schott
Oestr. Bot. Wochenbl. 3: 314. (1853).
Herb; rhizome horizontal, stout; leaf single, like a tattered umbrella; leaflets fish tail-shaped; petiole mottled, spiny; inflorescence c. 0.5–0.7m; peduncle spiny; spathe green, interior purple; styles curved downwards, warty, deep purple. Forest; 700m.
Distr.: Guineo-Congolian.
Ndanan 1: Cheek in MF 11102b 10/2002; **Ndanan 2:** Cheek 11102b fl., 10/2002.
Note: plot voucher MF365 is probably this taxon, Haigh ix.2010. *A. hookeri* was until recently treated as a synonym of *A. difformis* but the two are separable on differences in the styles. However, the distributions of these taxa have not been fully defined as yet; no conservation assessments can be made until this is clarified.

Cercestis camerunensis (Ntepe-Nyame) Bogner
Aroideana 8(3): 73 (1986).
Syn. *Rhektophyllum camerunense* Ntepe-Nyame Fl. Cameroun 31: 60 (1988).
Robust climber; stem rooting along whole length; leaves 1.5m long, scattered along whole stem; leaf blade pinnately divided, up to 50cm long; spathe up to 12cm long. Forest; 710m.
Distr.: Nigeria, Cameroon & Gabon [lower Guinea].
IUCN: LC
Ndanan 1: Cheek 11658 3/2004.
Note: locally numerous in Cameroon excluding the east of the country (Fl. Cameroun 31: 64, 1988).

Cercestis dinklagei Engl.
Fl. Cameroun 31: 66 (1988).
Syn. *Cercestis stigmaticus* N.E.Br.
Slender climber; rooting at same nodes as leaf-clusters, clusters spaced intermittently along stem; leaf-blade entire, lamina to 30cm, base deeply cordate, auricles oblong; spathe to 7cm long. Forest; 710m.

Distr.: Guinea (Conakry) to Congo (Brazzaville) [upper & lower Guinea].
IUCN: LC
Ndanan 1: Cheek 11516 3/2004.

Cercestis mirabilis (N.E.Br.) Bogner
Aroideana 8(3): 73 (1986).
Syn. *Rhektophyllum mirabile* N.E.Br.
Herb, initially terrestrial, with hastate marbled white leaves, becoming a climbing epiphyte, then losing the marbling and the leaf-blades becoming pinnately lobed, to 21 × 11cm; aerial roots warty. Forest; 700–710m.
Distr.: Benin to Bioko, Cameroon, Congo (Kinshasa), Angola and Uganda [Guineo-Congolian].
IUCN: LC
Mefou Proposed National Park: Cheek in MF 22 3/2004; 48 3/2004; 62 3/2004; Cheek in RP 74 3/2004; **Ndanan 1:** Cheek in MF 374 10/2002.
Note: plot vouchers MF22 & MF48 are probably this taxon, Haigh ix.2010.

Culcasia dinklagei Engl.
Fl. Cameroun 31: 80 (1988).
Herbaceous climber, or held erect by stilt roots; leaf blade to 12–35 × 4–15cm, obovate to oblanceolate, with transparent lines; petiole shorter than lamina, to 9cm; infructescence solitary or paired. Forest; 700m.
Distr.: Guinea (Conakry) to Congo (Brazzaville) [upper & lower Guinea].
IUCN: LC
Ndanan 1: Cheek 11262 10/2002; **Ndanan 2:** Cheek 11168 10/2002.

Culcasia parviflora N.E.Br.
Herbaceous climber; distinguished from other *Culcasia* spp. by the cordate lamina base and the petiole being c. 1/5 the length of the lamina. Forest; 700m.
Distr.: Guinea (Conakry) to Congo (Kinshasa) [Guineo-Congolian].
IUCN: LC
Ndanan 1: Cheek 11246 10/2002.
Note: Cheek 11246 is a tentative determination (probably *C. parviflora*) but need inflorescence, lacks the cordate blade base, possibly juvenile, Cheek ix.2010.

Culcasia sapinii De Wild.
Syn. *Culcasia seretii* De Wild.
Herbaceous climber; innovations (new shoots) copper-orange; petiole more than half the length but always shorter than lamina; leaves with translucent dots. Forest; 700m.
Distr.: W Africa to Congo (Kinshasa) [Guineo-Congolian].
IUCN: LC
Mefou Proposed National Park: Cheek in MF 105 3/2004; 71 3/2004; **Ndanan 1:** Cheek 11215 10/2002; Gosline 432 10/2002; Harvey 162 10/2002; 166 10/2002; 178 10/2002.
Note: uncommon through much of its range; only 5 previous collections at K, but 16 locations recorded in

Fl. Cameroun. Submontane forest habitat largely unthreatened. Gosline 432 and plot Vouchers MF71 and MF105 have tentatively placed in this taxon, Cheek ix.2010.

Culcasia simiarum Ntepe-Nyame
Fl. Cameroun 31: 92 (1988).
Terrestrial herb, to 25cm; lamina elliptic, 17.5 × 6.3cm, glabrous, apex acute to acuminate, base cuneate, margin entire; petiole to 4cm, sheathed; fruits orange-red, c. 1.7 × 1.5cm. Forest; 700m.
Distr.: Ivory Coast to Cameroon [Guineo-Congolian].
IUCN: LC
Ndanan 1: Harvey 174 10/2002.

Culcasia striolata Engl.
Terrestrial herb with stilt roots; lamina oblong, with numerous prominent short translucent lines. Forest; 650–700m.
Distr.: Guinea (Conakry) to Gabon [upper & lower Guinea].
IUCN: LC
Ndanan 1: Onana 2904 3/2004; Tadjouteu 607 3/2004; **Ndanan 2:** Darbyshire 258 fl., 3/2004.

Culcasia sp. of Mefou
Hemi-epiphyte, growing 3m above ground level; lamina elliptic-oblong, 29 × 8.6cm, glabrous, apex acuminate, base acute, margin entire; petiole 19cm, partially sheathed; sheath to 8cm. Lowland evergreen forest near swamp-edged water; 700m.
Mefou Proposed National Park: Cheek in MF 108 3/2004.

Lasimorpha senegalensis Schott
Mayo S. *et al.*, Gen. of Araceae: 138 (1997).
Syn. *Cyrtosperma senegalense* (Schott) Engl. Fl. Cameroun 31: 8 (1988).
Terrestrial herb, from rhizome; leaf-blades sagittate, 32–45 × 12–30cm; petioles erect, spiny, 6-angular, to 2cm diam.; inflorescence on peduncle 1–1.5cm; spathe thickly leathery, green streaked red, white and purple, erect, cylindric-ellipsoid, 40 × 8cm; spadix 8 × 1.5cm. Swamp forest; 600m.
Distr.: Senegal to Bioko, Cameroon and Congo (Kinshasa) [Guineo-Congolian].
IUCN: LC
Mefou Proposed National Park: Etuge 5178 3/2004.
Local Name: Engogwan (Etuge 5178).

Nephthytis poissonii (Engl.) N.E.Br.
Syn. *Nephthytis constricta* N.E.Br.
Syn. *Nephthytis gravenreuthii* (Engl.) Engl. Fl. Cameroun 31: 50 (1988).
Syn. *Nephthytis poissonii* (Engl.) N.E.Br. var. *constricta* (N.E.Br.) Ntepe-Nyame Fl. Cameroun 31: 56 (1988).
Terrestrial creeping herb; leaves triangular, posterior lobes considerably more developed than the anterior lobes; fruits orange-red subtended by a spreading,

persistent green spathe. Forest; 663m.
Distr.: Sierra Leone to Gabon [upper & lower Guinea].
IUCN: LC
Mefou Proposed National Park: Etuge 5304 3/2004.

Rhaphidophora bogneri ined. P.C.Boyce
Epiphyte, on *Ficus* c. 4m above ground; stems pendulous, almost to ground; lamina elliptic, c. 31 × 8.7cm, glabrous, apex acuminate, base acute, margin entire; petiole c. 18.3cm; spathe 9 × 2cm, black, elliptic to falcate in outline. Riverine forest; 640m.
Distr.: Cameroon and Gabon [Guineo-Congolian].
IUCN: DD
Ndanan 2: Darbyshire 314 fr., 3/2004.

Stylochaeton zenkeri Engl.
F.W.T.A. 3: 114 (1968); Fl. Cameroun 31: 43 (1988).
Terrestrial rhizomatous herb, to 45cm; lamina elliptic c. 15 × 7.5cm, acute, base shortly cordate, lower surface nearly white; petioles to 19cm, purple red; spathe inconspicuous, amongst petiole bases; fruits red, without spathe remains. Forest; 700m.
Distr.: Sierra Leone to Congo (Brazzaville) [upper & lower Guinea (montane)].
IUCN: LC
Ndanan 1: Cheek 11964 3/2004.

ASPARAGACEAE

Sebsebe D. (ETH), E. Fenton (K) & X. van der Burgt (K)

Asparagus drepanophyllus Welw. ex Baker
F.T.A. 7: 435 (1898).
Climbing shrub, to 5(–8)m; long root tubers in clusters; branches slender, flexuose, glabrous; branchlets numerous, spreading; spines 3–5mm long, flattened towards base; cladodia usually 3-nate, 5–45 × 1–3mm, unequal, spreading, linear, curved, bright green, distinctly 1-nerved; inflorescence raceme, 6–17(–23)cm; pedicels 3–6 in a cluster 2–3mm long, articulated in the middle; perianth campanulate, white to pale green, 2–3mm long; segments oblong; stamens shorter than the perianth; anthers yellow; fruits 5–7mm diam. Gallery, secondary forest and grassland; 700m.
Distr.: Cameroon, Equatorial Guinea, CAR, Gabon, Congo (Brazzaville), Congo (Kinshasa) & Angola [lower Guinea & Congolian].
IUCN: LC
Ndanan 1: Cheek in MF 212 10/2002; **Ndanan 2:** Cheek 11115 10/2002.
Note: both samples sterile (Cheek 11115, plot voucher MF212), flower and fruit details provided by Sebsebe Demissew, Harvey iv.2010. Although similar to *Asparagus warneckei*, *A. drepanophyllus* has shorter spines (3–5mm as opposed to 7–10mm), shorter pedicels that are articulated in the middle as opposed to the base (2–4mm as opposed to 5–6mm) and shorter tepals (2–3mm as opposed to 4–5mm), Sebsebe Demissew iv.2010.

Local Name: Ngnue (Cheek 11115).
Uses: MEDICINES - Injuries - roots crushed and applied to wounds (Cheek 11115).

CANNACEAE

J.-M. Onana (YA)

Canna indica L.

Cultivated ornamental, to c. 1m; stems erect, glabrous; leaves ovate-elliptic, broadly acuminate, abruptly cuneate, to 40 × 25cm, glabrous; inflorescences terminal, racemose, few-flowered; pedicels very short; bracts ovate, c. 1.3cm; flowers usually scarlet, with centres orange; petals linear-lanceolate, c. 4cm; fruit a capsule; pericarp muricate. Gardens, villages, roadsides; 700m.
Distr.: native to tropical America, introduced in Africa and other tropical parts [pantropical].
IUCN: LC
Ndanan 2: Cheek 11116 10/2002.
Local Name: Ekone Zog (Cheek 11116).
Uses: ENVIRONMENTAL USES — Ornamentals — planted for its flowers (Cheek 11116).

COMMELINACEAE

J.-M. Onana (K), Y.B. Harvey (K) & M. Xanthos (K)

Amischotolype tenuis (C.B.Clarke) R.S.Rao

Enum. Pl. Afr. Trop. 3: 21 (1995).
Syn. *Forrestia tenuis* (C.B.Clarke) Benth.
Decumbent herb, 30–60cm tall; leaves variegated white and pink above in juvenile plants, up to 5 × 3cm, in adults elliptic, purple below, c. 10 × 5cm, acuminate; petiole 1–2cm; inflorescence axillary, perforating leaf-sheath, sessile, bifurcate, branches 1–3cm long; flowers purple. Lowland forest; 710m.
Distr.: Nigeria to Congo (Kinshasa) [lower Guinea & Congolian (montane)].
IUCN: NT
Ndanan 1: Cheek 11840 3/2004.
Note: Cheek 11840 is tentatively placed in this taxon (cf. *Forestia tenuis*), Onana x.2009.

Aneilema beniniense (P.Beauv.) Kunth

Weak erect herb, to c. 1m; leaves elliptic, c. 15 × 5cm, acuminate, sessile or shortly petiolate; inflorescence terminal, dense, c. 3–4 × 3–4cm; flowers white or mauve; fruits longer than broad, apex rounded. Lowland farmbush, forest, stream banks and *Aframomum* thicket; 700m.
Distr.: tropical Africa.
IUCN: LC
Ndanan 1: Harvey 199 10/2002; **Ndanan 2:** Cheek 11100 10/2002.
Local Name: Soro (Cheek 11100).

Aneilema dispermum Brenan

Weak erect herb, to c. 1m; leaves elliptic, to 14 × 4cm, acuminate, sessile or shortly petiolate, margin ciliate; inflorescence terminal, dense, c. 4.5 × 3cm; flowers 4–5mm, white; capsules broader than long, 2-seeded. Montane forest edge, forest-grassland transition; 710m.
Distr.: Bioko, SW Cameroon, Malawi & Tanzania [afromontane].
IUCN: NT
Ndanan 1: Cheek 11770 3/2004; 11895 3/2004.

Aneilema umbrosum (Vahl) Kunth subsp. *umbrosum*

Straggling herb, to 1m; leaves elliptic or lanceolate, up to 13 × 4cm, acuminate, petiolate; sheath with rusty hairs at apex and sometimes on surface; inflorescence terminal, lax, up to 12 × 7cm with 8–30 branches; flowers white. Lowland farmbush & forest; 600–710m.
Distr.: Sierra Leone to Congo (Kinshasa) [Guineo-Congolian (montane)].
IUCN: LC
Mefou Proposed National Park: Etuge 5235 3/2004;
Ndanan 1: Cheek 11769 3/2004; Harvey 105 10/2002;
Ndanan 2: Cheek 11141 10/2002.
Local Name: Soro (Cheek 11141).

Commelina benghalensis L. var. *hirsuta* C.B.Clarke

Erect herb, c. 30(–150)cm tall; leaves ovate, to 6 × 3.5cm, subacuminate, truncate, petiolate; sheath with conspicuous rusty hairs all over outside; spathe c. 2 × 1cm; flowers bright blue, open 8.30–12am. Lower montane to submontane farmbush; 710m.
Distr.: Guinea (Conakry) to Malawi [afromontane].
IUCN: LC
Ndanan 1: Cheek 11855 3/2004; **Ndanan 2:** Cheek 11112 10/2002.

Commelina congesta C.B.Clarke

Erect herb, c. 60cm tall; leaves lanceolate, c. 9 × 3cm, acuminate, base obtuse, strongly asymmetric, subsessile, sheaths glabrous, rarely weakly hairy; spathes c. 3, crowded, c. 1.5 × 3cm, sessile, margins glabrous; flowers white to pale violet, open dawn to noon; fruit with 3, 1-seeded locules. Lowland forest, roadsides and villages; 710m.
Distr.: Guinea (Conakry) to Congo (Kinshasa) [Guineo-Congolian].
IUCN: LC
Ndanan 1: Cheek 11868 3/2004.

Commelina diffusa Burm.f. subsp. *diffusa*

Decumbent to straggling herb, 15–60(–120)cm; stems rooting freely at the nodes; leaves ovate-lanceolate, c. 5–8 × 0.8–2.5cm, acute, base rounded, subsessile; spathes 1, 1.5–3 × 0.8cm with unfused margins; peduncle c. 1cm; flowers blue to mauve; fruit with reticulate seeds. Forest edge, plantations & farmbush; 700m.
Distr.: pantropical.
IUCN: LC
Ndanan 1: Harvey 108 10/2002.

Commelina forskalaei Vahl

Annual, prostrate or widely straggling herb, 10–60cm; stems rooting at the nodes; leaf sheaths short, with dark red markings, glabrous or shortly pubescent; leaf-blades 37 × 13mm, ovate or elliptic, broadly acute to obtuse, sessile, roughly pubescent with short and long colourless hairs; spathes 12 × 9mm, margins fused, glabrous or pubescent, mucilage present; flowers c. 12mm diam.; petals rich pure blue. Grassland, open waste and cultivated ground, roadside, sand-dunes; 600m.
Distr.: tropical Africa, Madagascar, Arabia and India.
IUCN: LC
Mefou Proposed National Park: Etuge 5147 fl., 3/2004.

Commelina longicapsa C.B.Clarke

Decumbent herb, to c. 30cm; leaves 3–4 clustered at apex of stem, obovate, c. 13 × 6cm, shortly acuminate, base acute; petiole 3–4cm; spathes subleafy, pinkish, 2–3, crowded, c. 2 × 1.5cm; flowers white. Lowland forest, near streams; 700m.
Distr.: Liberia to Congo (Kinshasa) [Guineo-Congolian].
IUCN: NT
Mefou Proposed National Park: Cheek in MF 119 3/2004; **Ndanan 2:** Cheek 11147 10/2002.
Local Name: Etotogo (Ewondo) (Cheek 11147).
Uses: MEDICINES (Cheek 11147).

Floscopa africana (P.Beauv.) C.B.Clarke subsp. africana

Straggling herb, 20–60cm; stems & leaves red-purple; leaves lanceolate, to c. 8 × 1.5–2cm, acuminate; petiole 0.5cm; inflorescence paniculate, with up to 15 bracts, glandular-hairy; flowers purple; sepals 2–2.75mm; style not exserted; fruit pedicel 0.5–1.5mm. Lowland farmbush, often in damp areas; 710m.
Distr.: Gambia to Uganda [Guineo-Congolian].
IUCN: LC
Ndanan 1: Cheek 11518 3/2004.

Floscopa africana (P.Beauv.) C.B.Clarke subsp. petrophila J.K.Morton

Straggling herb, to 20cm; stems long creeping, lacking red pigment; leaves to three times as long as broad, 2.5–6cm long. Lowland farmbush & forest edge, sometimes on rocks; 650–710m.
Distr.: Liberia to Uganda [Guineo-Congolian].
IUCN: LC
Ndanan 1: Cheek 11517 3/2004; 11773 3/2004;
Ndanan 2: Darbyshire 216 fl., 3/2004.

Palisota ambigua (P.Beauv.) C.B.Clarke

Herb resembling *P. hirsuta* but inflorescence usually one per leaf rosette, only 3–10cm, puberulous, partial peduncles to only 3mm. Forest; 600–710m.
Distr.: Nigeria to Congo (Kinshasa) [Guineo-Congolian].
IUCN: LC
Mefou Proposed National Park: Etuge 5137 fl., 3/2004; Nana 37 3/2004; **Ndanan 1:** Cheek 11189

10/2002; 11508 3/2004; Harvey 207 10/2002; **Ndanan 2:** Cheek 11082 10/2002.
Local Name: Etotongo (Etuge 5137).

Palisota barteri Hook.

Herb, 20–80cm; lacking aerial stem; leaves forming a basal rosette, blades elliptic, c. 25–45 × 8–15cm, apex acute or acuminate, base cuneate to rounded, margin often black-hairy, lower surface green, glabrous; petiole 20–30cm; inflorescence with peduncle 10–50cm long, rarely bracteose; panicle subglobose to ovoid, 5–10 × 3–7cm; flowers white; fruits red. Lowland to lower montane forest; 700m.
Distr.: Sierra Leone to Bioko, Cameroon & Congo (Kinshasa) [Guineo-Congolian].
IUCN: LC
Ndanan 2: Cheek 11101 10/2002.

Palisota bracteosa C.B.Clarke

Acaulous herb; leaves to 80cm; leaf-blade elliptic, to 60 × 15cm; petiole winged; inflorescences 15cm; peduncles 10cm, with 2–3 reduced leaves 3cm; rachis with large, conspicuous bracts to 1.5 × 0.5cm. Forest; 658–700m.
Distr.: Guinea (Conakry) to Cameroon & São Tomé [upper & lower Guinea].
IUCN: LC
Mefou Proposed National Park: Etuge 5297 3/2004; **Ndanan 1:** Cheek in MF 345 10/2002; **Ndanan 2:** Cheek 11104 10/2002.
Note: plot voucher MF345 is tentatively placed in this taxon, Xanthos x.2010.

Palisota hirsuta (Thunb.) K.Schum.

Herb, to 3m; aerial stems present; leaves clustered at apices, oblanceolate, c. 15–30(–40) × 4–11.5cm, apex acuminate, base cuneate, midrib brown pubescent; sessile or petiole 2–3cm, brown hairy; inflorescences often clustered at apex, c. 30cm, terminal, lax, branches whitish to pinkish; peduncle to 10cm; partial peduncles to 2cm; flowers white to purplish; fruits glossy, blackish. Lowland forest; 700–710m.
Distr.: Senegal to Congo (Kinshasa) [Guineo-Congolian].
IUCN: LC
Mefou Proposed National Park: Cheek in MF 5 3/2004; Cheek in RP 79 3/2004; 85 3/2004; **Ndanan 1:** Cheek in MF 344 10/2002.
Note: the field identification of plot voucher MF344 has not been confirmed, Harvey viii.2010. Plot voucher RP79 is tentatively placed in this taxon, Xanthos x.2010.

Palisota mannii C.B.Clarke

Herb, 20–80cm; lacking aerial stem; leaves forming a basal rosette, lanceolate or lanceolate-obovate, 25–40 × 5–9cm, apex acuminate, base cuneate, margin hairy, lower surface white; inflorescence cylindrical, c. 12–18 × 3.5cm; peduncle 10–50cm long; pedicels longer than flowers; flowers white; fruits red. Forest & forest-grassland transition; 663–700m.

Distr.: S Nigeria to S Sudan, Uganda & W Tanzania [lower Guinea & Congolian].
IUCN: LC
Mefou Proposed National Park: Etuge 5302 3/2004; **Ndanan 1:** Cheek 11001 10/2002; 11234 10/2002.
Local Name: Etotogo (Ewongo) (Cheek 11234).

Pollia condensata C.B.Clarke
Erect herb with aerial stem, 0.3–1m; leaves narrowly elliptic as oblanceolate, c. 20 × 6cm, long acuminate, subsessile; inflorescence terminal, dense, ovoid, c. 2 × 1.5cm; flowers white, inconspicuous; fruits iridescent blue, hard, spherical, c. 4mm diam. Lowland forest & stream banks; 600–700m.
Distr.: Sierra Leone to Tanzania [Guineo-Congolian].
IUCN: LC
Mefou Proposed National Park: Etuge 5184 fl., 3/2004; **Ndanan 1:** Harvey 122 10/2002.

Pollia mannii C.B.Clarke
Sprawling herb, 30cm; leaves elliptic, c. 7 × 2.5cm, acuminate; petiole 1cm; inflorescence terminal, diffuse c. 6cm, 6–8 1-flowered branches; flowers white, actinomorphic; fruits hard, glossy, grey, 6mm. Lowland forest; 700m.
Distr.: Ivory Coast to Cameroon, São Tomé, Uganda & W Tanzania [Guineo-Congolian].
IUCN: LC
Ndanan 1: Gosline 426 10/2002.

Polyspatha paniculata Benth.
Erect herb, 10–30cm; leaves elliptic, to 12 × 6cm, acuminate; petiole 0.5cm; inflorescence terminal, single, elongate, 6–20cm; spathes 4–10, c. 7 × 5mm; flowers white. Forest; 700–710m.
Distr.: Guinea (Conakry) to Uganda [Guineo-Congolian].
IUCN: LC
Ndanan 1: Cheek 11264 10/2002; 11778 3/2004; Tadjouteu 588 3/2004; **Ndanan 2:** Cheek 11148 10/2002.
Local Name: Soro (Ewondo) (Cheek 11148).

Stanfieldiella imperforata (C.B.Clarke) Brenan var. *imperforata*
Straggling herb, 17–60cm tall; decumbent or stoloniferous below and rooting from lower nodes; leaves glabrous; leaf-blades 13 × 4cm, elliptic or narrowly so, ± purplish on lower surface, usually ± long-ciliate on "petiole" and margin of lamina near base; inflorescences lax, diffuse, 2–6cm long; outside of sepals and pedicels usually pubescent with gland-tipped hairs 0.5–1mm long; flowers with small, white petals; capsules exceeding sepals; seeds smooth. Lowland to submontane forest; 700–710m.
Distr.: Sierra Leone to Ethiopia, Uganda & Tanzania [Guineo-Congolian].
IUCN: LC
Ndanan 1: Cheek 11819 3/2004; Harvey 137 10/2002; **Ndanan 2:** Cheek 11099 10/2002.

Stanfieldiella oligantha (Mildbr.) Brenan var. *oligantha*
Straggling herb; leaves glabrous; inflorescence dense, branched, c. 1.5cm long; capsules exceeding sepals; seeds verrucose. Lowland & submontane forest; 605–710m.
Distr.: Liberia to Cameroon [upper & lower Guinea].
IUCN: LC
Mefou Proposed National Park: Cheek 13113 2/2006; Cheek in RP 63 3/2004; **Ndanan 1:** Cheek in MF 261 10/2002.
Note: all the above specimens have not been verified at Kew, Harvey viii.2010.

COSTACEAE

P.J.M. Maas (U), H. Maas (U), C. Specht (NY) & M. Cheek (K)

Fl. Cameroun 4 (1965).

Costus dinklagei K.Schum.
Fl. Gabon 9: 76 (1964); Fl. Cameroun 4: 76 (1965); Adansonia (Sér. 2) 7: 78 (1967).
Herb; inflorescences from ground; leafy stems erect, spirally twisting, 1.5–2.5m, pilose; leaf-blades alternate, elliptic-oblong, c. 14 × 4.5cm, apex long and tapering, base cuneate, softly hairy below; pseudopetiole 3mm, brown pilose; ligule sheath brown, 5mm, hairy; inflorescence cone 15cm; peduncle covered in bracts, 6–7cm; flowers pink with dark base; corolla 4.5 × 3.5cm. Forest; 600–700m.
Distr.: Nigeria and Cameroon [lower Guinea].
IUCN: NT
Mefou Proposed National Park: Etuge 5206 3/2004; **Ndanan 1:** Cheek 11961 3/2004; **Ndanan 2:** Darbyshire 276 3/2004.

Costus dubius (Afzel.) K.Schum.
Syn. *Costus albus* A.Chev. ex J.Koechlin Fl. Cameroun 4: 78 (1965).
Perennial herb, to 1.5(–2)m; inflorescence from ground; leafy stems spirally twisting, 1.5m, glabrous; leaves obovate-elliptic, c. 25 × 6cm, long-acuminate, base cuneate, appressed pubescent; pseudo-petiole 1cm; ligule sheath 5mm, brown, nerves raised; inflorescences terminal or on short lateral shoots at the base of the leafy stems, cones 20cm, flowers numerous; peduncle 15cm, bract-covered; flowers white with yellow throat. Forest; 700m.
Distr.: tropical Africa.
IUCN: LC
Ndanan 1: Cheek 11219 10/2002.

Costus lucanusianus J.Braun & K.Schum. var. *lucanusianus*
Terrestrial herb, 2–3m; inflorescence terminates the spirally leafy stems; leaves alternate, narrow elliptic, c. 25 × 10cm, apex long, tapering, base acute, lower

220

surface of leaves appressed pubescent, with hairy rim, hairs c. 5mm; pseudopetiole 1cm; ligule sheath brown, 2mm, glabrous, united to the leaf-sheath by a long-hairy raised ridge; capitulum 10cm, globose; calyces much longer than bracts; flowers orange, red and yellow, 4cm. Forest & farmbush, roadsides; 700m.
Distr.: Sierra Leone to Bioko, Cameroon, Gabon & Uganda [Guineo-Congolian (montane)].
IUCN: LC
Ndanan 1: Cheek 11025 10/2002; Cheek in MF 350 10/2002.

Paracostus englerianus (K.Schum) C.D.Sprecht
Taxon 55(1): 162 (2006).
Syn. *Costus englerianus* K.Schum.
Terrestrial herb, 5m; stoloniferous; leaves leathery, glossy, held at 45°, broadly elliptic, 12 × 8cm; flowers white, centre yellow. Forest; 600–710m.
Distr.: Sierra Leone to Gabon [upper & lower Guinea].
IUCN: LC
Mefou Proposed National Park: Etuge 5240 3/2004;
Ndanan 1: Cheek 11225 10/2002; 11893 3/2004.
Local Name: Soro (Ewongo) (Cheek 11225).

CYPERACEAE

M. Xanthos (K), D.A. Simpson (K) & B. Tolley (K)

Cyperus cylindrostachys Boeckeler
Linnaea 36: 383 (1870).
Green, or glaucescent rhizome, tuberous short or rarely horizontal; culms solitary, 15–30cm high, fairly rigid to slender, triangular, sometimes compressed, many leaved towards base; upper leaves mostly shorter, longer acuminate, 1.5–6.5mm wide with a margin, keel and adaxial surface scabrid; involucre 8–7 leafy bracts, rarely 5–7, patent, very equal; inflorescence an umbel, 7–14 rayed; rays fairly robust, sometimes setaceous, erect, patent, straight, 1.5–2.5cm long; spikes cylindrical, obtuse, 11–21 × 8–10.5mm; spikelets many, densely crowded, eventually very patent or reflexed, linear-oblong, acuminate, 2–3-flowered; bracts of spikelet minute, lanceolate, setaceous acuminate. Damp, grassy places; 600–700m.
Distr.: Senegal, Gambia, Mali, Sierra Leone, Liberia, Ivory Coast, Ghana, Togo, Nigeria, Cameroon [upper & lower Guinea].
IUCN: LC
Mefou Proposed National Park: Etuge 5155 3/2004;
Ndanan 1: Harvey 70 10/2002.

Cyperus dilatatus Schumach. & Thonn.
Erect herb, 0.5m, glabrous; culms in clumps; leaves 4; sheaths 4–5cm; blade c. 15cm × 2mm; inflorescence umbellate; bracts leafy, 3, unequal, longest 20cm × 3mm; spike clusters c. 8, stalks 0–8cm; spikes 2–6 per cluster, each cylindric to 7 × 1.5cm, scales concave, brown, midrib 2-ribbed, acumen hairy. Grassland; 600–710m.

Distr.: tropical Africa.
IUCN: LC
Mefou Proposed National Park: Etuge 5162 3/2004;
Ndanan 1: Cheek 11781 3/2004.

Cyperus mannii C.B.Clarke
Stout perennial to 2m; culms tufted or with a short rhizome; leaves often well-developed, to 1m or more; inflorescence anthelate, to c. 25 × 30cm, thrice-branched bearing small clusters of spikelets; inflorescence branches grooved, each with a conspicuous basal prophyll 1–2cm; involucral bracts leafy to 50 × 2cm; spikelets red-brown. Forest, forest edges, forest paths; 710m.
Distr.: Sierra Leone to Bioko, W Cameroon [upper & lower Guinea].
IUCN: LC
Ndanan 1: Cheek 11774 3/2004.

Cyperus reduncus Hochst. ex Boeckeler
Medium-sized annual, with solitary or few crowded stems, 12–35cm tall, triangular, scabrid; leaf blades 10–25cm long, flat with prominent longitudinal ribs, scabrid on margin and midrib, sheaths green to light brown; inflorescence an open anthela, 3–12cm wide, and with 4–6 leafy bracts; anthela with 4–12 main branches, each with one digitate cluster of spikelets and with or without additional spikelet-clusters on secondary peduncles, more rarely tertiary peduncles and spikelet clusters; spikelets 2–10 × 2.5–4mm, light brown with spreading curved glumes, 5–25 flowered; glumes 1.5–2.5mm long, strongly curved, reddish brown to greenish yellow with wide uncoloured margin, midrib 3-nerved, excurrent in a curved mucro. In seasonally wet grasslands and temporary pools and swamps; 679m.
Distr.: Senegal to S Nigeria, Congo (Kinshasa), Chad, Ethiopia, Uganda [tropical Africa].
IUCN: LC
Ndanan 1: Onana 2903 3/2004.

Cyperus tenuiculmis Boeckeler var. ***schweinfurthianus*** (Boeckeler) Hooper
Medium sized to robust perennial, with a rather thick creeping rhizome and swollen stem base; culms 30–80(–100)cm tall, triangular, usually scabrid; leaf-blades 10–40cm × 2–4(–8)mm, flat, scabrid on margin and major ribs, leaf-sheaths green or brown; inflorescence 5–20(–35)cm long with 3–15cm wide anthela consisting of one sessile and 2–6 stalked spikes, major peduncles (rays) up to 15cm long (in robust specimens up to 25cm long); inflorescence bracts leafy, erect or spreading, the largest 8–20cm × 1.5–4mm; spikes 1.5–3.5 × 1.5–4cm with 3–12 erect or spreading bracts; spikelets 8–30 × 1.3–2mm, linear-lanceolate, golden or yellowish brown, 6–20 flowered; glumes 3–4mm long, ovate, yellowish-brown, midrib green distinctly excurrent. In seasonally wet grassland, swamp-edges and on roadside banks; 680m.

Distr.: tropical Africa.
IUCN: LC
Ndanan 2: <u>Darbyshire 236</u> 3/2004.

Hypolytrum heteromorphum Nelmes

Stout, rhizomatous, perennial herb; culms slender, 25–40cm × 1–2mm; leaves densely set near base, 30–60 × 1.0–1.5cm, flat or ± plicate, margin slightly toothed; inflorescence a series of panicles, broader than tall; panicles of a few closely-set major branches arising from the axils of the main bracts; bracts greyish-brown, to 3.5cm; lowest branches to 2.5cm, with 2–6 sessile spikes at the tip; spikes 1.0–1.5 × 0.1–0.2mm, brown. Wet forest; 600–710m.
Distr.: tropical Africa [afromontane].
IUCN: LC
Mefou Proposed National Park: <u>Etuge 5175</u> 3/2004;
Ndanan 1: <u>Cheek 11938</u> 3/2004.

Hypolytrum heterophyllum Boeckeler

Beitr. Cyper. 1: 22 (1888).
Slender perennial, with thin, elongated rhizome; culms narrow, slender, 30cm × 2mm diam., triangular, scabrid; leaves of two forms; basal leaves many, adpressed to the sheath, short dense, lanceolate; upper leaves erect, herbaceous-membranous; sheaths linear, short acute, margins serrate; inflorescence a compound corymb, hemispherical, plurimose; spikelets shortly pedunculate, oblong-ovate obtuse when mature. Roadside; 710m.
Distr.: Liberia, Sierra Leone, Ivory Coast, Ghana, Nigeria, Cameroon, Equatorial Guinea, Gabon, Congo (Brazzaville), Congo (Kinshasa), Angola [upper & lower Guinea].
IUCN: LC
Ndanan 1: <u>Cheek 11520</u> 3/2004.

Hypolytrum pynaertii (De Wild.) Nelmes

Kew Bull. 10: 81 (1955).
Rather week perennial, to 75cm; culms erect to somewhat curved with brownish, strongly nerved, almost bladeless sheaths covering the base; leaf blades slightly to very much longer than the stems, 1–1.5cm wide, ± flat; inflorescence composed of numerous (25–35) spikes aggregated into a dense hemispherical bracteate head, 1–1.75 × 1.5–2.5cm, but looking ± globose in fruit and 2–2.5cm wide; bracts 3 or more, scarcely subfoliaceous, 2cm × 2–3mm, lanceolate or linear-lanceolate; glumes with apical membranous margins, about 3mm long, reddish-brown above, pale below with reddish-brown streaks. Swamp forest; 600m.
Distr.: Cameroon, Gabon & Congo (Kinshasa) [lower Guinea & Congolian].
IUCN: NT
Mefou Proposed National Park: <u>Etuge 5176</u> 3/2004.

Kyllinga nemoralis (J.R. Forst. & G. Forst.) Dandy ex Hutch. & Dalziel

Very leafy perennial, with slender, creeping rhizomes; culms usually distant, 10–25cm × 0.8 × 1.5mm, triangular, ridged, without swellings at its base; leaves 10–35 × 2–5mm, dark green, flaccid, flat, with small spine-like teeth on margin and midrib, on upper surface often with hairs on the secondary nerves as well; leaf sheaths brownish, the basal ones without blade; inflorescence a globose or slightly ovate head of a single spike, 3–8 × 3–8mm; involucral bracts 3–4, leafy, the longest 8–20cm; spikelets 2–2.5mm, with 1 flower only; glumes whitish but fading pale reddish-brown, acuminate, of equal length but not width, with 2–4 veins on each side of midrib. Deep or partly shaded forest floors; 710m.
Distr.: pantropical.
IUCN: LC
Ndanan 1: <u>Cheek 11818</u> 3/2004.
Note: *Kyllinga nemoralis* is a name accepted in the World Checklist of Cyperaceae (Govaerts *et al.* 2007) and F.W.T.A., but is a synonym of *Cyperus kyllingia* in The Sedges & Rushes of East Africa (Haines R. & Lye K., The Sedges & Rushes of E Africa: 247 (1983), from where this description has been taken, Xanthos ix.2010.

Mapania mannii C.B.Clarke subsp. *bieleri* (De Wild.) J.Raynal ex D.A.Simpson

Simpson D.A., Rev. of genus *Mapania*: 142 (1992).
Moderately robust, rhizomatous to stoloniferous perennial; culms solitary, erect, central, 16.5–49cm × 1.5–2.8mm, subtriqetrous, glabrous, green or reddish; leaf blades linear-elliptic, elliptic or oblong-elliptic, 10.5–37 × 2.2–6cm, apex narrowed, acute to broadly obtuse, base abruptly narrowed into pseudopetiole; involucral bracts 2(–3), 0.5–35 × 0.3–5cm, basal bract longest, ± sheathing, lower bracts foliaceous; inflorescence pseudolateral, globose, 2.1–4.5cm wide, composed of up to 30 spikes; spikes linear to elliptic, 0.8–1.2cm long, acute, distinct; spicoid bracts, lanceolate, 3.8–5.2 × 1.2–2mm, floral bracts 4, free 3.2–4.9 × 0.4–0.8mm, keel of lower two floral bracts wingless, hispid. Forest; 660m.
Distr.: Cameroon, Gabon, Congo (Kinshasa) [lower Guinea].
IUCN: NT
Ndanan 2: <u>Darbyshire 278</u> fl., 3/2004.

Rhynchospora corymbosa (L.) Britt.

A coarse, leafy, caespitose perennial; rhizome thick, creeping with numerous soft rootlets; culms 0.6–2.5m, to about 1cm across towards the base; leaves numerous, densely-crowded, the basal ones 50–100 × 0.1–0.2cm, very tough, with minute spinose teeth on the margin and midrib; inflorescence of one terminal and several lateral corymbs; involucral bracts leafy, 10–50 × 0.5–2.0cm; spikelets 6–10 × 1.5–2.5mm, in pedunculate clusters; glumes reddish brown. Lake shores, river-banks, and in shallow pools, tolerant of light shade; 600m.
Distr.: pantropical.
IUCN: LC
Mefou Proposed National Park: <u>Etuge 5177</u> 3/2004.

Scleria verrucosa Willd.

A tall, robust perennial, with thick creeping rhizome; culms 80–150cm tall, glabrous or scabrid; leaves 60 × 1–3cm, plicate, glabrous or hairy, scabrid on the margin and edges; inflorescence consisting of one terminal and 3–5 lateral panicles, all subtended by leafy bracts; male spikelets 4–4.5mm long, glumes reddish brown, usually minutely hairy on and near the obscure midrib; female spikelets 5–8mm long, glumes reddish brown or straw-coloured with numerous reddish brown steaks or patches, usually glabrous but minutely hairy on the margin, the glumes falling with the nutlet; nutlet about 3 × 2.5–3.45mm, sparsely to densely verrucose, often on longitudinal ridges, and with reddish hairs on the warts, whitish with a yellow tinge or pale brown. Swamp, riverine, or other wet forest; 600–710m.
Distr.: Senegal to Congo (Kinshasa), Uganda and Tanzania [Guineo-Congolian].
IUCN: LC
Mefou Proposed National Park: Etuge 5154 3/2004; **Ndanan 1:** Cheek 11771 3/2004; **Ndanan 2:** Darbyshire 297 fr., 3/2004.

DIOSCOREACEAE

P. Wilkin (K) & Seren Thomas (K)

Dioscorea bulbifera L.

Herbaceous climber, glabrous, to 3–7m; stems left-twining (sinistrorse); leaves alternate with a pair of membranous semicircular lateral projections clasping stem at petiole base, apex short-acuminate, not thickened. Farmbush; 700m.
Distr.: Senegal, Guinea (Conakry), Sierra Leone, Ivory Coast, Burkina Faso, Ghana, Nigeria, Cameroon [palaeotropics].
IUCN: LC
Ndanan 1: Cheek 11004 10/2002; 11074 10/2002; 11075 10/2002; Cheek in MF 356 10/2002.
Note: warty bulbils usually present.
Local Name: Bam (Ewondo) (Cheek 11074).

Dioscorea praehensilis Benth.

Sturdy climber, to 8m; single tuber per growing season; stems right-twining (dextrorse), prickly, glabrous; leaves opposite, chartaceous, not leathery, shortly cordate, acuminate to long-acuminate. Forest & forest edge; 700m.
Distr.: Sierra Leone to Cameroon & E Africa, S to Zambia, Zimbabwe, Malawi & Mozambique [tropical Africa].
IUCN: LC
Ndanan 1: Cheek 10999 10/2002; Cheek in MF 301 10/2002.

Dioscorea preussii Pax

Robust non-spiny climber, to 10m; stems left-twining, often 6-winged, subglabrous, with a few ± caducous, medifixed (T-shaped) hairs, also on the inflorescence and leaf apices; leaves alternate, broadly ovate, obliquely acuminate, deeply cordate, 10–30 × 8–35cm, villous-tomentose beneath. Forest and farmbush; 700m.
Distr.: Senegal to Cameroon, Gabon, Congo (Kinshasa), CAR, Uganda, Angola, Mozambique [tropical Africa].
IUCN: LC
Ndanan 1: Cheek 11072 10/2002; Cheek in MF 306 10/2002.

Dioscorea smilacifolia De Wild.

Perennial climber, to c. 4m; stems right-twining; rootstock with multiple tubers; leaves opposite, ovate, ovate-lanceolate or ovate-oblong, 6–10 × 4–7cm, leathery, 3-nerved, veins displaced towards margin. Forest gaps; 600–700m.
Distr.: Senegal to Uganda [Guineo-Congolian].
IUCN: LC
Mefou Proposed National Park: Nana 67 3/2004; **Ndanan 1:** Cheek 12003 3/2004.
Note: *D. smilacifolia* and *D. minutiflora* may overlap.

Dioscorea sp. aff. praehensilis Benth.

Vine, climber; stems right-twining, angular, light brown, glabrous; leaves opposite, sagittate in outline, 10.4–13.9 × 5–6.7cm, apex attenuate, base cordiform, basal lobes rounded, symmetrical and 2–2.5cm long, margin entire, 7–9 pairs of lateral veins, sparsely hairy on both surfaces; petiole 4–5cm long, glabrous. Forest; 700m.
Ndanan 1: Cheek in MF 360 10/2002.

DRACAENACEAE

G. Mwachala (EA)

Dracaena camerooniana Baker

Shrub, 0.3–8m; stems with pseudowhorls of leaves; leaves obovate, 5–33 × 1–8.5cm, base cuneate or attenuate, apex acuminate; petiole 1–4cm; inflorescence pendent, 5–50cm long; pedicels to 5mm, articulated above middle; flowers white (or with purple), 1.9–3cm; fruits orange or red, globose or depressed globose, 7–21mm diam.; seeds straw-coloured, hemispherical, 4–11 × 5–4 × 2–9mm. Forest; 600m.
Distr.: W Africa from Guinea to CAR, Congo (Kinshasa) and S to Angola and Zambia [tropical Africa].
IUCN: LC
Mefou Proposed National Park: Etuge 5170 3/2004.

Dracaena laxissima Engl.

Shrub, 1.5m, glabrous; stems 3–4mm diam., pale brown, striate; internodes 2–3cm; leaves evenly spread, alternate, elliptic, c. 15 × 5cm, acumen 1cm with a 5mm filament tip, base acute, lateral nerves c. 30 pairs, parallel; petiole 1cm, well-defined, base 9/10 amplexical; inflorescence terminal, paniculate, 15cm, diffuse, pendent, with c. 6 patent lateral branches each c. 5cm long. Forest; 710m.

Distr.: Nigeria to E Africa, Sudan, Zambia and Malawi [tropical Africa].
IUCN: LC
Ndanan 1: Cheek 11730 3/2004.
Note: can be scandent or scrambling, Mwachala xi.2010.

Dracaena mannii Baker

Tree, to 4m; stem and branches with prominent leaf scars; leaves clustered at ends of erect side branches; leaves narrowly oblong-elliptic, to 40 × 2cm, glabrous, apex acute, base attenuate, sessile; inflorescence a panicle of racemes, <0.5m, rachis yellow-green; pedicels to 4mm; flower buds to 4mm, green with red-brown tips; fruits not seen. Farmbush; 700m.
Distr.: Senegal to South Africa [tropical Africa].
IUCN: LC
Ndanan 2: Darbyshire 227 fl., 3/2004.

Dracaena ovata Ker Gawl.

Shrub, to 1.2m; stems with pseudowhorls of leaves; bracts 7–25mm, leaving a prominent circular scar around stem; leaves ovate-elliptic, 8–11 × 2–3cm, glabrous, apex acuminate, mucro to 1.5mm, base attenuate; petiole to 1cm; flowers crowded into a short subsessile terminal subspicate raceme, c. 12 flowers per cluster; pedicels to 3mm; flowers white, to 2.2cm in bud; fruits not seen. Riverine forest; 650m.
Distr.: Sierra Leone to Congo (Kinshasa) [Guineo-Congolian].
IUCN: LC
Ndanan 2: Darbyshire 313 fl., 3/2004.

Dracaena viridiflora Engl. & K.Krause

Shrub, to 3m; stems long, slender; leaves linear, c. 20 × 3cm, sheathing and amplexicaul at base, lateral nerves 8–10 pairs; inflorescence spikes terminal and axillary, to 10cm; flowers clustered; bracts ovate, inconspicuous; fruit globose, green. Forest; 700–710m.
Distr.: SE Nigeria & Cameroon [lower Guinea].
IUCN: NT
Ndanan 1: Cheek 11637 3/2004; **Ndanan 2:** Cheek 11170 10/2002.

GRAMINEAE

T. Cope (K), M. Xanthos (K), M. Cheek (K), E. Fenton (K) & X. van der Burgt (YA)

Acroceras amplectens Stapf

F.T.A. 9: 625 (1920); F.T.E.A. Gramineae pt. III: 565 (1982).
Annual herb, stoloniferous, with spindly branches, 30–60cm; leaves linear, 6–10cm by 5–10mm, base cordate; inflorescence in 2–5 racemose spikes, 3–10cm on an axis of 10–25cm; rachis glabrous; spikelets in pairs, one subsessile, the other on a pedicel 4–8mm, each pair overlaps the other, spikelets 5.5–6mm, glabrous; glumes not separated by an internode; lower glume 4mm, more or less acuminate; upper glume and lower lemma the same length as the spikelet, apex acuminate; upper lemma hard, the edges curled, clasping the palea. *Raphia* swamp, farmbush; 660–710m.
Distr.: Senegal, Gambia, Mali, Guinea (Conakry), Sierra Leone, Ivory Coast, Burkina Faso, Ghana, Togo, Niger, Nigeria, Cameroon, Congo (Kinshasa), Central African Republic, Burundi, Sudan, Uganda & Tanzania [tropical Africa].
IUCN: LC
Mefou Proposed National Park: Cheek in RP 6 3/2004; **Ndanan 2:** Darbyshire 321 fl., 4/2004.

Centotheca lappacea (L.) Desv.

Annual rhizomatous, with erect culms to 1m; leaf-blades lanceolate, asymmetric, 10–25 × 1.5–3cm, acuminate; hairs at base of leaf-blade and on sheath; inflorescence an open panicle, 15cm, lateral branches erect then spreading; spikelets lanceolate, 4–8mm, terete or slightly compressed; lemmas emarginate and mucronate at the tip, the lowest glabrous, the upper with reflexed tubercle-based bristles; short pedicels. Semi-forested land, undergrowth, fallow clearings in forest; 650m.
Distr.: pantropical but not E Africa.
IUCN: LC
Ndanan 1: Tadjouteu 632 4/2004.

Eleusine indica (L.) Gaertn.

Annual, prostrate at base, culms erect, up to 50–60cm; leaf-blades often folded, 5–35cm × 3–6mm; sheaths carinate; inflorescence spicate, 1–10 ascending racemes, 3–15cm × 3–7mm; spikelets in 2 rows, sessile, narrowly elliptic, 4–8mm, green or brownish-green; glumes unequal, lemmas 2.7–3mm long, narrow, membranous, keeled, longer than glumes; grain concealed within floret. Nitrophile of waste ground, roadsides; 660m.
Distr.: pantropical.
IUCN: LC
Ndanan 2: Darbyshire 320 fl., 4/2004.

Isachne kiyalaensis Robyns

Weakly ascending annual herb, to 45cm; leaf blades lanceolate, 3–6cm long, pubescent at the base and on the leaf sheath; ligule a line of hairs; inflorescence an open panicle (3–)6–7 × 1–1.5(–2.5)cm; spikelets solitary, fertile spikelets pedicelled, c. 1.5mm long, 2-flowered; florets disarticulating above the persistent glumes; glumes similar, lower glume 1.5mm long, membranous, elliptic, upper glume 1.5mm long, membranous, elliptic; lemmas dissimilar, lower lemma male, 1.3mm long, membranous, elliptic, lemma margins involute, palea membranous 0.9 the length of the lemma, upper lemma fertile 1–1.5mm long, coriaceous, obovate-broadly elliptic, puberulous; palea 1mm long, coriaceous; anthers 3, brown, 0.7mm long. *Raphia* swamp, forest; 700–710m.
Distr.: Guinea to Nigeria, Congo (Kinshasa), Angola [Guineo-Congolian].

IUCN: LC
Mefou Proposed National Park: Cheek in RP 21
3/2004; 68 3/2004; **Ndanan 1:** Cheek in MF 267
10/2002.

Leptaspis zeylanica Nees ex Steud.
Agric. Univ. Wag. Papers 92(1): 39 (1992).
Syn. *Leptaspis cochleata* Thw.
Perennial, to 1m; stoloniferous, trailing, rooting at lower
nodes, culms erect; leaf-blades oblong–oblanceolate,
10–35cm × 2.5–6cm, sharply pointed at tip,
asymmetrical; sheaths longer than internodes, imbricate;
inflorescence a panicle, 45–60cm; branches
subverticillate, lower branches up to 20cm, 2–3 branches
in a whorl, bearing short side branches with 1 male and
1 female spikelet, the male above; male spikelets
1–flowered, finely pubescent, the lemma 4mm long, the
glumes half as long; stamens 6; female spikelets
4–6mm, 1–flowered, reddish, lemma inflated,
conchiform, closed except for a small hole near the
apex, covered in hooked hairs. Undergrowth of dense,
humid forest, edges of fields & roads; 600–720m.
Distr.: tropical Africa & Asia [palaeotropics].
IUCN: LC
Mefou Proposed National Park: Cheek in MF 111
3/2004; Etuge 5153 3/2004; Nana 49 3/2004; **Ndanan
1:** Cheek 11538 3/2004; Cheek in MF 412 10/2002;
Onana 2820 3/2004; **Ndanan 2:** Cheek 11146 10/2002;
Darbyshire 251 fl., 3/2004.

Olyra latifolia L.
Perennial, 3–4 m, climbing or erect, often woody,
glabrous or pubescent at nodes; leaf-blades 5–20 ×
2.5–7cm, ovate, assymetric at base with short false
petioles, acuminate; short ligule; inflorescence a
panicle, 7–25cm, pyramidal; rachis scabrous; male
spikelets at base of female; male spikelets on filiform
pedicels, glumes rudimentary, lemma lanceolate, 4mm,
awned; female spikelets ovate, on claviform pedicels,
purplish; glumes 8–20mm; lemma and palea 4–6mm,
white, shiny; awns 3–4mm. Edges of clearings, slopes
in dense humid forest; 700–710m.
Distr.: tropical Africa and America [amphi-Atlantic].
IUCN: LC
Ndanan 1: Cheek 11689 3/2004; **Ndanan 2:** Cheek
11120 10/2002.
Note: Cheek 11689 is tentatively placed in this taxon,
specimen sterile, Cope ix.2005.
Local Name: Ekogze'h (Cheek 11120).

Panicum laxum Sw.
Annual herb, stoloniferous, 30–50cm, crimson at the
base; leaves 8cm by 2–5mm; panicle 15–17cm,
branches have simple, acute hairs or are glabrous,
secondary branches are very short, 3–7mm, with many
spikelets; spikelets are elliptic, glabrous, 1–1.2mm,
often crimson; lower glume 1/3 the length of the
spikelet, 1(–3)-nerved; upper glume 3(–5)-nerved;
lower lemma 3-nerved, palea the same length; upper

lemma finely streaked, tough; Forest; 550–660m.
Distr.: Mauritania, Guinea (Conakry), Sierra Leone,
Liberia, Ivory Coast, Burkina Faso, Ghana, Nigeria,
Cameroon, Gabon, Ascension Island, South Africa
(Cape), Mexico, Guatemala, British Honduras,
Honduras, El Salvador, Nicaragua, Costa Rica, Puerto
Rico, Cuba, Jamaica, Hispaniola, Trinidad, Guyana,
Surinam, Brazil, Paraguay, Panama, Colombia,
Venezuela, Ecuador, Peru, Bolivia, Galapagos,
Argentina, Uruguay [amphi-Atlantic].
IUCN: LC
Mefou Proposed National Park: Etuge 5201 3/2004;
Ndanan 2: Darbyshire 265 3/2004; 322 4/2004.

Paspalum conjugatum Berg
Perennial; stoloniferous; culms erect, 20–80cm; leaf-
blades linear to narrowly lanceolate, 4–20cm ×
5–12mm, slightly ciliate at edges; ligules hairy;
inflorescence 2 racemes, more or less joined, 5–15cm;
short pedicels; spikelets orbicular, 1.5–1.7mm, yellow-
green, closely appressed to the rachis of paired slender
racemes; ciliate fringe from margins of upper glume,
inner glume missing. Damp places in forest clearings,
grassland & rough ground; 660m.
Distr.: pantropical.
IUCN: LC
Ndanan 2: Darbyshire 323 fl., 4/2004.

Paspalum paniculatum L.
Perennial, rhizomatous, erect, 30–100cm; leaf-blades
10–45cm × 6–20mm, sheath carinate; ligule hyaline,
with hairs 5–10mm; inflorescence numerous fastigiate
racemes 4–11cm on 5–15cm rachis; long hairs on
branches of racemes; spikelets orbicular to broadly
elliptic, conspicuously plano–convex, dark brown,
1.3–1.4cm, pubescent, inner glume missing; anthers
white; stigmas crimson. Fallow in humid forests;
650–710m.
Distr.: Liberia to Angola and Uganda, also in
Polynesia, S America, New Guinea, Australia
[pantropical (montane)].
IUCN: LC
Ndanan 1: Cabral 105 3/2004; Cheek 11695 3/2004;
Darbyshire 282 fl., 3/2004; Harvey 201 10/2002;
Tadjouteu 625 4/2004.

Sorghum arundinaceum (Desv.) Stapf
Syn. *Sorghum vogelianum* (Piper) Stapf
Herb, annual or perennial, often robust, 0.3–5m, nodes
glabrous or softly pubescent; stilt-rooted; leaves often
large, 5–75 × 5–7cm, nerves often yellow; ligule
membranous, 2–3mm; panicle linear to open/loose,
10–60cm; grouped branches carry the articulated racemes,
fragile, with 2–7 pairs of spikelets; rachis filiform;
internodes and pedicels c. 4mm with dense tawny-
coloured hairs; spikelets in pairs, triads at apex; spikelets
yellowish or red-brown, sessile, lanceolate to oval, 4–9 ×
2–3mm, glabrous or with white pubescent hairs,
sometimes brown, lacking the filiform continuation of the

lemma or with a caudate awn (5–30mm); lower glume forked at the apex and hairy on the upper half, often black at maturity; spikelet pedicels linear to lanceolate, male or reduced, smaller if the spikelets are sessile, purplish-green. *Raphia* swamp, secondary forest; 600m.
Distr.: subsaharan Africa to Mauritius, Madagascar and the Comores [tropical & subtropical Africa].
IUCN: LC
Mefou Proposed National Park: Etuge 5253 3/2004.

Sporobolus tenuissimus (Mart. ex Schrank) Kuntze
Annual herb in small tufts; culms fragile, 15–20cm; leaves linear-ligulate; ligule ciliate, very short; panicle oblong, 5–25 × 2–6cm, delicate with pubescent branches & pedicels; spikelets 0.8–1mm, open at maturity, often dark grey; lower glume oblong; upper glume oval-oblong, sub-acute, 0.3–0.5mm; lemma oval, as long as the spikelet; anthers 3, 0.1–0.2mm; grain obovoid, 0.4–0.6mm. Village weed; 660m.
Distr.: Cameroon, Gabon, CAR, Congo (Kinshasa), Burundi, Uganda, Kenya, Tanzania, Zanzibar, Pemba, Mozambique [tropical & subtropical Africa].
IUCN: LC
Ndanan 2: Darbyshire 332 fl., 4/2004.

Streptogyna crinita P. Beauv.
F.T.E.A. Gramineae pt. 1: 23 (1970).
Perennial herb, clumps, 0.3–1.5m; leaves lanceolate, to 40 × 3.5cm, multinerved, shrunk at the base in a false petiole; ligule hairy, 2–3mm; raceme 10–40cm; pedicels 1–2mm; spikelet 2–3mm, all similar, stem spindly, contorted between the flowers; lower glume 3–6mm; upper glume 15–20mm; flowers 4–7 of which the lower are fertile; fertile lemmas softly pubescent at the base, oblong-lanceolate, 15–25mm; stigmas 2, 1–2cm. Dense humid forest, mostly secondary forest and on forest edges; 600–710m.
Distr.: W Africa to Congo (Kinshasa), Sudan, Uganda, Tanzania, India & Sri Lanka [tropical Africa, India & Sri Lanka].
IUCN: LC
Mefou Proposed National Park: Etuge 5223 3/2004; **Ndanan 1:** Cheek 11810 3/2004; Cheek in MF 300 10/2002; Tadjouteu 631 4/2004; **Ndanan 2:** Darbyshire 326 fl., 4/2004.

MARANTACEAE

J.-M. Onana (YA), H. Fortune-Hopkins (K) & M. Cheek (K)

Fl. Cameroun 4 (1965).

Ataenidia conferta (Benth.) Milne-Redh.
Erect herb, to c. 1.5m; branching; leaves to 40 × 20cm; calloused portion of petiole c. 2–4cm; inflorescence lateral, internodes about 1mm; bracts bearing sickles of cymes c. 2.5cm long; flowers white to pink; ovary densely long-pubescent; fruit ellipsoid, c. 10 × 6–7mm, pubescent towards the apex; perianth persistent. Forest; 700–710m.

Distr.: Ivory Coast to Cameroon, Gabon, CAR, Congo (Brazzaville), Congo (Kinshasa), Angola (Cabinda), CAR, Sudan and Uganda [Guineo-Congolian].
IUCN: LC
Ndanan 1: Cheek 11617 3/2004; Harvey 165 10/2002; **Ndanan 2:** Cheek 11130 10/2002.

Halopegia azurea (K.Schum.) K.Schum.
Erect herb to c. 3m, forming dense clumps; leaves several, to 50 × 15cm, linear-oblong, the sides ± parallel; apex long-acuminate; petiole to 1m, the calloused portion 2–5cm; inflorescence of 1–3 racemes, to 20–25cm long; axes pubescent; bracts to 2.5–3.5cm long, each bearing two cymes; flowers appearing 1 or 2 at a time, a decorative purple with blue and yellow staminodes exserted, 2.5cm long; ovary pubescent; fruits cylindrical, 11–12 × 3–4mm. Forest, by streams; 700–710m.
Distr.: Sierra Leone to Cameroon, Gabon, Congo (Kinshasa), Angola (Cabinda) [Guineo-Congolian].
IUCN: LC
Ndanan 1: Cheek 11768 3/2004; Harvey 99 10/2002.

Haumania danckelmaniana (J.Braun & K.Schum.) Milne-Redh.
Climbing herb, or scandent shrub, to 5m; stems slender, branching, with small prickles; leaf-blade ovate or oblong-elliptic, almost symmetric, 7–14.5 × 3–7cm, truncate or very broadly cuneate at base, acute to acuminate at apex, glabrous on both surfaces; petiole (above sheath, below calloused portion) absent, calloused portion 1–2cm; inflorescences pendulous, rachis zig-zag, to 10cm, pubescent; bracts subtending cymules sub-orbicular, 2 × 2cm, rounded at apex, white; corolla white; fruits 3-lobed, 2 × 3cm, ripening brown, covered by soft conical prickles. Semi-deciduous forest and by path; 700m.
Distr.: Cameroon, Equatorial Guinea, Gabon [lower Guinea].
IUCN: LC
Mefou Proposed National Park: Cheek in MF 38 3/2004; Nana 35 3/2004; **Ndanan 1:** Cheek 11030 10/2002.

Hypselodelphys scandens Louis & Mullend.
Lianescent herb to several metres long; branched; leaves elliptic or oblong-linear, 12–35 × 5–17cm, shortly acuminate, subtruncate; calloused portion of petiole above point of articulation to 3cm, conspicuously beaked at junction with midrib adaxially; inflorescence pendulous with a number of bifurcations, rachis glabrous or hirsute; branches usually zig-zag, c. 20cm long; internodes c. 1cm; bracts 3.5–4.5cm; flowers pale violet white and brown; fruit muricate, 3-lobed, c. 5 × 2–3cm; tubercles long, often curved, to 5mm; seeds not arillate. Secondary bush and *Raphia* swamp; 600–710m.
Distr.: Ivory Coast to Cameroon, Gabon, Congo (Kinhsasa) [Guineo-Congolian].

IUCN: LC
Mefou Proposed National Park: Etuge 5108 3/2003;
Ndanan 1: Cheek 11758 3/2004.

Marantochloa filipes (Benth.) Hutch.
A ± climbing herb, to 2.5m; leaves homotropic, ±
asymmetric, 7–15(–20) × 3–7cm; lamina often pruinose
abaxially, apex abruptly acuminate; petiole with
calloused portion c. 5mm long; inflorescence panicular,
very spindly, rachis simple or once branched; bracts
1.5–3.5cm; flowers white or pinkish; fruits smooth
(before drying), spherical with persistent perianth, c.
7mm diam., yellow, ripening red. Forest and stream
banks; 600–700m.
Distr.: Guinea (Conakry) to Bioko, Cameroon, Gabon,
Congo (Brazzaville), Congo (Kinshasa), Angola
[Guineo-Congolian].
IUCN: LC
Mefou Proposed National Park: Etuge 5138 fr.,
3/2004; **Ndanan 2:** Cheek 11139 10/2002.
Note: Cheek 11139 is tentatively placed in this taxon, it
could be *M. purpurea*, specimen missing, x.2010.

Marantochloa leucantha (K.Schum.) Milne-Redh.
Erect or climbing herb, to 5m; leaves homotropic,
15–20(–40) × 7–12(–25)cm, abruptly long-acuminate;
petiole with calloused portion c. 1cm long; inflorescence
of long loose panicles, much-branched, pendent, 30–40cm
long; flowers whitish; fruits smooth (before drying),
spherical, 1cm diam., reddish becoming yellowish on
drying; perianth not persistent. Forest; 600m.
Distr.: tropical Africa [afromontane].
IUCN: LC
Mefou Proposed National Park: Etuge 5293 3/2004.

Marantochloa purpurea (Ridl.) Milne-Redh.
Erect herb to 2.5m, branched; leaves 10–48 × 5–18cm;
petiole with calloused portion 1.5cm; inflorescence
loose, branched; bracts pale pink, subtending two, 2-
flowered cymes; peduncle 3–3.5cm; internodes of
rachis pubescent; flowers deep purple; perianth
persistent; ovary pubescent; fruits red, c. 7mm diam.
Wet areas in forest, gallery forest in savanna areas;
590–710m.
Distr.: Sierra Leone to Cameroon, Congo (Kinshasa),
Sudan, Uganda, Tanzania [Guineo-Congolian].
IUCN: LC
Mefou Proposed National Park: Cheek 13121 2/2006;
Etuge 5126 fl., 3/2004; **Ndanan 1:** Cheek 11511
3/2004; Harvey 102 10/2002; 155 10/2002; **Ndanan 2:**
Darbyshire 211 fl., fr., 3/2004.
Local Name: Emveal (Etuge 5126).

Megaphrynium macrostachyum (Benth.) Milne-
Redh.
Erect herb, to 2–3m, thicket-forming; stem simple
bearing a single terminal leaf; leaves elliptic, c. 30–60 ×
12–30cm; petiole with calloused portion to c. 15cm
long; inflorescence lateral, much-branched, each

raceme with up to 30 internodes of ± 5mm; flowers
whitish with red or purplish calyx; fruits glossy red,
smooth, with 3 conspicuous sutures; seed black, arillate.
Forest; 700–710m.
Distr.: Sierra Leone to Cameroon, Rio Muni, Gabon,
Congo (Kinhsasa), Angola (Cabinda), Sudan, Uganda
[Guineo-Congolian].
IUCN: LC
Ndanan 1: Cheek 11015 10/2002; 11739 3/2004;
Ndanan 2: Cheek 11134 10/2002.
Local Name: Okoane (Ewondo) (Cheek 11134).
Uses: MATERIALS — Unspecified Materials - used
for wrapping (Cheek 11134).

Sarcophrynium brachystachys (Benth.) K.Schum.
var. *brachystachys*
Erect herb, to 1.5m, unbranched; culms several, each
bearing a single terminal leaf; leaves ovate-elliptic,
slightly asymmetric 20–35 × 10–17cm, broadly cuneate at
base, cuspidate at apex, both surfaces glabrous, with
many closely parallel secondary nerves; petiole 30+cm,
calloused portion of petiole 4–6cm; inflorescence of cone-
like spikes from top of culm, 4–7cm long; with bracts
subtending flowers imbricate, distichous, chaffy, boat-
shaped, to 2cm long; corolla pale orang; fruits
subglobose, glossy red, c. 1.5cm diam. Forest; 650–710m.
Distr.: Senegal to Cameroon, CAR., Gabon and Congo
(Brazzaville) [Guineo-Congolian].
IUCN: LC
Ndanan 1: Cheek 11512 3/2004; 11618 3/2004; 11845
3/2004; **Ndanan 2:** Cheek 11132 10/2002; Darbyshire
213 fr., 3/2004.
Local Name: Ndutum (Cheek 11132).

Sarcophrynium schweinfurthianum (Kuntze)
Milne-Redh. var. *puberulifolium* Koechlin
Fl. Gabon 9: 151 (1964).
Differs from var. *schweinfurthianum* in having dense
pubescence on lower surface of leaf-blade. Forest;
700m.
Distr.: Cameroon [Cameroon endemic].
IUCN: NT
Ndanan 1: Cheek 11251 10/2002.

Sarcophrynium schweinfurthianum (Kuntze)
Milne-Redh. var. *schweinfurthianum*
FTEA Marantaceae: 5 (1952); Fl. Gabon 9: 150 (1964);
Fl. Cameroun 4: 150 (1965).
Upright herb, 2m; unbranched; leaf-blade ovate-elliptic,
slightly asymmetric, 35 × 15.5cm, cordate at base,
cuspidate at apex, both surfaces glabrous; sheath
pubescent; petiole long (25+cm), plus calloused portion
7cm; inflorescence produced half way up plant, lax,
rachis slender, zig-zag, to 13cm, glabrous; bracts
subtending cymules boat-shaped, 3cm long; corolla
whitish; fruits in 3s on pendent peduncles to 3.5cm
long, broadly obovoid, 1.5 × 1.5cm, indehiscent,
pericarp orange, fleshy; seeds embedded in white jelly.
Swamp forest; 700m.

Distr.: Cameroon, Gabon, Congo (Brazzaville), CAR, Sudan, Uganda [Guineo-Congolian].
IUCN: LC
Ndanan 2: Cheek 11133 10/2002.

Thalia welwitschii Ridl.

Erect herb, to 2m; tufted, unbranched; leaf-blade ovate, symmetric, c. 22 × 6–8cm, rounded towards base and very broadly cuneate, gradually acute with cuspidate apex, glabrous on both surfaces; petiole long (25+cm), plus calloused portion 0.3–1.5cm; inflorescence a lax panicle with glabrous rachis; bracts subtending cymules ovate or boat-shaped, 1.5cm long, glabrous; corolla purple; fruit a smooth 1-seeded capsule; seeds arillate. Swamps; 710m.
Distr.: Senegal to Sudan, Zambia and Angola [Guineo-Congolian].
IUCN: LC
Ndanan 2: Cheek 11930 3/2004.

Trachyphrynium braunianum (K.Schum.) Baker

Climbing herb, to 5m; stems branched, without prickles; leaves ± distichous; leaf-blade elliptic or ovate-oblong, slightly asymmetric, 7.5–12 × 2–4.3cm, cordate at base, acuminate at apex, both surfaces glabrous; petiole very short, 0.3mm above sheath and below articulation with calloused portion (0.5–1cm); inflorescence pendulous, sometimes branched near base, to 6cm, rachis relatively robust, minutely hairy; bracts subtending cymules boat-shaped, 2cm long; corolla white; fruit a 3-lobed capsule, 1.3 × 1.5cm, pericarp orange, covered by small triangular prickles; seeds glossy black, with basal aril. Forest, including river banks, secondary growth at roadside; 600–710m.
Distr.: Guinea to Cameroon, Gabon, Congo (Brazzaville), Congo (Kinshasa), Sudan and Uganda [Guineo-Congolian].
Mefou Proposed National Park: Etuge 5127 fl., 3/2004; **Ndanan 1:** Cheek 11876 3/2004; **Ndanan 2:** Cheek 11135 10/2002; 11919 3/2004.
Local Name: Eset (Ewondo) (Cheek 11135).

ORCHIDACEAE

P. Cribb (K), J.-M. Onana (YA), D. Roberts (K) & M. Cheek (K)

Fl. Cameroun 34 (1988), 35 (2001) & 36 (2001).

Aerangis stelligera Summerh.

Fl. Cameroun 36: 840 (2001).
Epiphyte; stems pendant, 5cm; leaves leathery, dark green, 3–5, oblong-lanceolate, 7–15 × 1.5–4cm, apex unequally bilobed; raceme pendant, 10–25cm, 3–6 flowered; flowers white; pedicel 3–6cm; dorsal sepal lanceolate, 3.5–5 × 0.7cm; petals lanceolate, acuminate, 2.5–4 × 0.6cm, labellum 3.3–4 × 0.5cm. Evergreen forest; 700m.
Distr.: Cameroon, CAR, Congo (Kinshasa) [lower Guinea].

IUCN: NT
Ndanan 1: Cheek 11259 10/2002.

Ancistrorhynchus capitatus (Lindl.) Summerh.

Epiphyte; stem 1–10cm; leaves 3–8, coriaceous, linear, 16–32 × 1.4–2.1cm, unequally bilobed at apex, each lobe 2–3-toothed, leaf-bases 2–2.7cm; inflorescence cauliflorous, capitate, densely many-flowered, to 1.5 × 2.5(–4)cm; flowers small, white or pale rose; labellum 4.5–5.7 × 4–5mm; spur 7–8.5mm. Forest; 700m.
Distr.: Sierra Leone, Liberia, Togo, Nigeria, Cameroon, Bioko, CAR, Congo (Kinshasa), Uganda [Guineo-Congolian].
IUCN: LC
Ndanan 1: Cheek 12015 3/2004.

Angraecopsis parviflora (Thouars) Schltr.

Epiphyte; stem 1–3(–10)cm; leaves 4–6(–9), falcate, linear-lanceolate, 6–16(–22) × 0.4–1.2cm; inflorescences 1–several, 5–17cm, densely 2–8-flowered; peduncle slender, wiry, 4–9cm; flowers white or greenish-white; spur longer than the lip, 0.6–0.9cm, very slightly or not at all swollen apically. Forest; 710m.
Distr.: Cameroon, Tanzania, Malawi, Mozambique, Zimbabwe, Madagascar, Mascarene Is. [afromontane].
IUCN: LC
Ndanan 2: Cheek 11942 3/2004.

Angraecum birrimense Rolfe

Epiphyte or epilith; stem to 100cm; leaves distichous, flat, more or less unequally bilobed at the apex, 7–14 × 1.5–3.5cm; flowers solitary on long peduncles or peduncles several-flowered; sepals and petals pale green, the sepals 3.5–5cm long; labellum white, centred green, ± orbicular, 4.0 × 2.5–3.5cm; spur 3.5–4.5cm. Forest, gallery forest, swampy places or terrestrial and climbing over mossy rocks; 700m.
Distr.: Sierra Leone to Cameroon [upper & lower Guinea].
IUCN: LC
Ndanan 1: Cheek 11968 3/2004.

Angraecum sacciferum Lindl.

Epiphyte; stems 5–20mm; densely branched; leaves leathery, 6–12, oblong-linear, 1.5–6 × 0.15–0.6cm, arranged in one plane, apex bilobed; inflorescence lax, 2–8cm, 1–6-flowered; flowers green to yellow, petals lanceolate, 2.4–3.2 × 0.5–1mm. Forest; 700m.
Distr.: Cameroon to E and South Africa [afromontane].
IUCN: LC
Ndanan 1: Cheek 12010 3/2004.

Ansellia africana Lindl.

Epiphyte; pseudobulbs cylindric, 10–50cm, covered in papery leaf-sheaths; leaves 8–10 at apex of pseudobulbs, narrowly lanceolate-oblanceolate, 15–30 × 1.5–6.5cm, acute to obtuse; inflorescence lax, to 40-flowered; flowers with sepals and petals subequal,

oblong, 15–30 × 5–10mm, yellow blotched with brown or red; labelleum trilobed, 1–2cm diam. Secondary forest undergrowth; 680m.
Distr.: tropical Africa.
IUCN: LC
Ndanan 2: Darbyshire 252 3/2004.

Auxopus macranthus Summerh.
Terrestrial heteromycotrophic (saprophytic) herb; stem 15–30cm with 1–several cataphylls; ± leafless; inflorescence with scape to 33cm, arising at apex of tubers; peduncle slender to 3cm; rachis up to 8cm, closely or loosely several–many-flowered; flowers orange, yellowish-brown or brown; tepals connate into an obscurely 2-lipped tube. Understorey leaf-mould in dense forest; 658m.
Distr.: Liberia, Ivory Coast, Ghana, Cameroon, Congo (Kinhsasa), Uganda [Guineo-Congolian].
IUCN: LC
Mefou Proposed National Park: Etuge 5299 3/2004.

Bolusiella talbotii (Rendle) Summerh.
Dwarf epiphyte; stem 0.3–1.5cm; leaves 3–6(–12), arranged in a fan, ensiform, (1.5–)2.5–6(–8) × 0.5–0.8(–1.2)cm; inflorescence lax, 3–11cm, 10–24-flowered; flowers white, centrally pink; labellum 1.7–2.5mm; spur 2mm, often green. Forest, coffee plantations; 700m.
Distr.: Sierra Leone to Cameroon, Bioko, Rio Muni, Annobón, Tanzania [afromontane].
IUCN: LC
Ndanan 1: Cheek 11949 3/2004.

Bulbophyllum falcatum (Lindl.) Rchb.f. var. *bufo* (Lindl.) J.J.Verm.
Bull. Jard. Bot. Nat. Belg. 56: 235 (1986); Orchid Monographs 2: 125 (1987).
Syn. *Bulbophyllum bufo* (Lindl.) Rchb.f.
Syn. *Bulbophyllum longibulbum* Schltr.
Epiphyte; pseudobulbs bifoliate, 3.5–8cm tall, 0.4–5cm apart; leaves broadly lanceolate to linear or oblanceolate, 8.5–21 × 1.3–4.5cm; inflorescence 11–40cm, 8–60-flowered, each one 8–30mm apart; peduncle 3–12cm; rachis flattened, blade-like, 7–28 × 0.21.8cm; flowers whitish, purple-spotted; ratio of length of median sepal to petal: 2.4–5; top part of petals not thickened, acute to acuminate, or subacute. Lowland and submontane forest; 700m.
Distr.: Guinea to Cameroon, Congo (Kinshasa) [Guineo-Congolian].
IUCN: LC
Mefou Proposed National Park: Darbyshire 449 4/2004.

Bulbophyllum intertextum Lindl.
Diminutive epiphyte; pseudobulbs 1-leafed, 0.2–2.5(–4)cm apart, 0.4–1.0 × 0.3–0.7cm; leaves elliptic to linear lanceolate, 0.7–10 × 0.3–1.1cm; inflorescence 2–30cm, 2–14(–20)-flowered; rachis

arching, nodding, terete, usually zigzag bent, 0.2–19cm; floral bracts 1.5–4 × 1–2mm; flowers very pale yellowish or greenish, often suffused red. Lowland to submontane forest, forest patches in grassland; 700m.
Distr.: tropical & subtropical Africa.
IUCN: LC
Ndanan 1: Cheek 11969 3/2004.

Bulbophyllum saltatorium Lindl. var. *saltatorium*
Orchid Monographs 2: 33 (1987).
Epiphyte; pseudobulbs unifoliate, 1.4–3cm apart, 1.2–2.7cm, distinctly flattened; leaves narrowly elliptic to linear-lanceolate, 3–11.5 × 1–2.3cm; inflorescence 3.5–8cm, 15–30-flowered; peduncle 1.5–4cm; rachis 1.5–4cm; floral bracts 4.5–6.5 × 1.8–2.7mm; flowers purple; labellum with marginal hairs × labellum width, dark purple-red with purple or brownish purple hairs. Primary forest; 700m.
Distr.: Sierra Leone, Liberia, Ivory Coast, Ghana, Nigeria, Cameroon, Rio Muni [upper & lower Guinea].
IUCN: LC
Ndanan 1: Cheek 12016 3/2004.

Calanthe sylvatica (Thouars) Lindl.
F.T.E.A. Orchidaceae: 282 (1984).
Syn. *Calanthe corymbosa* Lindl.
Terrestrial herb, to 70cm; pseudobulbs conical, 2–5 × 1.5cm; leaves ± rosulate, suberect, to 35 × 12cm, lanceolate; petiole 5–25cm; peduncle with 2 sheath-like leaves; inflorescence many flowered; flowers white or purple; petals 8–25 × 4–14mm, elliptic or oblanceolate; lip 11–15 × 6–25mm; spur slender, 1–4cm long. Forest; 700m.
Distr.: tropical and subtropical Africa [afromontane].
IUCN: LC
Ndanan 1: Cheek 12000 3/2004.

Calyptrochilum emarginatum (Sw.) Schltr.
Epiphyte; stem elongate, pendent, to 1(–3)m; leaves numerous, distichous, lanceolate-elliptic or oblong, 6–17 × 2.5–5cm; inflorescence many-flowered, very dense, up to 5cm; flowers white; labellum yellow or yellow-green, very fragrant, 7–10 × 7–10mm. Forest, gallery forest, plantations; 660m.
Distr.: Guinea (Conakry) to Cameroon, Bioko, Gabon, Congo (Kinshasa), CAR, Angola [Guineo-Congolian].
IUCN: LC
Ndanan 2: Darbyshire 339 fl., 3/2004.

Chamaeangis sarcophylla Schltr.
Fl. Cameroun 36: 758 (2001).
Epiphyte; rhizome horizontal, rooting 6cm; stem ascending to 10cm, bearing 4–9 leaves in 2 ranks; leaves strap-like, curved, 35 × 1cm, folding along midrib, leathery, base sheathing; inflorescence spike 20cm, arising below the leaves. Semi-deciduous forest and *Raphia* swamp; 600–700m.
Distr.: Cameroon, Congo (Kinshasa), Rwanda, Burundi, Uganda, Kenya, Tanzania, Malawi [tropical

Africa].
IUCN: LC
Mefou Proposed National Park: Etuge 5233 3/2004;
Ndanan 1: Cheek 12009 3/2004.

Corymborkis minima P.J.Cribb

Kew Bull. 51: 355 (1996).
Terrestrial herb; stems erect, 0.4–0.5m, from horizontal
rhizome, glabrous; leaves c. 6 per stem, alternate,
sheathing the stem, sheaths c. 8cm, ridged; blade
plicate, elliptic-oblong, to 22 × 4cm, gradually long
acuminate, acumen 3–4cm, base acute, pseudopetiole
poorly defined 5–10mm, longitudinal nerves 8, equal
conspicuous; inflorescence axillary, 2–5cm, few-
flowered; ovary 6mm; perianth white, 15mm. Forest,
undergrowth near river bank; 679–700m.
Distr.: Cameroon [Cameroon Endemic].
IUCN: EN
Ndanan 1: Cheek in MF 259 10/2002; Onana 2893
3/2004.

Cyrtorchis ringens (Rchb.f.) Summerh.

Epiphyte, stem arcuate, woody, to 30cm in old plants;
leaves linear, usually 6–7, thick and leathery, 7.5–12.5 ×
1.2–2.5cm; inflorescence 6–7(–16)cm, to 16-flowered;
flowers closely placed, creamy white, sweetly scented;
spur to 3cm. Evergreen forest; 700m.
Distr: Senegal, Sierra Leone to Cameroon, São Tomé,
Congo (Kinshasa), Burundi, Uganda, Tanzania, Zambia,
Malawi, Zimbabwe [tropical Africa].
IUCN: LC
Ndanan 1: Harvey, 107 10/2002.

Diaphananthe bidens (Sw.) Schltr.

Epiphyte; stem pendent to 50cm; leaves up to 22,
distichous, oblong-elliptic to ovate-elliptic, sometimes ±
oblique, 6.5–17 × 1.1–4.5cm; inflorescence 4–22cm, up to
25-flowered; flowers salmon-pink, yellowish-pink, flesh-
coloured or whitish; labellum 3.5–5mm. Forest; 700m.
Distr.: Guinea (Bissau) to Cameroon, E to Uganda, S to
Angola [Guineo-Congolian].
IUCN: LC
Ndanan 1: Cheek 11947 3/2004.

Diaphananthe pellucida (Lindl.) Schltr.

Robust epiphyte; stem short, up to 12 × 1cm; leaves
distichous, oblanceolate, 18–70 × 2–9cm;
inflorescences 1–7, pendent, 15–55cm, to 50–or more-
flowered; flowers white, translucent creamy yellow,
pale green or pinkish; labellum 0.8–1.1cm, shortly
fimbriate. Forest; 660–700m.
Distr.: Guinea (Conakry) to Cameroon, Congo
(Brazzaville), Congo (Kinshasa), Uganda [Guineo-
Congolian].
IUCN: LC
Ndanan 1: Cheek 12008 3/2004; **Ndanan 2:**
Darbyshire 337 fl., 3/2004.

Eurychone rothschildiana (O'Brien) Schltr.

Epiphyte; stem 2–8cm, with persistent leaf-sheaths;
leaves 3–8, 6–21 × 1.5–7cm, unequally and subacutely
bilobed, undulate; inflorescence laxly 2–6–12-flowered,
deflexed or pendulous, 3–10cm; flowers large, pleasantly
scented; tepals white or greenish, 18–25 × 6.5–10mm; lip
white, centrally green, with a purple-brown marking at
throat, 20–27 × 20–25mm. Forest; 600m.
Distr.: Guinea (Conakry) to Nigeria, Bioko, Cameroon,
Congo (Kinshasa), Uganda [Guineo-Congolian].
IUCN: LC
Mefou Proposed National Park: Etuge 5224 3/2004;
Ndanan 1: Onana 2879 3/2004.

Microcoelia caespitosa (Rolfe) Summerh.

Epiphyte; stem 1.0–3.1 × 0.25–5.0cm, almost leafless;
roots very long, to 90cm; scale-leaves acuminate, 6–7-
nerved, to 0.5–0.7cm; inflorescences up to 10
simultaneously, erect, spreading, to 3(–6.5)cm, each to
22-flowered; flowers to 1.8cm, whitish or greenish-
white. Forest, often near rivers; 710m.
Distr.: Sierra Leone to Uganda [Guineo-Congolian].
IUCN: LC
Ndanan 2: Cheek 11941 3/2004.
Note: Cheek 11941 is tentatively placed in this taxon,
sterile voucher, fertile material needed to identify fully,
Cribb iii.2010.

Polystachya affinis Lindl.

Epiphyte, to 50cm; pseudobulbs several, subspherical,
1–4.8cm across, 2–3-leaved; leaves petiolate,
oblanceolate, oblong-lanceolate or oblong-elliptic, 9–28
× 2.6–6cm, shortly acuminate; inflorescence lax, 8–60-
flowered, 6–40cm; flowers non-resupinate, white,
yellow or mustard-yellow, marked red or brown,
fragrant; labellum 6.3–8 × 4.7–5.7mm; spur to 6mm,
sacciform. Forest, riverine forest; 660–710m.
Distr.: Guinea (Conakry) to Cameroon, Gabon, CAR,
Congo (Kinhsasa), Uganda, Angola [Guineo-Congolian].
IUCN: LC
Ndanan 2: Cheek 11922 3/2004; Darbyshire 336 3/2004.

Polystachya mauritiana Spreng.

Syst. Veg. (ed. 16) [Sprengel] 3: 742 (1826).
Syn. *Polystachya tessellata* Lindl.
Epiphyte, (10–)20–60cm; pseudobulbs 15 × 0.5–0.7cm,
3–5-foliate; leaves oblanceolate or elliptic, (3–)10–30 ×
0.8–6.0cm; inflorescence paniculate, 10–50cm;
branches secund, distant, densely 20–200-flowered;
rachis and peduncle covered in sheaths; flowers small,
non-resupinate, cream, yellow, clear green or red-
purple. Forest, savanna, woodland; 660–720m.
Distr.: tropical Africa, South Africa & W Indian Ocean.
IUCN: LC
Ndanan 1: Cheek 12012 3/2004; Onana 2858 3/2004;
Ndanan 2: Darbyshire 338 fl., 3/2004.

Polystachya cf. *modesta* Rchb.f.

Epiphyte; stem erect, 1m; roots numerous, radiating to
6cm long; leaves arched, strap-like, leathery, folded

along midrib, sheathing peduncle base, 20–30 × 4mm; inflorescence spike 6cm; peduncle 4.5cm; fruit ellipsoid, 6 × 4mm. Secondary forest undergrowth; 680m.
Ndanan 2: Darbyshire 247 fr., 3/2004.
Note: flowering material needed to identify fully.

Rhipidoglossum kamerunense (Schltr.) Garay
Bot. Mus. Leafl. 23: 195 (1972).
Syn. *Diaphananthe kamerunensis* (Schltr.) Schltr.
Large pendent epiphyte; stem stout, 6–8 × 1.1–1.8cm; leaves 3–7, falcate, oblanceolate, 20–35 × 3–5.5cm; inflorescences pendent, 1–several, 10–12cm, laxly 8–10-flowered; peduncle 3.5–4.5cm; flowers pale translucent green. Secondary forest near *Raphia* swamp; 600m.
Distr.: Nigeria, Cameroon, Congo (Kinshasa), Uganda, Zambia [lower Guinea & Congolian].
IUCN: LC
Mefou Proposed National Park: Etuge 5225 3/2004.

Rhipidoglossum rutilum (Rchb.f.) Schltr.
Beih. Bot. Centralbl. 36(2): 81 (1918).
Syn. *Diaphananthe rutila* (Rchb.f.) Summerh.
Epiphyte, usually in canopy of tall forest; stems 3–40cm, occasionally branched; leaves linear to obovate or narrowly elliptic, 3–12 × 0.5–2.2cm; inflorescences 1–many, pendent, 5–19cm, densely up to 25-flowered or more; flowers translucent, pale green, pale yellow, often tinged rose, brown or purple. Riverine and montane forest, occasionally in old coffee plantations; 700m.
Distr.: Guinea (Conakry) to E Africa, S to Angola, Malawi, Mozambique, Zimbabwe [tropical and subtropical Africa].
IUCN: LC
Ndanan 1: Cheek 12014 3/2004.

Tridactyle anthomaniaca (Rchb.f.) Summerh.
Epiphyte; stems semi-pendent to 2m × 0.5–0.6cm; leaves numerous, fleshy-coriaceous, linear to narrowly elliptic-oblong, 3.5–11 × 0.6–1.9cm, shiny above, matt below; inflorescence very short, to 1cm, 2–4-flowered; flowers small, resupinate, green, pale green, yellow or white; labellum 3–6 × 1–2mm; spur filiform, (6–)11–16mm. Riverine, swamp and lower montane forest, plantations (coffee, cocoa, guava), often above water, or in sunny positions; 700–710m.
Distr.: Sierra Leone to Congo (Kinshasa), CAR, E Africa, S to Mozambique, Malawi, Zambia, Zimbabwe [afromontane].
IUCN: LC
Ndanan 1: Cheek 11755 3/2004; 12017 3/2004.
Note: Cheek 11755 is tentatively placed in this taxon, material poor, Cheek x.2010.

Vanilla cf. *acuminata* Rolfe
Climbing terrestrial herb, glabrous; each node with a root 5–12cm, and a leaf-blade narrowly oblong, 13 × 2cm, apex long-acuminate, base rounded; petiole 5mm. Forest; 720m.

Ndanan 1: Onana 2823 3/2004.
Note: flowering material needed to identify fully, Cheek x.2010.

Vanilla ramosa Rolfe
Climbing, or epiphytic herb; internodes 5.5–13 × 1.5–7cm; leaves shortly petiolate; lamina 8–17.5 × 1.5–6.5cm, oblong-elliptic, oblong-ovate to linear-lanceolate; petiole 1–1.5cm; inflorescence dense, usually axillary and branched basally, 2.5–8cm, 12–40-flowered; tepals ± 4 × longer than wide; labellum 14–25 × 22–27mm; median lobe oblong-elliptic. Forest; 680–720m.
Distr.: Ghana to Cameroon, Gabon, Congo (Brazzaville) & (Kinshasa), Tanzania [Guineo-Congolian].
IUCN: LC
Ndanan 1: Darbyshire 283 3/2004; Onana 2822 3/2004.

PALMAE
W. Baker (K), M. Trudgen (K) & M. Cheek (K)

Calamus deërratus G.Mann & H.Wendl.
Climber, to 7.5m or more, scrambling to tops of trees; stem with flat black spines; sheath papery, spiny; leaves to 1.5m, ovate patent; leaflets linear-lanceolate, c. 25–35 pairs, slender acuminate, 15–38cm long; petiole to 7.5cm; fruit ovoid. Evergreen forest; 710m.
Distr.: Senegal to Cameroon, Uganda and Angola [Guineo-Congolian].
IUCN: LC
Ndanan 1: Cheek 11655 3/2004.

Elaeis guineensis Jacq.
Single-stemmed tree, to 20m; leaves crowded, pinnately-compound, to 5m, arching, basal leaflets modified as spines; inflorescences partially hidden at leaf bases; fruits oblong-globose, angular by mutual compression, c. 4 × 3cm, ripening orange-red, marked brown. Plantations & farmbush, forest; 700–710m.
Distr.: tropical Africa, but cultivated throughout the tropics.
IUCN: LC
Mefou Proposed National Park: Cheek in RP 89 3/2004; **Ndanan 1:** Cheek in MF 247 10/2002.
Uses: FOOD ADDITIVES — cultivated for fruit oil.

Eremospatha macrocarpa (G.Mann & H.Wendl.) H.Wendl.
Climber, clustering, to c. 15m; leaf-sheaths unarmed; leaves with numerous basal, ovate, spiny, strongly-relexed leaflets, when adult (not seen); main leaflets lanceolate, c. 18 × 4cm; juvenile leaves fish-tailed. Forest; 700–710m.
Distr.: Sierra Leone to Cameroon [upper & lower Guinea].
IUCN: LC
Ndanan 1: Cheek 11640 3/2004; 11650 3/2004; 11831 3/2004; **Ndanan 2:** Cheek 11126 10/2002.

Laccosperma acutiflorum (Becc.) J.Dransf.
Kew Bull., 37(3): 456 (1982).
Climber, clustering; leaf-sheaths spiny, with ragged, ligule-like extension in leaf axil; leaflets 50+ pairs, irregularly lanceolate-sigmoid, held horizontally from rachis, c. 40 × 2.5cm. Forest; 700–710m.
Distr.: Sierre Leone, Ghana, Nigeria to Rio Muni, Congo (Kinshasa) [Guineo-Congolian].
IUCN: LC
Ndanan 1: Cheek 11675 3/2004; **Ndanan 2:** Cheek 11158 10/2002.
Local Name: Nkane (Ewondo) (Cheek 11158).
Uses: MATERIALS — Unspecified Materials — used to make panniers (Cheek 11158).

***Raphia* sp. of Mefou**
Unbranched palm, 10–15m tall; producing erect aerial roots from ground, c. 10cm long; aerial stem stout, covered by leaf bases; leaves with petiole c. 7m, rachis and leaflets c. 5m; leaflets subopposite, 70–80cm, rachis and petiolules angled with curved spines on ridges; inflorescence terminal, pendulous, 2–3m, side branches short and compact; fruits glossy, covered in scales, 8 × 5cm. 710m.
Mefou Proposed National Park: Cheek in RP 2 3/2004; **Ndanan 1:** Cheek 11780 3/2004.
Uses: MATERIALS — Unspecified Materials — leaflets used in making roof thatch, rachis and petiole for light furniture construction (Cheek pers. comm.).

PANDANACEAE

M. Cheek (K)

***Pandanus* sp. of Mefou**
Tree, to 8m; branches bending, crowned with a dense rosette of strap-like, pointed leaves c. 60 × 5cm, margins spiny; flowers and fruits unknown. River margin; 650m.
Ndanan 2: Darbyshire 262 3/2004.
Note: known at Mefou from a photographic record (Darbyshire 262). No specimen was collected since the plant was on the other side of the then deep Mefou river from the survey team. Ripe fruit are needed to identify to species. Although all W African species have been treated as *P. candelabrum* several different species have been recognised in recent decades, Cheek x.2010.

PONTEDERIACEAE

M. Cheek (K)

Eichhornia natans (P.Beauv.) Solms
Rosette forming herb, floating in muddy water; stem vertical, c. 3cm, glabrous; leaf-blade ovate, 20–25 × 15mm, apex acute to obtuse, base cordate, nerves all basal, c. 12 on each side of midrib, transverse nerves not present; petiole 3–6cm bearing a boat-shaped bract (inflorescence bud) 10 × 3mm on the upper surface 2cm

below the blade; stipules sheathing the stem oblong, 25 × 4mm, translucent, with longitudinal nerves; flowers blue, single, c. 7mm diam., petals 6. *Raphia* swamp and river edge; 710m.
Distr.: tropical Africa and tropical America.
IUCN: LC
Mefou Proposed National Park: Cheek in RP 90 3/2004; **Ndanan 1:** Cheek 11784 3/2004.

SMILACACEAE

J.-M. Onana (YA) & X. van der Burgt (K)

Smilax anceps Willd.
Meded. Land. Wag. 82(3): 219 (1982).
Syn. *Smilax kraussiana* Meisn.
Climber, to 7m; stem spiny; leaves coriaceous, alternate, elliptic, c. 14 × 8cm, mucro 0.5cm, base obtuse, nerve palmate, 3–5; petiole c. 2cm; inflorescence terminal, umbellate, 5cm diam. Forest; 700m.
Distr.: Senegal to S Africa [tropical & subtropical Africa].
IUCN: LC
Ndanan 1: Cheek in MF 359 10/2002; **Ndanan 2:** Cheek 11150 10/2002.
Local Name: Elsesali (Ewondo) (Cheek 11150).

ZINGIBERACEAE

D. Harris (E), M. Cheek (K) & J.-M. Onana (YA)

Fl. Cameroun 4 (1965).
Harris D.J. The vascular plants of the Dzanga-Sangha Reserve, Central African Republic. Scripta Botanica Belgica 23:1–274 (2002).

Aframomum flavum Lock
Bull. Jard. Bot. Nat. Belg. 48: 393 (1978).
Syn. *Aframomum hanburyi* sensu Koechlin Fl. Cameroun 4: 65 (1965).
Herb, to 4m; leaves narrowly elliptic, c. 45 × 8–11cm, acuminate, base cuneate, glabrous, ligule 5–8mm, suborbicular; inflorescence 4–6-flowered; peduncle 4–6(–20)cm; bracts broadly ovate, coriaceous, puberulent, 4.5 × 3.5cm; flowers yellow; fruit smooth, red; calyx persistent. Forest, clearings & thicket; 710m.
Distr.: Cameroon & Rio Muni to Gabon and CAR [lower Guinea].
IUCN: LC
Ndanan 1: Cheek 11549 3/2004.

Aframomum limbatum (Oliv. & Hanb.) K.Schum.
Herb, to 3–5m with creeping rhizomes; leaves oblong-lanceolate, c. 40 × 10cm, caudate-acuminate, base asymmetrically attenuate, pubescent on the lower midrib and margins, ligule rounded, 3mm; inflorescence 2–3-flowered on short spikes from the rhizome; bracts 2.5cm, submembranous; flowers purple, lobes c. 2.5cm, labellum obovate, undulate, 4.5–5cm diam.; fruit globose, smooth,

pale brown, (?) underground. Forest; 600–700m.
Distr.: Nigeria to Gabon & Uganda [lower Guinea].
IUCN: LC
Mefou Proposed National Park: <u>Nana 68</u> 3/2004;
Ndanan 1: <u>Cheek 11986</u> 3/2004.

Aframomum longiligulatum Koechlin
Fl. Cameroun 4: 46 (1965).
Leafy stem, 2.5m, glabrous; leaves alternate, sheath deeply ridged, ligule c. 8 × 5mm, pseudopetiole dark brown 3mm; blade oblong-elliptic, 21 × 5cm, acumen c. 3cm, base acute, midrib white; infructescence on stout erect peduncle, 28cm; fruits subtended by papery bracts c. 6 × 2cm; fruit orange, 5 × 1.5cm, red; calyx 2cm. Secondary forest; 700m.
Distr.: Cameroon, Gabon & CAR (lower Guinea).
IUCN: NT
Ndanan 1: <u>Onana 2202</u> 10/2002.
Uses: not valued for edibility (Onana 2202).

Aframomum pruinosum Gagnep.
Bull. Soc. Bot. France 55: 38 (1908).
Leafy culms, 2–4m, white waxy bloom; ligule inconspicuous, short, rounded, bifid, concealed by sessile cordate leaf-base; leaves oblong, widening 33 × 5cm, acumen 3cm, slender; inflorescence peduncle 10cm, bracteate; fruit ellipsoid-cylindric, 5 × 1.5cm, pale red, spotted purple, 3/4 concealed in papery oblong bracts 4 × 1.5cm; calyx persistent, 4cm. Evergreen forest; 700–710m.
Distr.: Cameroon, CAR, Gabon [lower Guinea].
IUCN: NT
Ndanan 1: <u>Cheek 11257</u> 10/2002; <u>11853</u> 3/2004.
Local Name: Esson-befam (Ewondo) (Cheek 11257).

Aframomum subsericeum (Oliv. & Hanb.) K.Schum. subsp. *subsericeum*
Herb, to 4m with long rhizomes; leaves oblong-lanceolate, c. 35 × 6.5cm, caudate-acuminate, base attenuate, lower surface pubescent between the nerves, hairs not appressed, spreading perpendicular to veins, ligule coriaceous, 5mm; inflorescence a 2–3-flowered spike; peduncle to 10cm; bracts coriaceous, emarginate, 3cm; corolla pink, lobes 3.5–7cm, labellum purple with yellow throat, obovate, 6cm diam., margin undulate; fruit 6 × 3cm, smooth, red. Forest; 600–710m.
Distr.: Ivory Coast, Nigeria to Congo (Kinshasa) [Guineo-Congolian].
IUCN: LC
Mefou Proposed National Park: <u>Etuge 5209</u> 3/2004;
Ndanan 1: <u>Cheek 11633</u> 3/2004; **Ndanan 2:** <u>Cheek 11156</u> 10/2002.
Local Name: Odzom (Ewondo) (Cheek 11156).
Uses: FOOD ADDITIVES — Infructescences — dried fruit crushed and put in soup (Cheek 11156). MEDICINES — Unspecified Medicinal Disorders - medicine for children (Cheek 11156). FOOD — Infructescences — fruits edible (Etuge 5209).

Aframomum sp. 1 of Mefou
Leafy stem, 2.5–5m; leaf sheath longitudinally ridged with transverse bars, glabrous; ligule rounded, 3mm, bifurcate to the blade; leaf-blade oblong, 35–38 × 7–8cm, acumen 1.5cm, slender, base acute to obtuse, lower surface sparsely hairy, secondary nerves fairly well differentiated, separated by 4–5 intersecondaries; petiole 5–7mm; peduncle near culm, 6cm, bearing 3 branches; fruit narrowly ovoid, smooth, red, 6 × 2.5cm, part enclosed in large bracts; calyx 2–3cm. Riverside forest; 700–710m.
Distr.: Cameroon (Mefou) [].
Ndanan 1: <u>Cheek 11017</u> 10/2002; <u>11897</u> 3/2004.
Note: a new species to be published by David Harris, Cheek x.2010.

Aulotandra kamerunensis Loes.
Bot. Jahrb. Syst. xliii: 389 (1909).
Herb, 30cm, glabrous; lacking aerial stem; leaves several, erect, from stout horizontal rhizome; sheath cylindric; erect 7–8 × 0.6cm, papery; blade lanceolate, 20–26 × 6.5–8.8cm, apex acute to acuminate, mucron 2mm, base obtuse with midrib concealed by blade, secondary nerves parallel, arising at a steep angle from the midrib and discernible as ridges in the midrib for 1–2cm before departing at an angle of c. 40°, slowly curving up to the apex and margin, inter-secondary nerves 3, weaker, parallel, with transverse nerves forming square or oblong compartments, each compartment with 7–15 black dots on the lower surface, in transmitted light, translucent; pseudopetiole 12–35 × 0.15–0.2cm, finely ridged; inflorescence arising from rhizome, 2–3-flowered; peduncle erect, 5–6cm, covered in sheathing papery oblong bracts 30–40 × 5–7mm, internodes 1.5–2.7cm; pedicels 2cm, subtended by bracts; ovary 9 × 3mm; calyx tube 9 × 4mm; tepal tube 1.5cm, outer tepal lobes 3, oblong-elliptic, 3.2 × 0.7cm, inner tepals purple and white, forming a tube c. 7 × 2cm. Forest understorey; 650m.
Distr.: Cameroon endemic.
IUCN: EN
Ndanan 1: <u>Tadjouteu 630</u> 4/2004.
Note: the only species of the genus in Africa, extremely rare, known previously from only two other collections. Cheek x.2010.

Renealmia africana (K.Schum.) Benth.
Herb, from short thick rhizome; leaves with a distinct false petiole; blades elliptic, to 30 × 8cm or more, apex acuminate, base narrowly cuneate, veins prominent on both surfaces; inflorescence arising from the rhizome, near the leaves; flowers small, delicate, whitish-translucent; lateral inflorescence branches spreading to upright; fruits spherical to ellipsoid, c. 8mm diam., reddish becoming black. Forest; 660–700m.
Distr.: Nigeria to Congo (Kinshasa) [lower Guinea & Congolian (montane)].
IUCN: LC
Ndanan 1: <u>Cheek 11250</u> 10/2002; <u>Harvey 157</u> 10/2002; **Ndanan 2:** <u>Darbyshire 273</u> fl., 3/2004.

GYMNOSPERMAE

GNETOPSIDA

GNETACEAE

E. Fenton (K), Seren Thomas (K) & X. van der Burgt (K)

Gnetum buchholzianum Engl.

Woody climber; stem grey-brown, cylindrical; leaves opposite, elliptic, 3.6–14 × 2.5–5.6cm, apex cuspidate, base obtuse, 4–8 lateral veins, glabrous; petiole 7mm; spikes axillary, to 5.2cm; fruit brown when young, green when mature. Forest; 600–700m.
Distr.: Cameroon [lower Guinea].
IUCN: LC
Mefou Proposed National Park: <u>Nana 32</u> 3/2004;
Ndanan 1: <u>Etuge 5110</u> 3/2003; <u>Tadjouteu 608</u> 3/2004.
Local Name: Okok (Etuge 5110).
Uses: FOOD — Leaves — Vegetable "Eru" (Etuge 5110).

PTERIDOPHYTA

P.J. Edwards (K), Seren Thomas (K), L. Fay, J.-M. Onana (YA) & Y.B. Harvey (K)

Fl. Cameroun 3 (1964)

Note: here, we are not currently following Roux, J.P. (2009). Synopsis of the Lycopodiophyta and Pteridophyta of Africa, Madagascar and neighbouring islands. Strelitzia 23: 1–296

LYCOPSIDA

SELAGINELLACEAE

Selaginella myosurus (Sw.) Alston

Scrambling herb; stem straw yellow with long creeping stolons. Forest regrowth, roadsides and farmbush; 700–710m.
Distr.: Senegal to Bioko, Cameroon, Gabon, Congo (Brazzaville), Congo (Kinshasa), Angola and Kenya [Guineo-Congolian].
IUCN: LC
Ndanan 1: <u>Cheek 11574</u> 3/2004; <u>Darbyshire 281</u> fl., 3/2004; <u>Fay 4701</u> 10/2002.

FILICOPSIDA

P.J. Edwards (K)

ADIANTACEAE

Adiantum vogelii Mett. ex Keyserl.

Small fern; rhizome creeping; stipe hairy, angular, black; fronds bipinnate, c. 15 × 11.5 × 5–14.6cm, hastate to elliptic in outline, 7–8 opposite to alternate pinna on each frond; pinna lanceolate, 5.6–10.5 × 2.3–2.6cm, c. 12 pairs of alternate ultimate segments on each pinna, segments trapeziform, apex obtuse, dentate, base attenuate; sori marginal. Forest, plantations and stream banks; 650–700m.
Distr.: Senegal to Bioko, Cameroon, CAR, Gabon, Congo (Brazzaville), Congo (Kinshasa), Angola and Zanzibar [Guineo-Congolian].
IUCN: LC
Mefou Proposed National Park: <u>Fay 4824</u> 3/2004; <u>4832</u> 3/2004; **Ndanan 2:** <u>Darbyshire 295</u> fl., 3/2004.

Doryopteris concolor (Langsd. & Fisch.) Kuhn

Roux, J.P., Synopsis of the Lycopodiophyta and Pteridophyta: 187 (2009).
Syn. Doryopteris kirkii (Hook.) Alston
Small fern; rhizome hairy, creeping; stipe glossy, dark brown, scaly at base; fronds pinnatisect, broadly hastate in outline, c. 8.1 × 9.5cm, apex attenuate, base truncate, margin crenate; sori continuously marginal, spherical, light brown. Forest understorey; 680–700m.
Distr.: tropical Africa.
IUCN: LC
Mefou Proposed National Park: <u>Fay 4799</u> 3/2004; **Ndanan 2:** <u>Darbyshire 302</u> 3/2004.

ASPLENIACEAE

Asplenium emarginatum P.Beauv.

Terrestrial fern; rhizome curved, grey-brown; stiple sparsely covered in hairs, angular; fronds broadly ovate, pinnate; pinna 10.5–15.3 × 3.6–4.5cm, lanceolate, apex emarginate, base cuneate, margin denticulate; sori linear and along venation either side of costa. Forest; 700m.
Distr.: Guinea (Conakry) to Bioko, Cameroon, Gabon, Congo (Brazzaville), Congo (Kinshasa), Angola, CAR, Sudan and Uganda [Guineo-Congolian].
IUCN: LC
Mefou Proposed National Park: <u>Fay 4810</u> 3/2004.

Asplenium gemmascens Alston

Epiphyte or terrestrial herb; rhizome long-creeping, to 7cm diam.; fronds 60–80cm long, stipe dark brown or matt grey, 15–30cm long, lamina dark green, lanceolate, 1-pinnate; pinnae in 10–16 pairs, elongate-deltoid, 6–7.5 × 2–3cm; sori many, on each side of the costa along the veins at ± 45° angle Forest; 700m.
Distr.: Nigeria, Cameroon, Congo (Kinshasa), Rwanda, Uganda, Kenya, Tanzania [Guineo-Congolian].
IUCN: LC
Mefou Proposed National Park: <u>Fay 4815</u> 3/2004.

Asplenium paucijugum Ballard

Bull. Jard. Bot. Nat. Belg. 55: 147 (1985); Acta Botanica Barcinonensia 40: 8 (1991).

Syn. *Asplenium variabile* Hook. var. *paucijugum* (Ballard) Alston
Terrestrial, low epiphyte or on rocks; rhizome creeping, 3–6mm diam.; fronds tufted, erect, 20–60cm long, ± 1-pinnate; stipe pale brown or green; lamina dark green with 1–2 pairs of subopposite pinnae; sori closely parallel long the oblique veins, 7–29mm long. Forest, sometimes in stream beds; 700m.
Distr.: tropical Africa.
IUCN: LC
Mefou Proposed National Park: Fay 4838 4/2004.

CYATHEACEAE

Cyathea camerooniana Hook. var. *camerooniana*
Syn. *Alsophila camerooniana* (Hook.) R.M.Tryon var. *camerooniana* Acta Botanica Barcinonensia 31: 26 (1978).
Tree fern, to 4m; trunk slender, 0.6–3m tall, not spiny; fronds 1.2 × 0.4 to 2.5 × 0.55m; stipe c. 10cm; pinnae sessile, gradually reducing in size in lower 1/4 of frond; sori near costules, at forking of nerve. Forest & streambanks; 660–700m.
Distr.: Guinea (Conakry) to Bioko, Cameroon, Congo (Brazzaville) & Gabon [Guineo-Congolian (montane)].
IUCN: LC
Mefou Proposed National Park: Fay 4836 3/2004; **Ndanan 2:** Darbyshire 266 fl., 3/2004.

Cyathea manniana Hook.
Syn. *Alsophila manniana* (Hook.) R.M.Tryon Acta Botanica Barcinonensia 31: 27 (1978).
Tree fern, to 10m, erect, slender; stipe sharply spinose; fronds arching to horizontal, to 2.5 × 1m, 3-pinnate, old fronds not persistent. Forest & farmbush; 660m.
Distr.: tropical Africa [afromontane].
IUCN: LC
Ndanan 1: Fay 4732 10/2002; **Ndanan 2:** Darbyshire 267 3/2004.

DAVALLIACEAE

Davallia chaerophylloides (Poir.) Steud.
Epiphytic fern; rhizome c. 1cm diam., creeping, hairy; stipe angular, golden brown, glabrous; fronds olive green, 4-pinnate pinnatisect, ovate, apex attenuate, base oblique; pinna alternate, 1–7 × 0.6–3.7cm, ovate, apex acute to attenuate, base oblique; pinnules alternate, up to 6 pairs per pinnae, triangular to trapeziform, ultimate segments deeply pinnatisect; sori solitary in the lobes on vein apex, subtended by blunt or sharp teeth. River edge; 700m.
Distr.: tropical Africa, South Africa, Comoros, Madagascar, Seychelles.
IUCN: LC
Mefou Proposed National Park: Fay 4829 3/2004.

DENNSTAEDTIACEAE

Blotiella currori (Hook.) R.M.Tryon
Fl. Zamb. Pteridophyta: 84 (1970); Bull. Jard. Bot. Nat. Belg. 53: 265 (1983); Acta Botanica Barcinonensia 38: 29 (1988).
Syn. *Lonchitis currorii* (Hook.) Mett. ex Kuhn
Herb; rhizome erect or short-creeping, thick and woody, dense ferruginous or golden hairs; fronds tufted, 1.2–3(–3/5)m; stipe straw-coloured, to 1m, with dense felt of hairs at base, ± glabrous above; lamina triangular to lanceolate, 0.5–2.5 × 0.45m, 1–3-pinnate; rachis winged, yellow or brownish pubescent; terminal pinna broad, hastate and pinnatifid; pinnae large, 15–75 × 4–12cm, acute; sori narrow, c. 0.6mm wide, continuous or interrupted. Forest; 700m.
Distr.: tropical Africa.
IUCN: LC
Mefou Proposed National Park: Fay 4806 3/2004.

Blotiella mannii (Baker) Pic.Serm.
Acta Botanica Barcinonensia 38: 30 (1988).
Large terrestrial, to 2.5m; rhizome erect to suberect; stipe with long dark brown hairs near the base; lamina densely and coarsely short hairy, to 2 × 0.5m; pinnae up to 35cm long, narrowly oblong-acute, increasingly deeply incised from the apex towards the base into mostly adnate narrowly triangular acuminate weakly sinuate lobes with wide rounded sinuses between them; sori linear, almost continuous except around the apices of the pinna segments; indusia very narrow, membranous. Forest, *Raphia* swamp; 700m.
Distr.: Guinea, Liberia, Nigeria, Bioko, Cameroon and São Tomé [Guineo-Congolian (montane)].
IUCN: LC
Mefou Proposed National Park: Fay 4805 3/2004; 4807 3/2004; 4833 3/2004.

Microlepia speluncae (L.) T.Moore var. *speluncae*
Terrestrial fern; stipe straw yellow, angular, covered in short brown hairs; lamina 3-pinnate, pinnatisect, alternate, to 28.5 × 11.7cm, ovate to falcate, to 15 pairs of alternate pinna on each frond; pinna ovate, 1.4–6.3 × 0.6–2.8cm, apex of pinna attenuate, base oblique, ultimate segments deeply pinnatisect to crenate at apex; sori 1–3 per segment, intramarginal, round, 1mm diam., light brown. Forest and farmbush; 710m.
Distr.: pantropical.
IUCN: LC
Mefou Proposed National Park: Cheek in RP 15 3/2004; 65 3/2004.

Pteridium aquilinum (L.) Kuhn subsp. *aquilinum*
Terrestrial fern; thicket forming; rhizome long-creeping, subterranean; fronds to 1.5m tall; stipe erect, the base black (remainder brown); lamina 3-pinnate pinnatisect; sori marginal with fimbriate indusia on both sides. Grassland and forest; 700m.

Distr.: cosmopolitan [montane].
IUCN: LC
Mefou Proposed National Park: Fay 4826 3/2004;
Ndanan 2: Darbyshire 228 3/2004.

DRYOPTERIDACEAE

Lastreopsis currori (Mett. ex Kuhn) Tindale
Bull. Jard. Bot. Nat. Belg. 55: 178 (1985); Acta
Botanica Barcinonensia 40: 51 (1991).
Tall fern; rhizome erect, creeping; stipe straw yellow
glabrous, scaly at base; fronds c. 9 in rosette, varies
from tripinnate at base to bipinnate and once pinnate,
pinnatisect at apex, c. 22 pairs of pinna; pinna alternate,
to subopposite, ovate to oblong in outline, 1.2–17.5 ×
0.5–13cm, apex attenuate, base asymmetrical, up to 12
pairs of alternate segments on each pinna, ultimate
segments ovate to lanceolate, 0.7–2.1 × 0.4–0.8cm,
apex acute, base asymmetrical, margin serrate to
crenate in places; sori intramarginal, spherical, in two
rows either side of midrib, usually one on each
segment. Evergreen forest; 700m.
Distr.: Guinea (Conakry) to Bioko, Cameroon, Gabon,
Congo (Brazzaville), Congo (Kinshasa), Angola,
Uganda, Burundi and Madagascar [Guineo-Congolian].
IUCN: LC
Mefou Proposed National Park: Fay 4804 3/2004;
4823 3/2004; **Ndanan 1:** Cheek 11254 10/2002.

Lastreopsis subsimilis (Hook.) Tindale
Fl. Cameroun 3: 278 (1964); Acta Botanica
Barcinonensia 40: 50 (1991).
Syn. *Ctenitis subsimilis* (Hook.) Tardieu
Tall fern; rhizome erect, scaly and creeping; stipe
striate, glabrous; fronds bipinnate pinnatisect; pinna
1.6–21.6 × 0.4–18.4cm, apex attenuate, crenate to
serrate, base suboblique, c. 12 pairs of opposite to
alternate segments; ultimate segments ovate to
lanceolate, deeply pinnatisect, 2.2–13.7 × 0.5–2.4cm,
apex attenuate, base suboblique, margin crenate; sori
round, up to 6 per segment in pairs or in linear rows.
Forest, edge of *Raphia* swamp; 700m.
Distr.: Liberia to Bioko, Cameroon and Gabon
[Guineo-Congolian].
IUCN: LC
Mefou Proposed National Park: Fay 4802 3/2004.

Tectaria angelicifolia (Schum.) Copel.
Terrestrial; fronds 1-pinnate, olive-green, ovate; pinna
deeply pinnatisect in places, lanceolate, 3.3–14.1 ×
1.7–7cm, apex attenuate, base suboblique, margin
crenate, rachis black and glossy adaxially, pubescent
abaxially; sori round, on the adaxial surface, up to 15
sori per segment. Lowland evergreen forest; 700m.
Distr.: Guinea to Bioko, Cameroon, Gabon, Congo
(Brazzaville), Congo (Kinshasa), Angola, Uganda and
Sudan [Guineo-Congolian].
IUCN: LC

Mefou Proposed National Park: Fay 4831 3/2004.

Triplophyllum fraternum (Mett.) Holttum var.
elongatum (Hook.) Holttum
Kew Bull. 41(2): 254 (1986).
Terrestrial fern; similar to var. *fraternum* but differing in
the more widely spread out fronds (to 6cm) and the
larger ultimate segments (to 8 × 3cm). *Raphia* swamp;
700m.
Distr.: Cameroon and Príncipe [lower Guinea].
IUCN: LC
Mefou Proposed National Park: Fay 4818 3/2004.
Note: this variety was formerly considered to be
confined to Príncipe Island, Edwards x.2005.

Triplophyllum fraternum (Mett.) Holttum var.
fraternum
Kew Bull. 41(2): 253 (1986).
Terrestrial; rhizome erect, dark brown, pubescent; stipe
hairy, to 60cm; fronds spaced to 2cm apart, olive green,
varies from tripinnate to bipinnate and once pinnate,
pinnatisect; pinna ovate to lanceolate, 1.5–21.7 ×
0.6–14.5cm, c. 10 pairs of alternate ultimate segments
on pinna, apex attenuate, base asymmetrical, ultimate
segments pinnatisect in places, elliptic to oblong in
outline, 1.5–2.3 × 0.9–1cm, apex obtuse, base
asymmetrical, margin entire to sinuate in places; sori
intramarginal, up to 11 per segment, round spherical.
Raphia swamp and forest; 700m.
Distr.: Guinea (Conakry), Liberia, Ivory Coast, Sierra
Leone, Cameroon, Bioko, Equatorial Guinea, Congo
(Kinshasa), Angola, Tanzania, Madagascar [tropical
Africa].
IUCN: LC
Mefou Proposed National Park: Fay 4822 3/2004.

Triplophyllum protensum (Sw.) Holttum
Kew Bull. 41: 247 (1986); Acta Botanica Barcinonensia
40: 45 (1991).
Syn. *Ctenitis protensa* (Afzel. ex Sw.) Ching Fl.
Cameroun 3: 272 (1964).
Rhizome very hairy c. 4mm diam.; scales to 3mm,
brown; stipe pubescent, golden brown; fronds
bipinnate, deeply pinnatisect in areas; pinna dark
green, c. 18.5 × 13.6–15.9cm, ovate to trianguar in
plan, apex narrowly attenuate, base asymmetrical,
margin varies from predominately repand to crenate
and entire; pinnules oblong to ovate, 1–11.5 ×
0.5–3cm, apex acute to attenuate, base asymmetrical;
segments 0.3–1.3 × 0.4–0.9cm; sori spherical and
either side of pinnule midrib, up to 26 sori per pinnule.
Forest and plantations, sometimes on rocks, roadside
weed; 710m.
Distr.: Senegal to Bioko, CAR, Gabon, Congo
(Brazzaville), Congo (Kinshasa), Angola and Uganda
[Guineo-Congolian].
IUCN: LC
Ndanan 1: Cheek 11502 3/2004.

HYMENOPHYLLACEAE

Trichomanes africanum Christ

Epiphytic trailing shrub-like fern; fronds pinnate and deeply pinnatisect, pinna ovate to lanceolate in outline, 4–10.5 × 2–2.8cm, apex lobed, base narrowly attenuate; pinnules pinnatisect, up to 9cm, each pinnule, 0.6–2 × 0.3–0.4cm, somewhat forked and lobed; ultimate segments ligulate to oblong in plan, apex rounded to obtuse. Forest; 700m.
Distr.: Guinea (Conakry) to Bioko, Cameroon, São Tomé, CAR, Gabon, Congo (Brazzaville), Congo (Kinshasa), Angola and Sudan [Guineo-Congolian].
IUCN: LC
Mefou Proposed National Park: Fay 4834 3/2004.
Note: Fay in Darbyshire 4834 is a mixed collection of both *Trichomanes africanum* and *Trichomanes erosum*, Edwards x.2005.

Trichomanes erosum Willd. var. erosum
Syn. *Trichomanes aerugineum* Bosch
Syn. *Trichomanes chamaedrys* Taton
Syn. *Microgonium chamaedrys* (Taton) Pic.Serm. Webbia 23: 181 (1968); Webbia 35: 254 (1982).
Syn. *Trichomanes erosum* Willd. var. *aerugineum* (Bosch) Bonap. Fl. Zamb. Pteridophyta: 76 (1970); Acta Botanica Barcinonensia 32: 17 (1980).
Epiphyte; filmy trailing herb to 1m; stipe densely hairy, dark brown; pinna outline varies from obovate to oblanceolate, to elliptic and lanceolate, 0.8–1.7 × 0.3–1.4cm, filiform-cylindrical carpels on apex of certain pinna, apex usually rounded or obtuse, base attenuate or somewhat rounded, glabrous. Forest, sometimes on rocks, swamp; 710m.
Distr.: tropical and subtropical Africa.
IUCN: LC
Ndanan 1: Cheek 11521 3/2004.
Note: see also Fay in Darbyshire 4834 which is a mixed collection of both *Trichomanes africanum* and *Trichomanes erosum*, Edwards x.2005.

LOMARIOPSIDACEAE

Bolbitis sp. of Mefou

Terrestrial herb; rhizome horizontal, long, creeping, c. 5mm; stipe sparsely hairy, c. 10.2–21.5cm; fronds with up to 9 alternate pinna 7–24.4 × 8.8–22.5cm, pinnate; pinna lanceolate to ovate in shape, 7.6–12.7 × 2–3.6cm, apex attenuate, base asymmetrical to obtuse in places, margin ± entire, glabrous. Lowland evergreen forest; 700m.
Mefou Proposed National Park: Cheek in MF 30 3/2004; 94 3/2004.

Lomariopsis guineensis (Underw.) Alston

Epiphytic climber; rhizome hairy, c. 8mm, scales have long hairy margins; stipe hairy grey brown; sterile fronds pinnate and alternate, c. 15 pairs of alternate pinna; pinna oblong, 6.5–9.3 × 1.8–2.3cm, apex

acuminate, base rounded, margin entire; fertile fronds linear, 2.5–5.2 × 0.3–0.4cm, apex narrowly acute, base truncate, margin entire; sori completely covering underside of pinna. *Raphia* swamp, forest; 600–710m.
Distr.: Guinea (Conakry) to Bioko, São Tomé, CAR, Gabon, Congo (Brazzaville), Congo (Kinshasa), Angola and Sudan [Guineo-Congolian].
IUCN: LC
Mefou Proposed National Park: Etuge 5210 3/2004; Fay 4801 3/2004; **Ndanan 1:** Cheek 11525 3/2004; 11957 3/2004.

MARATTIACEAE

Marattia fraxinea J.Sm. var. fraxinea

Very large terrestrial fern; rhizome erect, to 40 × 30cm; fronds tufted to 4m, stiff, fleshy; stipe with brown flushing and long white- or green-streaks; swollen base with a pair of green to dark brown, thick, fleshy stipules; lamina ovate in outline, 2-pinnate, to 2 × 1m. Forest; 680–700m.
Distr.: palaeotropical [montane].
IUCN: LC
Ndanan 1: Fay 4720 10/2002; Harvey 205 10/2002; **Ndanan 2:** Darbyshire 253 fl., 3/2004.

NEPHROLEPIDACEAE

Nephrolepis biserrata (Sw.) Schott

Small fern; rhizome long, hairy; stipe hairy, striated; fronds pinnate; pinna alternate, 1.3–16.6 × 0.7–2.5cm, oblong, apex obtuse, base truncate, margin serrate particularly at the apex; sori intramarginal in evenly spaced lines from base to apex of pinna. Forest, plantations, *Raphia* swamp; 700–710m.
Distr.: pantropical.
IUCN: LC
Mefou Proposed National Park: Cheek in RP 23 3/2004; 27 3/2004; 41 3/2004; **Ndanan 1:** Cheek in MF 371 10/2002; Fay 4704 10/2002.

Nephrolepis cordifolia (L.) Presl
Kew Mag. 8: 112–118, t. 175 (1991).
Herb; rhizome erect, c. 10cm long; stolons up to 2m, bearing scaly tubers to 3.5cm; fronds 5–20; stiple dark olive-brown, 5–25cm long; lamina linear-oblong, 35–72 × 2.5–7.5cm; pinnae 4–100 on each side, sterile pinnae oblong, to 4.5 × 0.5–0.9cm, fertile pinnae abruptly narrowed at apex; sori placed about halfway between margin and midrib. Forest; 700m.
Distr.: probably in Africa only as an escape from cultivation; very widespread in New World tropics and Old World tropics, subtropics from India to Japan and NE Australia; perhaps native in Mauritius and Seychelles [pantropical].
IUCN: LC
Mefou Proposed National Park: Fay 4827 3/2004.

PARKERIACEAE

Ceratopteris thalictroides (L.) Brongn.

Bull. Sci. Soc. Philom. 1821: 186 (1821).

Herb, to c. 35cm; rhizome dark brown, hairy, creeping; stipe 7.5–13.5cm × 2mm, striated and golden brown; fronds pale green, 2–3 pinnate, pinnatifid in places, 5–13.6 × 4.2–8.5cm, ovate in outline, apex of fronds retuse, c. 3 pairs of pinna on each frond, alternate or opposite, ultimate lobes in more finely divided fronds; sori spherical, marginal. *Raphia* swamp; 700–710m.

Distr.: tropical Africa and Asia.

IUCN: LC

Mefou Proposed National Park: <u>Cheek in RP 72</u> 3/2004; <u>Fay 4809</u> 3/2004.

POLYPODIACEAE

Drynaria laurentii (Christ ex De Wild. & T. Durand) Hieron.

Epiphyte; rhizome creeping, scales ferruginous; sterile fronds elliptic, 13–40 × 5–16cm, base cordate, deeply lobed for $^1/_3$–$^2/_3$ of lamina, lobes acute or obtuse; fertile fronds horizontal or drooping, 20–70 × 10–40cm, divided to the rachis into linear lobes 8–25cm long, apex acute or rounded; petiole 5–35cm, winged; sori thick, to 3mm diam., round, in a row on each side of the costa. Forest; 700m.

Distr.: Angola, Bioko, Burundi, Cameroon, Ivory Coast, Congo (Kinshasa), Equatorial Guinea, Gabon, Ghana, Guinea, Kenya, Liberia, Nigeria, Príncipe, São Tomé, Sierra Leone, Tanzania, Togo, Uganda [tropical Africa].

IUCN: LC

Ndanan 1: <u>Fay 4723</u> 10/2002.

Note: Fay 4723 not seen at Kew, field determination, Harvey viii.2010.

Microsorum punctatum (L.) Copel.

Epiphyte; rhizome c. 8 x0.6mm, embedded in a thick felt of roots with no visible stipe; pinna falcate, narrowly elliptic, apex acuminate, margin entire, glabrous, ± 6cm on widest section of the pinna; costa 2mm wide, prominent both abaxially and adaxially, brown; sori densely packed on abaxial side for half of the pinna length (apical side), venation not prominent. Forest; 700m.

Distr.: palaeotropical.

IUCN: LC

Mefou Proposed National Park: <u>Onana 2811</u> 3/2004.

Platycerium elephantotis Schweinf.

Fl. Zamb. Pteridophyta: 145 (1970); Bull. Jard. Bot. Nat. Belg. 53: 206 (1983).

Syn. *Platycerium angolense* Welw. ex Hook.

Epiphyte; stipe dark chestnut brown, elliptic in shape, angular, 8mm × 3mm; lamina large and split centrally into two pinna, U-shaped groove where the pinna are separated; pinna ± 51cm long, symmetrical, margin entire, apex fork-shaped, acuminate, venation parallel, linear secondary hairs, up to 1.8cm long, base decurrent; sori on the abaxial side of lamina on the margins of the fork endings (apices), up to 4cm thick. Lowland evergreen forest; 700m.

Distr.: tropical Africa.

IUCN: LC

Mefou Proposed National Park: <u>Cheek in MF 24</u> 3/2004.

PTERIDACEAE

Pteris atrovirens Willd.

Rhizome erect, creeping; stipe straw-coloured, often reddish at base; fronds with pairs of subalternate pinna; pinna elliptic to lanceolate, 8.1–14 × 3.5–4.7cm, apex acute, base truncate; ultimate segment oblong to falcate in shape, 7–15 pairs, 1.2–2.2 × 0.5–0.8cm, obtuse, acute or dentate at apex, base pinnatifid, veins anastomising in the lobes; linear sori nearly reaching to the apex of the lobes, the apices sterile. Evergreen forest near river; 700m.

Distr.: Guinea (Conakry) to S Nigeria, Príncipe, São Tomé, Bioko, Cameroon, Gabon, Congo (Kinshasa), Sudan, Angola, Uganda, Kenya [tropical Africa].

IUCN: LC

Mefou Proposed National Park: <u>Fay 4821</u> 3/2004; <u>4828</u> 3/2004.

Pteris burtonii Baker

Rhizome erect, dark brown, with linear-lanceolate scales 3–4mm long; stipe brown, angular; rachis covered with minute hairs; sterile pinna lanceolate, 12.5–22.5 × 3.9–4.9cm, 5–7 pinnae on each frond, apex cuspidate, base alternate to cuneate, rachis narrowly winged by decurrent bases, margin entire, apex serrate, veins anastomasing; fertile fronds with 7 pairs of pinnae; pinna falcate, 9.8–22 × 1.2–6.5cm, apex narrowly acute, rachis narrowly winged by decurrent pinna bases, veins in segments usually densely anastomasing, pinna entire; sori linear, running continuously along margin of pinna nearly reaching the sterile crenate tops. Stream banks in forest; 700m.

Distr.: Guinea (Conakry) to Bioko, Cameroon, CAR, Gabon, Congo (Brazzaville), Congo (Kinshasa), Angola, Burundi and Tanzania [Guineo-Congolian (montane)].

IUCN: LC

Mefou Proposed National Park: <u>Fay 4825</u> 3/2004.

Pteris intricata C.H.Wright

Rhizome woody, erect or ascending with dark brown linear-lanceolate entire scales up to 1cm long; rachis castaneous with small spines; fronds almost bipinnate; pinna lanceolate, 15.8 × 5.5cm, apex attenuate, base asymmetical; ultimate segments 0.4–2.7 × 0.2–0.5cm, apex acute to obtuse, base pinnatifid, margin entire; sori at margins with serrate sterile tips. *Raphia* swamp; 700m.

Distr.: Guinea (Conakry), Sierra Leone, Mali, Nigeria, Cameroon, Equatorial Guinea, Congo (Kinshasa), Rwanda, Uganda, Kenya, Tanzania, Mozambique, Zambia and Angola [tropical Africa].

IUCN: LC

Mefou Proposed National Park: Fay 4820 3/2004.

Pteris mildbraedii Hieron.

Rhizome short, erect, with dark lanceolate scales, hairy, creeping; stipe 20–70cm long, straw yellow to brown; fronds oblong, simply pinnate with 2–6 pairs of pinnae; pinna oblong to lanceolate, c. 1.5 × 4cm, c. 21 pairs of oblong to falcate ultimate segments on each; ultimate segments, 1–2.2 × 0.2–0.6cm, rounded to subacute at the apex, sterile tips dentate; apical segment triangular-lanceolate, c .1.3 × 0.5cm; linear sori continuous at the margins, absent at the dentate apex. Forest, often on stream banks. 700m.

Distr.: tropical Africa.

IUCN: LC

Ndanan 1: Cheek in MF 208 10/2002; 342 10/2002.

Pteris similis Kuhn

Fern; stipe straw coloured; frond nearly bipinnate; pinna lanceolate, 11.5–15 × 2.2–4.9cm, c. 11 pairs of ultimate segments on each pinna, apex acute, base asymmetic; ultimate segments sub-opposite, falcate to oblong in outline, 0.5–3 × 0.6–0.9cm, apex obtuse and serrate, base pinnatifid, veins anastomasing in the lobes; prominent ribs with spines on the underside; Forest, often on stream banks; 710m.

Distr.: Guinea to Bioko, Cameroon, Gabon, Congo (Kinshasa), Angola, Uganda, Sudan and Tanzania [Guineo-Congolian].

IUCN: LC

Mefou Proposed National Park: Cheek in RP 31 3/2004.

SCHIZAEACEAE

Lygodium microphyllum (Cav.) R.Br.

Climber; stipe dark brown, twining, 1mm diam.; fronds with up to 5 pairs of pinnae; pinnae alternate, 1.5–3.1 × 0.7–1.5cm, sterile pinna lanceolate, apex acute, base truncate to cuneate, margin crenulate, fertile pinna lanceolate, ± triangular, margins lobed with clusters of sori on the underside. *Raphia* swamp; 700m.

Distr.: tropical Africa to South Africa, S Asia and introduced to USA.

IUCN: LC

Mefou Proposed National Park: Fay 4803 3/2004.

THELYPTERIDACEAE

Christella hispidula (Decne.) Holttum

Acta Botanica Barcinonensia 38: 51 (1988).

Rhizome scaly, erect; fronds 1-pinnate, pinnatifid; stipe straw yellow, angular, hairy on upper surface; c. 19 pairs of sessile pinnae, 4–10 × 1.1–1.5cm, falcate in outline, apex attenuate, base truncate, both surfaces of pinna covered in appressed hairs; ultimate segments c. 0.5 × 0.4cm, pinnatifid, oblong in outline, apex subacute, apex of terminal segments at apex minutely crenate; sori round, 6–10 per segment, brown. Forest; 700m.

Distr.: pantropical.

IUCN: LC

Mefou Proposed National Park: Fay 4814 3/2004.

Cyclosorus striatus (Schumach.) Ching

Syn. *Thelypteris striata* (Schumach.) Schelpe Fl. Zamb. Pteridophyta: 199 (1970).

Rhizome dark brown-black, 0.4mm diam., long-creeping; stipe straw brown, angular; pinna dark green, c. 23–30cm long, narrowly elliptic, apex narrowly attenuate, base truncate, glabrous; ultimate segments 0.7–2.3 × 0.3–0.5cm, sub-opposite, oblong, apex obtuse, pinnatifid; sori in two linear lines either side of midvein, c. 18 pairs of sori on each segment. Forest margins and swampy ground; 710m.

Distr.: tropical Africa.

IUCN: LC

Mefou Proposed National Park: Cheek in RP 1 3/2004; 16 3/2004; 28 3/2004; **Ndanan 1:** Cheek 11772 3/2004; Fay 4711 10/2002; 4734 10/2002.

Pneumatopteris afra (Christ) Holttum

Bull. Jard. Bot. Nat. Belg. 53: 283 (1983); Acta Botanica Barcinonensia 38: 57 (1988).

Syn. *Cyclosorus afer* (Christ) Ching

Terrestrial; rhizome wide-creeping; fronds 3 × 18–120 × 36cm; stipe c. × length of the lamina, 1-pinnate; pinnae crenate, 1–2 pairs of basal pinnae much reduced; sori medial; indusia hairy. Plantations and open areas in forest; 700m.

Distr.: tropical Africa.

IUCN: LC

Mefou Proposed National Park: Fay 4808 3/2004.

VITTARIACEAE

Antrophyum mannianum Hook.

Epiphytic fern; trunks/lower branches epilithic; rhizome short-creeping; fertile fronds 8 × 5–25 × 17cm, entire, obovate-orbicular, acuminate; stipe length similar to frond length; sori elongate along the veins all over undersurface. Forest, sometimes on rocks; 700m.

Distr.: tropical Africa [afromontane].

IUCN: LC

Mefou Proposed National Park: Fay 4816 3/2004.

WOODSIACEAE

Diplazium sammatii (Kuhn) C.Chr.

Terrestrial herb; rhizome erect; stipe thick, black, 15–30(–60)cm long; fronds lanceolate, 40–50(–70) × 25–30cm; pinna subopposite at the base, alternate towards apex, 13–15 × 2–2.5cm, base truncate, margin serrulate, terminal pinna lanceolate; sori elongated, not varying in size, near the costa. River banks in forest; 700m.

Distr.: Guinea (Conakry) to Cameroon, Equatorial Guinea, Congo (Kinshasa), Sudan, Madagascar [Guineo-Congolian].

IUCN: LC

Mefou Proposed National Park: Fay 4819 3/2004.

Diplazium welwitschii (Hook.) Diels

Acta Botanica Barcinonensia 38: 41 (1988).

Terrestrial herb; rhizome erect, scaly; stipe angular, glabrous; fronds elliptic in outline, pinnate, mildly pinnatisect; pinna alternate, sometimes sub-opposite, 2.9–16.6 × 0.8–2.8cm, falcate to lanceolate, apex attenuate, base cuneate to truncate, margin ranges from incised to serrate; sori arranged symmetrically in lines along venation either side of costa of fertile pinna. Forest, *Raphia* swamp; 700m.

Distr.: Sierra Leone to Bioko, Cameroon, Gabon, Congo (Brazzaville), Congo (Kinshasa), Angola, Uganda and Tanzania [Guineo-Congolian].

IUCN: LC

Mefou Proposed National Park: Fay 4800 3/2004; **Ndanan 1:** Onana 2204 10/2002.

INDEX

Accepted names in roman, synonyms in *italic*. **Bold** text indicates figures.